Quantitative
Investigations
in the Biosciences
using
MINITAB™

Quantitative Investigations in the Biosciences using MINITAB™

John Eddison

Senior Lecturer
Department of Agriculture and Food Studies
University of Plymouth
United Kingdom

CRC Press
Taylor & Francis Group
Boca Raton London New York

CRC Press is an imprint of the
Taylor & Francis Group, an **informa** business

A CHAPMAN & HALL BOOK

CRC Press
Taylor & Francis Group
6000 Broken Sound Parkway NW, Suite 300
Boca Raton, FL 33487-2742

First issued in hardback 2017

© 2000 by Chapman & Hall/CRC
CRC Press is an imprint of Taylor & Francis Group, an Informa business

No claim to original U.S. Government works

ISBN 13: 978-1-5848-8033-2 (pbk)
ISBN 13: 978-1-1384-6986-0 (hbk)

Visit the Taylor & Francis Web site at
http://www.taylorandfrancis.com

and the CRC Press Web site at
http://www.crcpress.com

Library of Congress Cataloging-in-Publication Data

Eddison, John (John C.)
        Quantitative investigations in the biosciences using Minitab / John Eddison.
            p.   cm.
        Includes bibliographical references.
        ISBN 1-584-8803303 (alk. paper)
        I. Biometry.   2. Minitab.   I. Title
    QH323.5.E32   1999
    570'.1'5195 — dc21
                                                                                99-043693
                                                                                CIP

Contents

Preface

As biologists, our primary task is to seek out biological answers to biological questions. This process of discovery comprises three interdependent elements of equal importance: (i) defining the question and planning the investigation; (ii) carrying out the investigation, analysing the results, and discovering an answer; and (iii) communicating the answer to the appropriate audience(s) in a language that they can understand. There are many textbooks that provide excellent accounts of the analytical techniques involved in the second stage, but far fewer that give due prominence to the planning and reporting elements of research in the biosciences. The aim of this book is to bring together these three important elements of *Design, Analysis,* and *Communication*.

Throughout the book, I focus on the *biological question*. In the past, students have tended to spend a disproportionately large amount of time analysing data, leaving insufficient time to think about the results in a biological context. In this era of statistics packages, this excuse is no longer valid. Any device that creates time for researchers and students to think more about the design and conclusions of an investigation and reduces the time devoted to number-crunching must be welcomed and embraced. For this reason, all the analytical techniques that I have described here are illustrated using the Minitab™ statistical package.

Having enthused about the advantages of statistics packages, I must add a note of caution. There is a great danger in rushing through analyses using such packages without a real understanding of the underlying processes. For this reason, I have not only illustrated analyses by full-worked examples, but also devoted some time to discussing the type of biological question to which each technique should be applied. I am convinced that by knowing *why* and *how* such analyses are performed, one acquires the skill to choose appropriate methodologies. Furthermore, one is also in a much better position to recognise the wild errors that can be generated easily on the computer through operator error.

I have observed that the justifiable sense of achievement provided by a successful calculation all too often leads to an eagerness on the part of students to find out the *statistical* significance of a result while forgetting to consider its relevance to the *real* problem. In an effort to compensate for this tendency, I have included formal conclusions in all the worked examples and I would urge a similar practice when working through the exercises.

I hope that this book provides a basic handbook of investigative methods for biologists and that it also serves as a useful starting point for statistics students who need to integrate their studies with Minitab. The analytical techniques and examples have been drawn from a variety of subject areas in the biosciences, and, given this diversity of subject matter and processes that are found in biology, both parametric and non-parametric techniques are included.

The book is organised into sections that relate to different types of questions. Chapter One introduces the key elements of an investigation: defining the question, categorising four different types of problems, and describing how we construct testable hypotheses. It also outlines the reasons why we need to use quantitative methods in order to obtain answers, and discusses the important issue of drawing conclusions and reporting. The remaining chapters are grouped into four sections. Part I (Chapters 2–4) deals with handling, summarising, and presenting data; basic statistical principles; and sampling techniques. Part II (Chapters 5–7) describes how we can answer *Questions of Comparison*. For example: Is the blood cholesterol level of healthy men different from those who have suffered a heart attack? It also describes more sophisticated investigations where the effects of several factors are examined simultaneously. For example: How do food type and ambient temperature influence the number of eggs laid by *Drosophila* fruit flies, and do these factors interact with each other? Part III (Chapter 8) considers problems that examine *Sequential Relationships* (those that can be plotted on a graph). For example: What is the relationship between drug dosage and blood pressure? Part IV (Chapter 9) describes how we can address problems of *Associations* and *Agreement* that are best examined in the form of tables, such as: Is there any association between disease status and blood group? or Does the ratio of phenotypes reflect the predictions of a genetic model?

I am very conscious of the fact that non-statistics students find bare mathematical treatments impenetrable, and so, while it would be impossible to write a statistics textbook without including any formulae, I have made every effort to discuss principles in words before introducing algebraic terms; and the level of arithmetic does not exceed elementary algebra.

There are many individuals and groups to whom I am extremely grateful for assistance during the writing of this book. I should like to thank Minitab Inc. for supporting me through the Author Assistance Programme. Portions of MINITAB Statistical Software input, output, and data sets contained in this book are printed with the permission of Minitab Inc. MINITAB is a trademark of Minitab, Inc. and is used herein with the owner's permission (www.minitab.com). I also thank Academic Press for allowing me to reprint the brain and body weight data (Question 8, Chapter 2), and Oxford University Press and the Biometrika Trust for the data on cuckoo eggs (Question 10, Chapter 3). I am very grateful to Achi Dosanj, Mark Pollard, and Stephanie Harding at Chapman & Hall/CRC who have provided considerable encouragement over the long gestation of this book. I owe a great debt to my colleague Hayley Randle and an anonymous reviewer whose detailed comments on early drafts have improved the text enormously. (Any remaining errors are, of course, my responsiblity, and I shall be pleased to receive notification of any that have slipped through unnoticed.) Finally, I must acknowledge the support and encouragement of my wife, Catherine, and son, Paul, who have kept me going while writing.

John Eddison

CHAPTER 1

Introduction

1.1   Introduction

In all areas of the sciences, questions are posed and answers are pursued. While some popular fiction would have us believe in inspired and charismatic scientists making discoveries that can change mankind virtually overnight, real examples of inspirational or accidental discoveries (e.g., Röntgen's discovery of the effects of X-rays on photographic plates) are extremely rare. In fact, the converse is nearer to the truth, most answers being obtained through methodical, lengthy, and, sometimes, exciting endeavour. In short, science does not differ from any other occupation in that it involves its fair share of hard work.

Given that most investigators obtain their results through systematic enquiries, the question arises as to the nature of the system or systems that these 'normal' scientists employ. Is there one basic method or are there many? For example, are there any areas of common ground in the way physicists, physiologists, and ecologists approach investigations in their respective spheres of science? And, if there are any general principles of scientific method, what are they? These and similar questions have occupied scientific philosophers for many years (e.g., Popper, 1959; Kuhn, 1970; Hull, 1974). Fortunately, accounts of the work performed by practical scientists, combined with the thoughts and writings of scientific philosophers, have provided us with a methodological framework within which we can conduct rigorous biological investigations. The purpose of this chapter is twofold. The first aim is to introduce these basic principles of investigative methods with their various processes and components, illustrating their application in the biosciences. The second objective of this chapter is to demonstrate the importance within this methodological framework of quantifying our observations.

1.2   The process of conducting an investigation

The big question

Whatever the nature of an enquiry, whether it be general, specific, in the form of an experiment conducted rigorously under laboratory conditions, or general observations in an inhospitable wilderness, the investigator must be aware of the purpose of his or her work. This objective has sometimes been described as the "big question." Without some known direction or defined aims for the information-gathering process (whether it be experimental or observational), the quality or appropriateness of the data obtained may be inadequate, or possibly even so poor that they are useless. There is also the important, but often ignored, consideration of the motivation of the investigator, which can be influenced considerably by the presence or absence of clearly-defined goals. For example, the prospect of leaving a warm building to spend a day observing animals in the wild with the certainty of returning later thoroughly soaked by rain and quite chilled may instill a certain reluctance in the investigator. While a clear knowledge of the ultimate goal will not make the rain any drier nor the wind any warmer, it may help to increase the observer's motivation and maintain attention so that data quality is more assured.

The formalisation of key questions is not only confined to studies where the investigator has an element of choice in the subject of the project. In applied investigations, objectives are often determined by a sponsor, but the initial remit may be couched in terms that are not precise enough to be translated directly into an experimental design.

The first principle of any investigation, therefore, is to be able to formalise the aims and objectives in a clear, concise, and unambiguous manner. This may seem to be common sense; it is! However, there is a great temptation to spend insufficient time on this aspect of the project and to proceed with some sort of physical activity. So, remember that thought *prior* to action will generally not only minimise the energy required to find the answers, but will also reduce the potential for fundamental errors that, in turn, may decrease the value of the project. It is worth adding at this point that all investigations comprise many stages, many of which precede data collection, the latter generally comprising the main physical activity. The requirement for intellectual activity is not confined solely to the development of ultimate goals, and a considerable amount of thought is required before and after any physical action.

The form that questions may take is almost endless. They may be extremely general, such as: What factors influence the vegetational composition of an area of moorland? Or, what changes, if any, take place in blood chemistry when an individual is stressed? Alternatively, the questions could be quite specific. For example, what is the sequence of behaviours that culminates in mating between mallards (*Anas platyrhynchos* L.)? Or, what is the relationship between grain yield of maize and the concentration of nitrogen in the fertiliser?

All of the above questions are perfectly adequate starting points for an investigation, but they all require further clarification, generating further questions that have to be answered before proceeding on to the next stage.

As an example of the need for clarification, consider the first question concerning vegetational distribution. At what level are we to conduct this investigation? What factors are we going to consider? Are we to confine ourselves to physical factors such as soil characteristics (e.g., pH, soil water content), topography, and climatic conditions? Or might we include an examination of grazing pressure, the effect of soil invertebrates, or trampling by human visitors? To what ultimate ends are the results to be applied (sometimes determined by the financial sponsor of the work)? Do we intend to confine our conclusions to the specific section of the moorland that we have studied? To the entire location? Or to all areas that bear some resemblance to our study area? Clearly, answers to these simple questions will have direct implications on the nature of the data that we will need to collect and also to the manpower, expertise, and equipment required to carry out this project.

The fact that wide-ranging questions need to be made more specific before a project can be started, is no surprise. However, even the more specific questions may need to be reconsidered and clarified to some degree.

Consider the example of mallard pre-copulatory behaviour, or any other sequence of behaviours for that matter. Tinbergen (1963) described four key questions that have to be answered about the level at which we can examine the behaviour patterns.

1. What are the mechanisms (external and internal) that have stimulated the individual to perform the behaviour? What cues initiate the behavioural sequence?
2. How did the behaviour develop in the individual? That is to say, what environmental and genetic factors combined for this behaviour to develop within the individual?

3.   What use or survival value does this behaviour serve the individual?
4.   How did this behaviour evolve within the species?

These four questions or problems were developed in the context of behavioural research and they illustrate very clearly the need to define the objectives of an investigation at the very outset, showing that a particular phenomenon may be examined at different levels and the central questions of the investigation vary according to the level of interest. Of course, this principle is parallelled in all scientific disciplines.

The level of the problem will be an important factor in determining the detail of the data that will have to be gathered in order to answer the key questions. The mallard example illustrates this point well. If we were interested in the evolutionary development of the displays of this duck, we would not need to obtain a detailed knowledge of the neural anatomy of the visual system even though vision would be an important sensory system in the courtship displays.

Without well-defined aims, the investigator will not be able to determine what data to gather, and this may lead to investigations being conducted that either may not be able to provide the answers that are sought, or that might provide perfectly correct answers to quite different questions. In answering questions, all scientific investigators try to be as objective and as unambiguous as possible. In order to achieve these aims, it is necessary to couch both the problem and the solution in quantitative terms. This necessarily requires the scientist to become conversant with the different types of problems and quantitative measurements that can be applied to his or her field of study. Therefore, there is a need to define, in quantitative terms, the particular problem under examination because it is only with such a specification that we can ensure that a study is performed successfully. Moreover, if it were possible to produce a simple categorisation of basic problem types, this would be a very useful tool that could be employed at the start of any investigation.

Problem types

At first sight, classifying the great diversity of biological problems into a small number of categories may seem to be an overwhelming task. How could we devise general headings across such different topics as molecular biology, physiology, biomechanics, and ecology? In fact, a

basic classification of problem types can be constructed that cover
many scientific disciplines (not just biology) and this is just one exam
ple of all branches of science being unified. Later in this chapter, th
principles of hypothesis testing, which also apply throughout scienc
will be introduced.

Throughout the biological and medical sciences, researchers nee
to perform *comparisons*. For example, are the hairs of reindeer thicke
than those of tropical deer? Are the red blood cell counts of Peruviar
inhabiting the high Andes different from those living at lower alt
tudes? Is the level of cortisol produced by the adrenal glands of sow
in one husbandry system different from that of similar sows kept i
an alternative housing regime? Furthermore, the comparisons do nc
have to be restricted to two groups. We might wish to compare thre
or more similar groups simultaneously, such as the effect of severa
different pest control measures on the yield of crops infested by aphid
(the groups might include several concentrations of a pesticide as we
as an organic method of aphid control), or the effectiveness of severa
drugs or treatment regimes in combatting a specific disease. Alterna
tively, we may have a single set of measurements that we wish t
compare against an independent prediction. For example, we migh
wish to establish whether the mean dissolved oxygen content of ;
series of water samples falls within a specified acceptable range. Ii
all of these comparisons, whatever variables are compared, the prin
ciples of the questions posed are the same. Is X bigger than Y? Or i
there any difference between a, b, and c?

Many investigations require the simultaneous measurement of tw
variables on a series of individuals (or *subjects*). For example, we ma
wish to examine the effect of the concentration (or *dose*) of a drug or
the blood pressure (*response*) of human patients, or perhaps the rela
tionship between bone length and age in a species of mammal. Both
of these examples incorporate the notion of cause and effect: drug
concentration exerting an effect on blood pressure, and body size as
measured by bone length being dependent upon age (at least in early
life). Allied to this sort of problem are those that also examine the
simultaneous variation in two or more variables on a series of subjects
but do not contain any reference to a cause and effect relationship
Examples of such relationships include the height and weight of
humans, and the bill length and depth of fulmars (*Fulmarus glacialis*
L.). The latter relationship is used to differentiate between the sexes
of this Northern Hemisphere seabird using a technique called discrim-
inant analysis. These types of problems, with or without any implicit
notion of cause and effect, can be summarised by the following

		Species A		
		Present	Absent	Total
	Present	16	40	56
Species B	Absent	35	15	50
	Total	51	55	106

Table 1.1. Presence/absence of two species: measured as frequency of occurrence in a quadrat survey.

question: Is a high value of X associated with values of Y that are high, low, or is there no consistent pattern to the relationship between X and Y? These problems examine relationships over a sequence or range of values, and, for want of a better term, I shall refer to them as problems of *sequential relationships*.

Both groups of problems described so far are based upon numerical measurements of various kinds. Qualitative data, on the other hand, are often concerned in problems of *association*. For example, animal and plant ecologists are often interested in the coexistence or mutual exclusivity of pairs or groups of species. Questions that they might ask could include: Is Species A often found with Species B? or Are the two species never found together? or Is there no pattern to their respective distributions? The researcher would approach this problem by making many observations of different habitats and recording the presence or absence of the species of interest. The data would take the form of a set of *frequencies*: the number of habitats in which both species were present, in which only A was observed, in which only B was present, and the number where neither Species A nor B was observed (as shown in Table 1.1). Having answered such questions, the ecologist would then proceed to search for the reasons underlying the observed pattern.

Similarly, in anatomical studies one might ask whether a particular characteristic is found in both males and females equally? Or is there a sex-related difference? By observing the respective frequencies of the characteristic in both males and females, one can determine whether or not there is a sex-related association, and, if so, with which gender the characteristic is linked.

Another sort of problem based upon qualitative data and frequencies is encountered when one wishes to compare a set of observed frequencies with a pattern of frequencies independently predicted by some other source. The analyses of such problems are closely related to those of association. The latter forms are termed *goodness-of-fit* problems and answer the following type of question: How well do the observed frequencies fit the predicted pattern? This type of analysis is found in many areas of biology and is particularly common in genetics. For example, in simple breeding experiments, Mendel's laws might predict a ratio of 1:2:1 in the three possible variants of a particular phenotypic characteristic in the offspring of a particular cross. A goodness-of-fit analysis would be performed to compare the observed frequencies with those predicted under the laws of Mendelian genetics.

In essence, these are the four major groups of problems (comparisons, sequential relationships, associations, and goodness-of-fit) that biologists encounter most frequently. This is a very simple classification and there are several variations within each group that depend to a large extent on the nature of the data being studied. Furthermore, this grouping is not all-embracing. Other more complex types of problem do exist, but, as a starting point, an ability to recognise these common types and an understanding of how to conduct investigations that fall into these categories will provide a firm basis upon which to embark upon biological research.

The hypothesis

Given that our ultimate objective in any experiment or investigation is to increase our understanding of a particular system (e.g., moorland management or reproductive biology of mallards), we need to devise refined questions, the answers to which will indeed increase our understanding of the wider system under examination. As such, the interpretation of any results should be related back to the "big question," thereby helping us to explain the behaviour of the whole system in a more general way. Some general principles of the way in which we devise investigations have been developed that help us to achieve this objective. These principles have become the cornerstone of scientific method. The existence of such principles indicates that there is a single scientific method that is common to all science.

By way of illustration, consider an investigation into the human pulse rate. If we measured the pulse rates of several healthy individuals, we would probably find that their pulse rates were grouped around 72 beats.min^{-1}. We would then be able to construct a general statement or a *hypothesis* that the normal pulse rate of a healthy individual is 72 beats.min^{-1}. This is a general description of our knowledge of the human pulse rate. Subsequently, if we repeated the measurements after all of the individuals had performed a specific physical exercise, we would probably find that the pulse rates had increased. This second observation is at variance with the earlier one, forcing us to reject the original hypothesis and to construct a new, more generally-applicable hypothesis that explains both results. The results can be explained by relating the increased need for oxygen on the part of skeletal muscles during exercise to the response of the heart to that need by pumping more blood (and hence oxygen). A new hypothesis might be that the normal, *resting* pulse rate of a healthy individual is 72 beats.min^{-1}. We could proceed further to perform more complex tests that included exercises of different durations to discover how the heart responded to different levels of oxygen demand.

There are several methodological points to understand concerning hypothesis construction and testing. The first is that to reject or falsify a hypothesis forces us to formulate a new, more generally-applicable theory to explain our observations. This is a very productive method of increasing our understanding, and this process has become known as the hypothetico-deductive method and was proposed by Sir Karl Popper.

A second point is that a single, inconsistent observation is all that is required to disprove a hypothesis; equally, repeated observations consistent with a hypothesis do not make that hypothesis any more true. This point can be demonstrated by an ornithological example.

Suppose we walked alongside a river and observed a single white swan. From that single observation, we could legitimately construct the hypothesis: All swans are white. Further along the river, we might encounter several more white swans. However, these additional observations do not add anything to our hypothesis. They may reinforce in our mind the idea of swans being white, but they add nothing to the theory. If we then visit a wildfowl collection or travel to Australasia, we will encounter black swans, the first observation of which will disprove our hypothesis, thereby requiring us to explain the contradictory observation and to formulate a new testable hypothesis.

Once we have seen black swans, we must then move on to explain the observations. This explanation would be in the form of a testable hypothesis, for example, members of one species of swan can be either white or black. We could then proceed to conduct breeding experiments or biochemical analyses to discover if there are genetic differences between white and black swans. These experiments could disprove the hypothesis, showing that black and white swans are not from the same species. Moreover, this analysis, depending on the swans sampled, should also show that there are several different species of white swan (e.g., Mute, Whooper, and Bewick).

Clearly, a single observation contrary to the hypothesis is far more productive in terms of producing a general understanding of a system than many confirmatory observations. This example highlights the fact that it is impossible to *prove* a hypothesis. It doesn't matter how many white swans we see before we find the first black one; they don't prove that all swans are ·white. We can only *reject* hypotheses, and only one black swan is needed to do that.

The process of scientific investigation is one of continually erecting hypotheses that are, in turn, challenged and rejected, giving rise to new hypotheses, which, in their turn, are challenged, and so on … By challenging each hypothesis in turn, we move forward to a much more general set of scientific laws describing a particular system. This cycle of hypothesis construction, testing and rejection, explanation, and formulation of new, more general hypotheses, is at the root of scientific method and is applicable to all areas of science. In fact, not only has it been argued that if a hypothesis is not testable then it is not a scientific hypothesis, but also that if we are not constructing testable hypotheses then we are not performing good science (Popper, 1959). The construction of testable hypotheses is, therefore, a fundamental stage in the development of an investigation and precedes any data gathering. It is this hypothetico-deductive method that is regarded as appropriate for all branches of scientific investigation.

In our classical approach to scientific method, we actually erect two hypotheses every time we conduct an experiment or investigation: a *null hypothesis* and an *alternative hypothesis*.

The null hypothesis is challenged and, if it is rejected, the alternative hypothesis is accepted. If the outcome of the investigation is that the null hypothesis is not rejected, then the null hypothesis still stands, but it is not proven; we accept it as the currently best description of the system under examination. We will return to hypothesis testing in later chapters in the context of a variety of problems.

The swan example is one of great simplicity because there is no difficulty in deciding whether a swan is white or black (as long as we chose to ignore the occurrence of immature grey or brown swans). However, the examination of the pulse rate provides an altogether more complex problem for the investigator. .

In measuring a characteristic such as pulse rate that cannot be categorised into black/white or similar discrete groups, but can vary, we are confronted by the question: How different must an observation be from the hypothesised value before we reject the hypothesis? For example, would we accept the null hypothesis that the resting pulse rate of a healthy human is 72 beats.min^{-1} if a series of measurements were clustered around 70 beats.min^{-1}, 60 beats.min^{-1}, or 50 beats.min^{-1}? Expressed more generally, this question becomes; are deviations from the hypothesised value due to natural variation in the characteristic? Or are they due to another particular causal factor that needs to be identified if we are to understand more clearly the system under examination?

Such variability is not confined to biological measurements but is found in any subject area where quantification occurs. Variability is of great importance in the biological sciences where no two individuals are exactly the same in every respect even if they are genetically identical and have been subjected to similar experiences. In fact, one area of behavioural research that is currently receiving considerable attention is the topic of individual variation and how different individuals respond to particular experiences.

The branch of statistics called inferential statistics deals with the difficulties of having to draw conclusions based upon measurements of characteristics that vary. With particular regard to the question of the pulse rates, a statistician would conclude that there was a tendency of resting pulse rates to fall around 72 beats.min^{-1}, but that there was a certain amount of variability around that value. If the statistician were then asked whether a particular series of pulse rate measurements were consistent with a null hypothesis of 72 beats.min^{-1}, he or she would take into consideration the natural variation as well as the closeness to the actual hypothesised value. A conclusion could then be made as to whether the values under consideration were consistent with the null hypothesis.

If we are to follow the hypothetico-deductive method of approaching scientific investigation, we must be able to test hypotheses and draw conclusions that are based upon observations of characteristics that vary to some degree. Statistical inference provides techniques that enable us to attach some measure of confidence to our conclusions. Statistical methods help us to describe our observations unambiguously, and thereby allow us to challenge hypotheses in a rigorous and scientific manner. The majority of this book is devoted to explaining how different types of biological problems may be addressed by particular quantitative methods.

The data and choice of investigative method

Having established the central question of a study and constructed one or more testable hypotheses, the investigator must determine what data should be collected in order to test those hypotheses. It is only by collecting data that similarities, differences, trends, or associations can be identified objectively. Earlier examples have already demon strated that biological data can take several forms, including bot quantitative measurements (e.g., pulse rates) and qualitative charac teristics (e.g., colour of swans).

Within the quantitative data types, some variables, or attribute may take the form of simple counts or *frequencies*. Common example would include the number of red blood cells on a haematocrit, or plant in a quadrat or grid square. Geneticists often have to census th frequency of a specific phenotypic characteristic among the offsprin of a breeding experiment. Ornithologists and entomologists migh count the number of eggs laid by females of the species that they ar studying. Count data are very common throughout biology and ar *discrete* or discontinuous variables, that is to say they are always whol numbers or *integers*. The number of eggs found in a nest might, fo example, be 4 or 5, but never 4.5.

Other quantitative biological variables are measured on a *contin uous* scale and take the form of *real* or decimal numbers. The numbe of decimal places is dependent to a large extent upon the sophisticatio or *precision* of the measuring equipment. Between any two points o a continuous scale there are an infinite number of possible measure ments, but it is the recording device and the eyesight of the user tha determine how many places of decimals are available. Typical exam ples of these continuous measurements include the height or weigh of individual organisms; the length of bones, teeth, wings, or hairs body temperature or blood pressure; the concentration of hormones i the blood; or the duration of a particular behaviour. The biologist therefore, has to become conversant with the range of data types tha he or she may encounter. Furthermore, it is also important to remem ber that the investigation of a particular problem may require th collection of a diverse selection of measurements.

The parts of the planning process that concern data acquisitio will include many stages such as specifying both the nature of th data and the methods by which they will be collected, and possibl even how to identify the data if the investigator is not familiar with the subject organisms. Given that biological material is intrinsicall variable, the specification of data requirements and methods of dat collection will be critical if we are to be able to draw unambiguou conclusions that discriminate between the effects of natural variatio and those of specific causal factors.

Perhaps we should first consider exactly what is meant by data. A *datum* (the plural is *data*) is a value of a particular characteristic suc as the thickness of a shell, the lower critical temperature of a pig, ar index of ecological diversity, or the colour of a swan. The property or characteristic being measured or observed is referred to as the *variable*. There are four basic types of data and, since the nature of the problem under examination and the hypotheses that we test will determine the data that must be collected, it is important to be aware of these data types.

1. *Nominal:* Some variables can only be described qualitatively. They are often described by the term, "attribute" (as opposed to *measurement,* which implies a numerical value) and are known as nominal data (from the Latin word *nomen*, meaning name). Examples of such data are widespread throughout biology and include colour, gender, and behaviours (e.g., standing, walking, eating), in short any sort of categorisation (even *yes/no*). While these attributes cannot themselves be reduced to numbers (although one could measure colour in terms of wavelength of light), they can be combined with frequencies of their occurrence and thus be examined quantitatively. An individual observation can be allocated to a particular category according to the "value" of a nominal variable, such as gender (male or female) or colour of a swan (black or white). For example, in elementary genetics, one often performs simple breeding experiments, observing the outcome of particular crosses, such as counting the number of red-eyed and white-eyed *Drosophila* in a hybrid generation. In such an instance, eye colour is the nominal variable which can take two forms or "values" (red or white).

2. *Ordinal:* This is the simplest form of quantitative data which allows us to place a series of observations in a sequence or order (hence the term *ordinal*) according to a particular characteristic. This sequence is often referred to as a *rank* order, and several analytical techniques that can be applied to this type of data include the word "rank" in their names. A classic example of this type of scale is the dominance ranking in animal behaviour. (Dominance is assessed by the outcome of contests between individuals; winners are classed as dominant to losers.) In this type of ordering, an animal with a rank of 1 is classified as superior to the second ranking individual, which, in turn, is dominant to the third ranking animal, and so on ...

In this example, we are able to state that one individual is more dominant than the next, but we cannot make any inferences concerning the magnitude of the differences between pairs of individuals on the scale. So, the difference in dominance between animals 1 and 2 is not necessarily the same as between individuals 3 and 4; only the *order* is known.

Ranks are also used very effectively when the attribute under examination is difficult to quantify precisely. A typical example, taken from plant pathology, is where an assessment of leaf rust might be made on a group of plants by means of a scoring system where 0 = no infection, 1 = slight, 2 = moderate, and 3 = heavy. While any plant scored as 3 is more heavily affected than a plant with a score of 2, one cannot say that the difference between scores 3 and 2 is the same as between 2 and 1 or between 1 and 0.

Another situation where a scoring system can be applied is where the variable of interest takes a variety of forms that are difficult to reduce to a common scale. For example, if we wished to relate the resting pulse rates of a group of individuals to their normal levels of exercise, we would be faced with the problem of trying to equate levels of one form of exercise with another. Is it possible to decide whether swimming 1500 m each day comprises more or less activity than running 3 mi/day? We can overcome this problem by creating a simple ranked scale of exercise: *a lot, moderate,* and *very little,* ascribing scores of 3, 2, and 1, respectively, to the different levels of exercise. By doing this we are then able to ask questions such as: Do individuals differ with respect to their resting pulse rates according to the level of regular exercise that they perform?

Ordinal scales are also often found where characteristics are scored or measured on an index, such as with behavioural measures, condition scores in livestock, or some diversity indices.

The two remaining data types are truly quantitative and are measured on *interval* and *ratio* scales. They differ from each other quite distinctly as their descriptions below will show. However, they also share properties that distinguish them quite markedly from both nominal and ordinal data.

3.   *Interval:* This data type is measured on a numerical scale and so data of this sort can be used not only to order individuals according to a characteristic, but also enable us to draw inferences about the differences (or *intervals*) between individuals. On an interval scale, the distances or differences between points on the scale have meaning and differences of equal magnitude are equivalent. Temperature is an example of a variable measured on an interval scale, and so the difference between 90°C and 100°C is equivalent, in terms of size, to the similar difference between 50°C and 60°C. In this respect, ratio and interval data are the same.

Where interval scales differ from ratio data is that the former are measured on arbitrary scales, that is the scales are defined in terms that are independent of the property that they measure, and, as a consequence, these scales do not possess true zero values. Taking the Celsius and Fahrenheit temperature scales as examples once more, 0°C is defined as the freezing point of pure water and 0°F is the freezing point of sea water (and remember there is also absolute zero, 0°K). Not one of these zero points actually means no heat.

One property of interval data that does need to be noted is that ratios of differences between points on scales are similar irrespective of the units of measurement. This property is most readily demonstrated by an example. If we take two pairs of temperatures (e.g., 80°C and 50°C, and 60°C and 40°C) and calculate their respective differences using both Celsius and Fahrenheit scales, we will find that the ratios are identical:

80°C (176°F) − 50°C (122°F) = 30°C *or* 54°F

60°C (140°F) − 40°C (104°F) = 20°C *or* 36°F

giving the following ratios:

30:20 = 1.5 and 54:36 = 1.5

Another interval scale is time of day or year, where the zero points (midnight or December 31) are entirely arbitrary.

4. *Ratio:* This is a very common type of quantitative data
    and includes such variables as height, weight, pressure,
    and rates of occurrence (e.g., pulse rate). The fundamental
    difference between these data and those measured on an
    interval scale is that ratio data are measured on non-
    arbitrary scales that possess zero values which have a
    meaning in the context of the scales themselves. For exam-
    ple, values of 0 m and 0 yd both mean zero height, length,
    or width and 0 kg and 0 lb both mean no weight; contrast
    these with the meanings of 0°C, 0°F, or 0°K.

    We saw earlier, in the description of interval data, that
    the ratios of *differences* between pairs of values are similar
    irrespective of the units of measurement. This property is
    also shared by ratio data. However, as a consequence of
    having a true scale of measurement, ratio data possess one
    further property that separates them from interval data.
    The ratio of two *measurements* on a ratio scale will be the
    same irrespective of the units of the scale. Using both
    metric and imperial units, try calculating the ratio of the
    heights of two individuals, one measuring 1.85 m (6.01 ft)
    and the other measuring 1.65 m (5.36 ft). Do the same
    with the ratio between two temperatures measured on the
    Celsius and Fahrenheit (interval) scales. Use the ratio
    between 80°C (160°F) and 45°C (104°F)*.

In many instances, differentiating between ratio and interval measurement is not crucial, but, on some occasions, it is essential and
therefore one should understand this distinction.

The various data types are summarised in Table 1.2.

The establishment of the nature of data types associated with a
particular study is a key component in the preliminary stages of any
investigation. The simple classification outlined above is a very useful
tool in this process.

One word of caution, however, before proceeding. Many projects
will not fall neatly into one category of problem type, nor will they
involve the use of a single data type. Often different analyses utilising
various data sets will be required. Sometimes a researcher may gather
only a small proportion of the total data and rely upon others to supply
the rest. Therefore, before embarking on data collection, one must

---

* The ratios 1.85:1.65 and 6.01:5.36 both equal 1.12; the temperature ratios do
not give identical values, 80:45 = 1.78 and 160:104 = 1.54.

Data Type	Qualitative/ Quantitative	Discrete/ Continuous	Examples
Nominal	Qualitative	Discrete	Gender, colour
Ordinal	Quantitative	Discrete	Dominance rank
Interval	Quantitative	Discrete or continuous	Temperature, time of day or year
Ratio	Quantitative	Discrete or continuous	Height, weight, length

Table 1.2. The four basic data types.

establish in great detail the requirements and experimental protocol that will lead to a successful outcome of the study. Fortunately, the availability of many statistical packages on computers now means that the time required for numerical calculations is very small relative to the time spent thinking and drawing conclusions. In other words, more time is available now for the biologist to apply his or her biological understanding and skills rather than having to practice mathematical expertise.

Returning now to the process of conducting an investigation, we have to achieve the transition from having constructed a testable hypothesis to collecting the data that can be used to test that hypothesis. Decisions have to be made regarding how the hypothesis is to be tested, for example, by means of a structured, controlled experiment, by observing natural phenomena or a combination of the two, and what data in particular are required. The first stage of this substantial process is to examine the published literature. Studying the literature will enable the scientist to obtain valuable background information in order to answer the following questions:

What is already known about both the key questions and the subject material under examination?

How convincing are the published findings?

(If the key questions have already been answered satisfactorily, then report as such and move on to other stages of the investigation.)

What methods were employed to perform this earlier work?

Were experimental or observational methods used?

What methods were effective?

What methods were not effective?

What questions still remain unresolved?

What data are required to answer these unresolved questions?

The literature may also provide information concerning two other aspects of an investigation. The first relates to the variability of the subject species or material. Is the material with which we are working very variable or will two specimens selected at random be fairly similar? As discussed earlier, the examination of variability lies at the very root of statistical analysis and will, therefore, be discussed in some detail throughout later chapters of this book.

The second area on which the literature may provide guidance relates in particular to experimental work. Previous workers may have identified ranges of particular experimental variables within which experiments could be conducted profitably. For example, if we were to examine experimentally the effect on moorland vegetation of trampling by sheep, we might create fenced enclosures of similar size each of which contained a different number of sheep. Published work might indicate that above certain stocking densities (sheep per hectare) the sheep destroyed vegetational communities through overgrazing and trampling. Or agronomists may have demonstrated that during the winter months the soil was so wet that even low densities of sheep created severe damage if they were confined to a small area.

The importance of studying the literature prior to engaging in practical work cannot be stressed enough. Pointlessly repeating work carried out earlier or repeating mistakes revealed by others is neither worthwhile nor satisfying. Moreover, the additional benefits to be gained from learning more about your subject material through the experiences of others are very great.

Once the decision on the data requirements has been made, the collection methods have to be specified. Practical considerations invariably prevent us from performing a complete census of the population or study area under investigation because the resource constraints of time, manpower, and the availability of analytical equipment will limit us to sampling a proportion of the biological material of interest. For example, if we used radio telemetry to study the ranging behaviour of foxes (*Vulpes vulpes*), we would be limited by both the number of radio transmitters and the manpower required to monitor a large number of individuals. In fact, there are very few examples of studies where it has been possible to take measurements from an entire population. (A postgraduate contemporary of mine studying magpie robins did ring the entire world population, 40 individuals on the Seychelles.)

There are equally compelling ethical and methodological reasons, in addition to resource limitations, that require us to use representative samples rather than entire populations. If we are investigating a development process or conducting any other time-based study, we will need to take sequential measurements at set intervals during the investigation in order to monitor progress. In some cases, such measurements will be destructive, as in the case of dry matter estimates in botanical studies, necessitating us to take fresh samples on each occasion. Experiments of this type have to be planned such that the initial number of plants is adequate to support the non-replacement sampling régime and that the remaining population is not influenced by the increased space caused by our removing samples. Similar considerations prevail in animal-based studies, for example, where the development of carcinomas in rats or mice is monitored over a period of time.

The need to pay due regard to ethical considerations in animal research is quite clear. One must not subject any animal to unnecessary suffering and it follows, therefore, that in any experiment one must use only the minimum number of individuals that are required to answer the question for which the experiment has been designed. Many universities have an ethical committee that considers proposals for animal research and many learned societies and scientific journals concerned with the reporting of research using animals have similar groups that review manuscripts prior to publication.

In addition to the direct suffering caused to the subjects of an experiment, one must also be conscious that in field studies it is possible to cause distress indirectly by disturbing the natural habitats of animals that are not themselves central to the investigation. Moreover, one must pay due regard to habitat conservation in general. For this reason, resource and ethical considerations combine to provide the general rule of using as few animals and plants and as small a study area as is necessary to perform an investigation properly. Statistical techniques are available that enable us to calculate the minimum sample size that would be required to detect an effect of a specified magnitude. These techniques will be described in Chapter 4 (Sampling).

The terms *population* and *sample* have been introduced, but we must define carefully what is actually meant by the two words. A population is the group of animals or plants which we are investigating and about which we draw conclusions and make statements, while a sample is a subset of that group which we study and use as a representation of the larger population. This statistical definition of *popu-*

lation should not be confused with the more general understanding of the term as a group of individuals inhabiting a specific country or area. The distinction between population and sample and the implications of this distinction to analytical methods will be discussed in greater detail in Chapter 3.

In taking samples, we have to ensure that they are a true reflection of the population from which they were drawn so that the hypotheses that we have constructed can be tested rigorously. Our sampling strategy must also take into account other factors that may impinge upon our subjects. This will require the provision of adequate control measures.

For example, if we were to examine the effect of exercise on pulse rates, we might take a random selection of individuals, measure their pulses before and after exercise, and then attribute any observed differences to the effects of exercise. There is a major flaw in this method. We cannot be certain that the observed effects are due to exercise rather than some other, unrecorded factor that acted at the same time as the exercise. In order to overcome this problem, we should divide the subjects into two equivalent groups, one of which exercises while the other rests. The former is termed the *experimental group* while the latter is called the *control group*. We would have to allocate subjects to the two groups randomly in an effort to ensure that we did not have all the super fit individuals in the exercise group and the less fit individuals in the control group (or *vice versa*). The process of randomisation is extremely important in experimental studies and will appear in many later examples. The resting pulse rates of both groups are measured and then the individuals in the experimental group perform their exercises while the controls rest. After completing the exercises, all subjects have their pulse rates recorded. Hopefully, the initial pulse rates of both groups will be similar, indicating that the two groups definitely were equivalent. In order to determine if exercise had any effect on pulse rate, we would have to compare the pulse rates of the two groups. If there is a difference, we can attribute that difference to exercise. Without the control group, we would be unable to relate observed effects to exercise. When designing experiments, we must be careful to construct control groups in such a way that they differ from experimental groups *only* in terms of the experimental variable that we are investigating, which, in this example, is exercise.

To obtain a representative sample is not always a straightforward process. A correct sampling strategy requires a knowledge of both the population being studied and also the environment in which the investigation is being conducted. Many factors influence the choice of sampling techniques and some of the key issues that need to be considered are listed below:

> The nature of the investigation: experimental or observational?
>
> Age structure of the population: discrete year classes; eggs; larvae; pupae; adults?
>
> Seasonal effects on the population: recruitment into adult population; breeding cycle; mortality?
>
> Differential mortality between age classes?
>
> Spatial organisation: regularly-spaced crops; location of breeding sites?
>
> Social organisation: territorial; solitary; social; sexual segregation; gregarious?
>
> Reproductive biology: vegetative; transmission of seeds; sexual?
>
> Can we be sure that our sampling strategy does not modify the subject population?

Sampling techniques are extremely important and are discussed in detail in Chapter 4. Remember that samples provide our best estimates of the population in which we are interested and to which our null hypothesis refers. If the samples are inappropriate, then we will not be able to test our null hypothesis. Therefore, we should never be embarrassed to ask a professional statistician for advice concerning sampling procedures (or any other aspect of investigative design) prior to starting a study.

Before leaving data collection and sampling, we should remember that even the best planned sampling strategy is subject to the vagaries of chance and accidents. While we might not be able to avoid catastrophic events, we can prepare ourselves for some by being meticulous in recording *everything* that happens during the course of an investigation in a log book. Long-forgotten incidents may enable us to explain anomalous results at the conclusion of an investigation.

The analysis

This section could be subtitled, "How do we use the data to test the null hypothesis?" Clearly, the response to this question depends upon the objectives of the investigation, the null hypothesis itself, and the nature of the data. Not surprisingly, this has provided the basis of many books as well as forming the subject of a good deal of the remaining chapters of this one. Therefore, I shall not devote much space here to a discussion of individual analytical methods since they can be described more clearly in the context of specific problems. There are, however, a few basic points that are common to all types of investigation that need to be mentioned. These are best outlined here, although they will reappear in a variety of contexts in later chapters.

In an ideal world, one should decide upon an appropriate statistical analysis *prior* to collecting any data. Classically, the sequence of the investigative process is as follows: the null hypothesis is formulated so that, in its testing, one can answer the key question; then the investigator decides upon the data required to test that hypothesis, and, at the same time, an appropriate analytical technique is chosen to test the null hypothesis using the data. This process will be discussed in considerable detail in later chapters.

Following the collection of data, but prior to performing any *inferential* statistics, it is important to summarise the data using *descriptive* statistics. This area of exploratory data analysis (EDA) or initial data examination (IDE) has been revolutionised and has been the subject of enormous expansion in recent years because of the arrival of modern computers. These methods include both graphical and numerical techniques, several of which will be introduced in the next chapter on data familiarisation. This descriptive phase of the analysis fulfills several extremely important functions.

First of all, descriptive statistics enable us to obtain a "feel" for the data revealing patterns that might exist within the data set. Moreover, any reports that we may be required to produce will, almost certainly, have to include a summary of results which can be based, in part, upon this description.

The use of simple graphical techniques can sometimes reveal items of data that are widely at variance with the rest of the data set. These outliers may be indicative of data collection or transcription errors. Once identified, they can be checked and corrected, if necessary.

All analytical techniques make assumptions about the nature of the data to which they will be applied. There is no benefit to be gained by discussing them here in detail, but they will be outlined and discussed in later chapters as the various techniques are introduced. Descriptive statistics fulfill the useful function of enabling us to verify that such assumptions are valid.

The final general point to be made concerning analysis is a warning not to push data too hard. If, for any reason, the quality of the data that you have collected is not good enough to answer the question that

you have posed, be prepared to give an approximate answer with qual-
ifications, or even no answer at all. Perhaps, with the benefit of hind-
sight, you might be able to identify improvements to the methodology
that could be made in the future or offer other constructive suggestions
for further work that could lead to more conclusive answers.

The conclusions and communication

The statistical analysis can be regarded as complete once a decision
has been made either to reject or accept the null hypothesis. One
should also remember that many investigations are multi-faceted,
having several different data sets and hypotheses. The decisions concerning hypothesis rejection and acceptance do not, however, signal
the completion of a whole investigation. Before a scientist can start to
address subsequent questions, he or she must draw conclusions that
place the statistical results into perspective with respect to the original
objectives of the investigation and also within the wider context of the
published findings of earlier work. Remember that, in the biological
sciences, statistical analysis is only one stage in the process of making
biological statements. It is only by discussing results in biological
terms that one can really judge the true worth of a study and be able
to make reasoned decisions about the direction of future work. From
this, it is obvious that an ability to communicate clearly is an important
characteristic that every scientist should possess.

The conclusions from a study will show whether the original objectives have been achieved and will identify the main findings of the
project. By referring back to the initial objectives, the conclusions
should act rather like a mirror to the introduction, with the contribution of the results of the investigation being discussed in the context
of previous knowledge. It is in the conclusions that evidence or suggestions can be introduced as to the mechanisms through which the
results may have arisen. Indeed, one of the primary objectives of any
subsequent experiment should, wherever possible, be the identification
of underlying mechanisms. Observational studies may conclude with
speculations about causal mechanisms. Experimental investigations
can be designed specifically to test the veracity of those speculations.

In projects carried out with commercial or management objectives,
recommendations for action based upon the results should be proposed.
Reports of investigations of a less applied nature would be concluded
with suggestions for the next questions to be addressed and, perhaps,
proposals for the methods that could be employed.

The communication of both the results and their implications are obviously very important. Scientists must, therefore, be good and willing communicators. There may be an intellectual satisfaction of a personal nature to be gained from solving a problem, but if the results are not transmitted to a wider audience the project will have helped no one. Editorial committees of learned journals are generally quite specific about the structure of a scientific paper that they will publish. When preparing project reports for sponsoring bodies, you may not be provided with so much guidance, so it always pays to ask a sponsor, prior to writing, not only about the format of a report, but also about the readership for which it is aimed. For example, the audience may not have any statistical knowledge and require only a brief summary of results and recommendations.

So far, the structure of an investigation has been described in some detail and the various components (questions, problem types, hypothesis construction, data types and data collection, analysis, conclusions, and communication) have all been introduced. The various examples have demonstrated the intrinsic variability of biological data. It is this variability that requires us to quantify our observations wherever possible so that we can draw unambiguous conclusions and answer questions with some degree of confidence. The remaining chapters describe how we can collect and characterise our data, and then use them to answer biological questions as rigorously as possible.

1.3   Summary

1. The general principles of scientific method have been introduced.
2. A categorisation of biological question has been proposed with four basic types of problems: comparison, sequential relationship, association and goodness-of-fit.
3. Four basic data types were introduced: nominal, ordinal, interval, and ratio, as well as the concept of variability.
4. The basic principles of hypothesis testing were introduced.
5. The process of drawing conclusions about populations from samples was described, and the consequential importance of sampling techniques was emphasised.
6. The importance of scientists being good communicators was stressed.

PART I

Data Familiarisation and Presentation

CHAPTER 2

Exploring, Summarising, and Presenting Data

2.1 Introduction

The introductory chapter described, in general terms, the ideal form that investigations should take. In that description, I stressed the need to make detailed plans *prior* to data collection. In theory, at least, those plans should include a specification of the analytical techniques that would be required during the course of the investigation. In an ideal world, the scientist would then have the straightforward task of returning to the office, entering the data into the computer, performing the prescribed analysis, drawing the appropriate conclusions, and writing the final report. Nothing could be simpler. However, the real world is not ideal and so there are several reasons why one should spend some time exploring a data set and employing various numerical and pictorial descriptive techniques prior to performing the planned analytical statistics. Some of these methods were mentioned in the previous chapter and they will be discussed in more detail here.

The first function of data description is one of familiarisation. In complex investigations, we may have to synthesise a variety of data types from one or more sources and so, by exploring the data, we will be able to gain a better understanding of the relationships that may exist between the different data sets. Furthermore, such a familiarisation process also helps us to conceptualise more clearly the objectives of the investigation. With this better understanding of the objectives and relationships, the final stage of the study (drawing conclusions) becomes much easier.

Secondly, we have seen that scientific data, and biological data especially, can be extremely heterogeneous. Sometimes this variability is a natural phenomenon intrinsic to the subject matter with which we are concerned. On other occasions extreme variability may be due to errors that occurred during data collection or while entering the data into the computer. For example, an individual may not have correctly mastered the data collection techniques and committed consistent or inconsistent errors; or measuring equipment may not have been calibrated correctly or was malfunctioning; and there will always be scope for transcription errors occurring during the transfer of data from written notes to computer files. Such problems might have given rise to individual data items that stand out as widely different from the rest of the data set. These particular values, whether they arise naturally or through errors, are referred to as *outliers*. One should always make as much effort as possible to explain outliers, and one of the most effective methods by which they can be detected is through the various data description techniques that are available. In the last chapter, I recommended keeping a diary for an investigation which can provide an invaluable instrument in the explanation of outliers. Entries in the diary may act as reminders of possible explanatory events. An example of the usefulness of maintaining a project diary was related to me by a colleague who was investigating various aspects of grassland management. As part of his study, he regularly took samples of soil from an experimental enclosure and later analysed them for nitrogen content. On one occasion, one particular sample was found to have an extremely high nitrogen concentration. Referring back to the diary for the relevant dates of sampling, an entry was found that recorded that cattle from a neighbouring field had invaded the experimental plots through a broken fence. A very plausible explanation of the high nitrogen recording would be that the soil sample had been taken inadvertently from a urine patch created by one of the bovine intruders.

Thirdly, we shall see later that many analytical techniques are based upon assumptions about the data to which they are applied. Clearly, therefore, before conducting any analysis, we have to ensure that such assumptions have not been violated. An important part of this validation procedure is achieved by examining the data descriptively.

Finally, the task of scientists is to ask questions, and no scientist is likely to analyse a data set according to a predetermined plan and then discard it without conducting further explorations to discover any other interesting relationships that might exist within it. Exploratory data analysis is therefore an essential first step immediately after data entry.

This process of data exploration, sometimes referred to as exploratory data analysis (EDA) or initial data examination (IDE), has received much attention in recent years, particularly since the arrival of desktop computers (Tukey, 1977; Velleman and Hoaglin, 1981; Chatfield, 1988). As a consequence of this attention, new techniques have been developed that aid us in our explorations. Essentially, data exploration requires us to look at data sets, or sections of them, in a variety of ways, summarising our observations, attempting to distil the essential *wood* from the peripheral *trees*. Therefore, exploring and summarising data are, in many respects, very similar. They both reduce what may be large diffuse collections of observations into a series of diagrams, tables, and summary numbers that enable us to get a *feel* for the data. They focus our attention on the essential elements of our investigation and they can reveal phenomena that are not immediately obvious from the whole set of raw measurements.

In addition to helping us increase our familiarity with our data, producing summaries also has another function. After we have analysed the data and drawn our conclusions, we need to communicate those findings to a wider audience. We must present our results and conclusions in as simple a form as possible, which will necessarily require succinct summaries. Therefore, many of the techniques of summarising data are common both to data exploration and to data presentation. The three activities that comprise the title of this chapter, exploring, summarising, and presenting data, are therefore inextricably linked together. The objective of this chapter is to introduce the basic techniques available to biologists to perform these tasks, and, in doing so, I hope to demonstrate the strong links that exist between data types, problem types, and methods of data description. This chapter is organised into three main sections. In Section 2.2, I shall describe some of the basic methods that Minitab provides for data organisation and manipulation. In the two following sections, I will introduce numerical and graphical methods of data description, explaining the circumstances where each technique is most appropriately employed.

2.2 Organising data using Minitab

This section introduces some basic manipulative skills that will enable you to organise data on a computer. The link between data organisation and the structure of an investigation, particularly the mechanics of data collection, is very important. The design of recording sheets for use in the field can be tailored to fit a particular method of data entry. And, in these days of automatic data loggers, the link between data collection and data entry is even closer. Once the data are stored, we may wish to select particular subsets for analysis, such as separating data referring to males from those relating to females. Or we may need to perform simple calculations on the raw data, such as finding the square root of each value or adding together the contents of two columns. By mastering these and other basic skills, it is possible to make the process between data collection and analysis extremely swift and efficient.

Some of the techniques introduced here, such as those concerned with data entry, are clearly essential; others may not appear to be quite so important. However, I have found that time spent practising simple manipulative techniques fulfills two very useful functions. First of all, it provides a means of introducing users to Minitab, or even to computers, without simultaneously asking them to absorb a deluge of new statistical principles and operations. Many students find difficulty with statistics and/or the use of statistical packages and I have found it useful to separate the introduction of novel computing from new statistical principles wherever possible. Secondly, many simple manipulative and arithmetic operations are part and parcel of an investigation and so these simple exercises provide an informative review of some of the many techniques that are available within Minitab (and most other statistics programs as well). I do recognise that the most effective way to learn a new technique is by practice, particularly on data that are relevant to the learner; rest assured, you will encounter all of the following techniques in the context of real problems in subsequent chapters. My hope is that by working through these brief exercises you will remember that a particular operation is possible, or remember that several useful techniques are described in this chapter, and, therefore, know where to look as a first source of information.

In order to perform the exercises in this book you will need to have access to a computer on which Minitab is installed. This statistical program is available on many different types of computers from small personal computers (such as Apple Macintosh and IBM compatibles) up to very large multi-user machines. Whatever computer is available to you, familiarise yourself with its basic operations such as switching it on, gaining access to stored programs and data (including username and password procedures where appropriate), and how to terminate a computer session. You should also find out where the printers associated with your computer are located. If you are not very experienced with the operation of your computer, then you should seek assistance from a computing service staff if you work in a business or educational establishment and ask for copies of any appropriate introductory guides that may be available. If you are using your own computer, you will need to consult the owner's handbook provided by the manufacturer.

Before introducing Minitab itself, we need to review some basic concepts of computing in general. We need to know how information is stored on a computer and how to identify our various sets of data so that we can recall them without difficulty whenever we wish. Fortunately, the principles of data storage are fairly similar for all computers.

Stored information is organised into portions of the disks called *files* or *documents* that fulfill a similar function to files in a normal office filing cabinet. In a similar fashion to when you create a new file for storing in a filing cabinet, a computer file must be given a name which identifies it uniquely. There are commands on all computers that will produce, on the computer screen or printer, a list of files that have been created and are accessible, and you will need to be conversant with this command on your particular computer.

Filenames generally comprise two sections: a first section that is invented by the person who creates the file, and a second portion or *file-ending* (sometimes referred to as the *filetype*) that, generally, is produced automatically by the program through which the file is created. On many computers, the two sections are separated by a full stop or period, which is an integral part of the file-ending on many computers. Various types of files can be created by Minitab and each type has a different ending representing a different function. We shall encounter some of these filetypes in this book and each will be described as it appears. An example of a filename would be **MYDATA.MTW**, where **MYDATA** is the filename chosen by the computer user, and **.MTW** is generated by the computer program.

Initially, you should be aware of two basic types of files: *data* (the computerised version of our observations), and *programs* (sets of instructions that the computer will obey slavishly). The major example of a program that concerns this book is Minitab, which performs data handling and analysis functions. Within Minitab (and many other programs), data files are often referred to as *worksheets* and Minitab worksheets have the file-ending .MTW, which is derived from the words, "MiniTab Worksheet." We shall encounter many examples in this book where data are manipulated within Minitab to produce graphs, summary statistics, and analyses. These related elements of an investigation can all be grouped together by Minitab in a single file called a *project*, which is identified by the file ending **.MPJ**.

If we continue with the filing cabinet analogy, the computer equivalent to a drawer in a filing cabinet is called a *directory* or *folder* (the use of these names varies according to computer system), which holds a number of files. In fact, directories can also contain sub-directories, which, in turn, can contain further sub-directories, all classified in an hierarchical system. On many computers you will find a main Minitab directory that contains both the program files and also a data subdirectory. This subdirectory holds many individual data files, some of which will be used in later examples.

Once you have gained access to your computer, you will need to start Minitab. The way you do this will depend upon the computer available to you and its operating system. However, there are two general methods that will be found on many different types of machines and the difference between them lies in the way that the user interacts with the computer, either by entering commands from a keyboard or by pointing to *icons* with a mouse.

The objective of this book is to help you design and carry out biological investigations using Minitab and this section introduces some simple Minitab commands; later chapters include commands to perform specific analyses. For extended detail of all commands, you should consult the main *Minitab User Guide and Reference Manual* and on-line **HELP** command. If you are in an educational establishment your computing service may also have produced simplified manuals that may be useful.

Generally speaking, users of Minitab (or any other statistics package) need to provide two types of input, namely *commands* and *data*, in order to perform their required analysis. Commands can be passed to the program by two possible methods: either typed directly via the keyboard, or by selecting commands from menus using the mouse; the two methods are quite interchangeable. The keyboard commands are the same whatever computer is being used. However, there are some differences in the menus on different computers. To reduce confusion, where I have depicted menus, only those that would appear on IBM-PC compatible computers are shown. The opening screen, which appears when Minitab is started is shown in Exhibit 2.1 and the menu headings are to be seen at the top of the screen. Keyboard commands are entered at the **MTB >** prompt, which can also be seen in Exhibit 2.1. The vertical line to the right of the prompt is the current cursor position that means Minitab is awaiting a command from the user.

Data are simply entered into the appropriate cell of the worksheet using a combination of the keyboard, mouse, and/or arrow keys. Data can easily be edited or replaced by retyping with the new data. The **Edit** menu also enables you to insert or delete columns, rows, or cells.

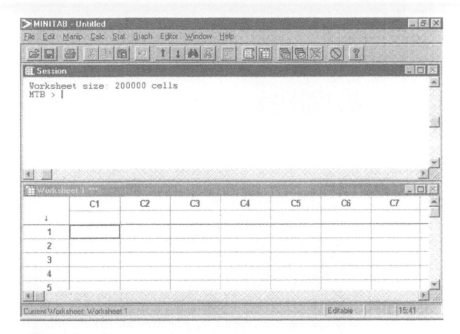

Exhibit 2.1. Minitab startup screen: main session and data windows. Note the menus at the top of the window and the **MTB** > prompt at which keyboard commands may be entered. A Minitab project may include several worksheets.

Although a number of exhibits in this book will illustrate how to select various commands through accessing menus, to show full screen displays on all occasions would create an excessively long book, which would be dominated by pictures of Minitab menus. Therefore, in later chapters, menu choices will generally be described in a shorthand form. For example, the process of opening or retrieving a file, which is already stored is achieved by accessing the **Open Worksheet** option in the FILE menu (as illustrated in Exhibit 2.2.). The shorthand style followed in later examples would show this operation as:

FILE> Open Worksheet

Even if you use menu-based commands, Minitab generates the corresponding keyboard commands on the screen. This enables you to save commands into files, which can be used to repeat a sequence of often-used commands. The output of any operations will be displayed in full and annotated to provide a full explanation. An example of how the Minitab screen appears when a file is opened is shown in Exhibit

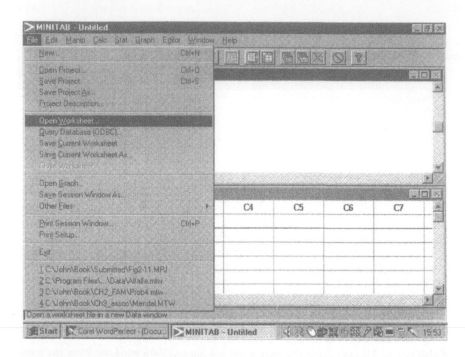

Exhibit 2.2. The **FILE** menu, indicating the option to open a previously stored Minitab data file or worksheet.

2.3. The exhibit also shows the keyboard command, **RETRIEVE**, to open a stored data file (referred to by Minitab as a *Worksheet*). This command is followed by the filename enclosed in single quotes. The reverse operation, to store data in a file on disk, takes a similar form using the command, **SAVE**. The menu item that performs this operation is shown in Exhibit 2.2. as **Save Current Worksheet**.

The worksheet information listed in Exhibit 2.3. shows the details of eight columns, each of which has been named. Data can be stored within Minitab as either columns, constants (single values), or matrices. Each data storage type is identified by a letter and a number; **C** for columns, **K** for constants, and **M** for matrices. Names can be added to all data types and this should be done to make the results of analyses more easily understood. A column can be given a name by entering it at the head of the column in the worksheet immediately below the column number.

Many Minitab commands can be modified or refined by subsidiary instructions or **SUBCOMMANDS**. Examples would include commands acting only on specific rows in a column, or, in graphics com-

```
MTB > Retrieve  "C:\Program Files\MTBWIN\Data\Pulse.mtw".
Retrieving worksheet from file: C:\Program Files\MTBWIN\Data\Pulse.mtw
Worksheet was saved on 12/23/97 12:42:06


Current worksheet: Pulse.mtw

MTB > Info

Information on the Worksheet


Column  Count  Name
C1         92  Pulse1
C2         92  Pulse2
C3         92  Ran
C4         92  Smokes
C5         92  Sex
C6         92  Height
C7         92  Weight
C8         92  Activity

MTB >
```

Exhibit 2.3. Worksheet information as displayed by Minitab after a file, *PULSE.MTW*, had been opened. The command **INFO** lists the names and numbers of all columns, constants, and matrices that are stored in the file.

mands, choosing specific fonts for axis labels. While most users will access these subcommands via dialog boxes, it is useful to see a simple example in order to understand the listing of commands that Minitab generates when a command is given by the menu route. A simple example based upon the **COPY** command illustrates the general principles of subcommands.

Imagine that we have a column of data that ranges in value from 0 to 10 and we wish to carry out an operation on those values less than 5. We can copy the data from the original column to a new column, but the copy command can be modified so that only the required data are transferred. The whole process is illustrated in Exhibit 2.4. Notice in this exhibit the semi-colon and full stop in the copy command. These are essential components of any command with subcommands. The semi-colon indicates that a subcommand is to follow (a command may be followed by several subcommands), and the period terminates the composite command (main command and associated subcommands). The colon between the 0 and 4 specifies a range of values, in this case 0 to 4, inclusive; this is the conventional way that ranges of values

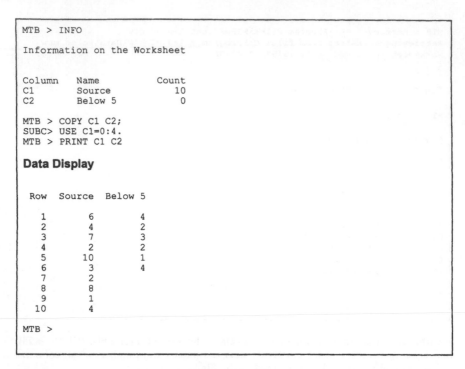

```
MTB > INFO

Information on the Worksheet


Column    Name              Count
C1        Source               10
C2        Below 5               0

MTB > COPY C1 C2;
SUBC> USE C1=0:4.
MTB > PRINT C1 C2

Data Display


 Row   Source   Below 5

  1        6        4
  2        4        2
  3        7        3
  4        2        2
  5       10        1
  6        3        4
  7        2
  8        8
  9        1
 10        4

MTB >
```

Exhibit 2.4. The **COPY** command with its subcommand **USE** copies the data stored in Column 1 to Column 2, transferring only the data that fall into the range 0–4.

(or rows of data) are specified in Minitab. In addition to **COPY**, this exhibit also includes the **PRINT** command, which causes the data specified in that command (in this case columns 1 and 2) to be listed on the screen.

The alternative way of executing the copying operation is to use the menu-based method as illustrated in Exhibits 2.5., 2.6., and 2.7. This menu-based method is shown here in full in order to show the form of dialog boxes that provide an alternative way of entering subcommands. While these dialog boxes may seem long-winded, they do prompt the user for the correct inputs. Therefore, if you are unsure of which subcommand is appropriate, they are probably the most efficient way of constructing complex commands for the inexperienced Minitab user. Remember also that all Minitab instructions, irrespective of how they have been entered into the computer, appear on the screen in their text-based form as an accurate record of the commands issued.

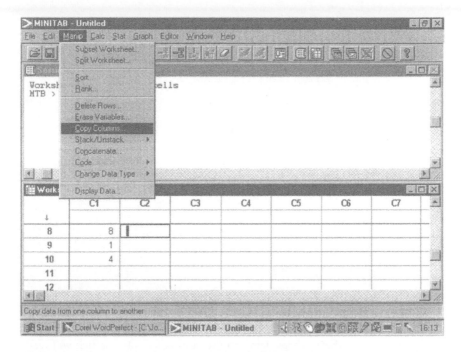

Exhibit 2.5. Menu-based system of using the **COPY** command.

As with all major computer programs, Minitab includes an on-line **HELP** system that provides detailed instructions about the various commands. It can be accessed through the main **HELP** menu (see Exhibits 2.1., 2.2., and 2.5.). Furthermore, there is a context-sensitive **HELP** facility that appears in every dialog box. This latter feature provides assistance with the procedures associated with the command currently being used. It is accessed by clicking on the button marked **HELP** in a dialog box (see Exhibits 2.6. and 2.7.).

The beauty of computers is that once data have been entered and stored, they do not need to be entered again. If data have to be presented in a variety of formats, all further manipulations can be performed by computer commands. The **COPY** command is one such manipulative command. The pair of Minitab commands, **STACK** and **UNSTACK**, to be introduced in this chapter also fulfill manipulative functions, which are frequently required. As their names suggest, these two commands perform opposite functions. **STACK** will combine two columns (or sets of columns), while **UNSTACK** separates columns into subsets of data. Their respective functions are illustrated in the

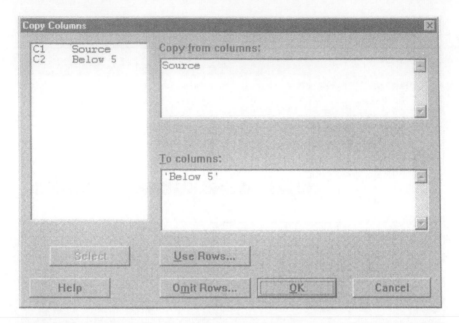

Exhibit 2.6. The first dialog box of the **COPY** command in which the user selects the source columns from which to copy data and the destination columns that will receive data. Notice the **USE ROWS** and **OMIT** rows options. Column 2, named *Below* 5, requires inverted commas in the **To Columns** box because it contains a number in the name.

following example. As part of a physiological experiment, the serum glucose levels of ten rats (5 males and 5 females) were measured and entered into Minitab in a single column (**C1**) with a second column holding codes that indicated whether a particular measurement was from a male or female (1 = male, 2 = female). Subsequent to data entry, we might wish to rearrange the observations in two separate columns, one for each sex. The command **UNSTACK** is used to do this as shown in Exhibits 2.8 and 2.9.

Exhibit 2.10. shows the commands that would appear on the screen, and it also shows the reverse process, using **STACK**. As its name suggests, this command combines columns. It can also take a subcommand that creates a subscript column in which is held codes signifying the subgroups from which each data item originated. Referring to the on-line **HELP** and the exhibits shown here, generate the display shown in Exhibit 2.10. using the menu-based **STACK** commands.

The final group of operations to be introduced here falls under the general heading of simple arithmetic. Frequently, we may need

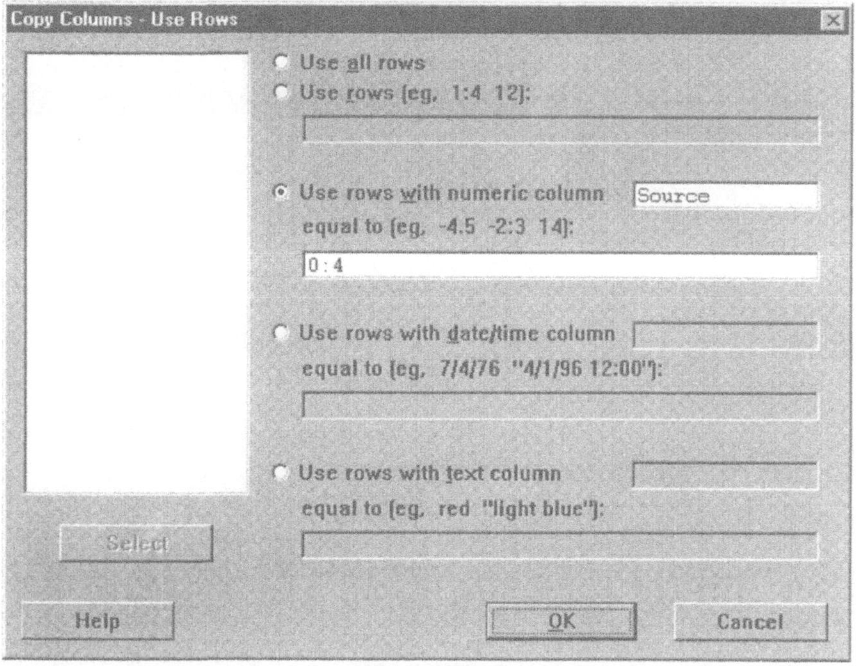

Exhibit 2.7. The final dialog box associated with the **COPY** command in which the user can specify the selection criteria to determine which values are copied.

to manipulate data arithmetically and, like all statistics packages, Minitab provides a straightforward method of performing these tasks. The command used is **LET** and it takes a very simple form as shown in Exhibit 2.11. In this example, the contents of columns 1 and 2 are added together creating a third column (**C3**). There is a wide variety of built-in mathematical functions, and you should take the time both to explore what is available and also to discover how to perform this sort of operation. The various built-in functions can be found in the calculator function that can be accessed via the following menu route:

CALC > Calculator

The exercises at the end of this chapter provide you with some practice at manipulating data in this way.

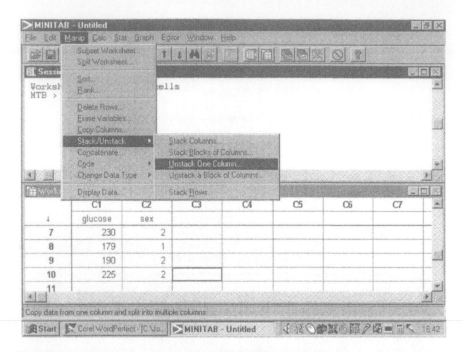

Exhibit 2.8. The menu to access the **UNSTACK** command.

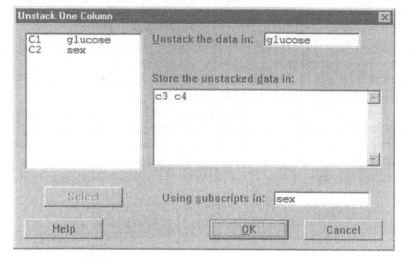

Exhibit 2.9. The Minitab dialog box in which we specify the columns (C3 and C4) into which we wish to unstack the glucose data. Column 2 (*sex*) acts as a code, indicating from which gender each glucose measurement has been obtained.

```
MTB > Unstack 'glucose' c3 c4;
SUBC>    Subscripts 'sex'.
MTB > Stack C3 C4 c5;
SUBC>    Subscripts c6.
MTB > print c1-c6
```

Data Display

Row	glucose	sex	C3	C4	C5	C6
1	185	1	185	192	185	1
2	206	1	206	198	206	1
3	210	1	210	230	210	1
4	192	2	192	190	192	1
5	192	1	179	225	179	1
6	198	2			192	2
7	230	2			198	2
8	179	1			230	2
9	190	2			190	2
10	225	2			225	2

```
MTB >
```

Exhibit 2.10. An example of the **UNSTACK** and **STACK** commands. The former splits Column 1 into two new columns, using the values in Column 2 to determine into which destination column each datum in Column 1 should be copied. The reverse process is performed by **STACK**, and, in this case, a new subscript column (C6) is created to indicate the origin of each value in the new Column 5.

```
MTB > LET C3 = C1 + C2
MTB > PRINT C1-C3
```

Data Display

Row	C1	C2	C3
1	25	1	26
2	26	4	30
3	24	5	29
4	23	7	30
5	25	8	33
6	28	6	34
7	27	3	30

```
MTB >
```

Exhibit 2.11. A simple example of the **LET** command. More complex operations can be entered directly or achieved using the **CALCULATOR** function.

2.3   Numerical methods of data description

One of the principal tasks of scientists is to ask questions in an objective manner, and so we must at all times attempt to minimise the potential for misinterpreting results. Later in this chapter, we shall see various graphical methods that we can use to describe and present data. On many occasions pictorial methods provide an extremely clear and unambiguous description. However, we may encounter problems where small, but biologically significant differences, trends, or associations may exist, but, because of their subtlety, such relationships are not readily discernible from a picture. In such circumstances, numerical methods of data description provide a more unambiguous method.

Listing and tabulating data

At the simplest level, we could merely produce lists of raw data on a computer screen or on paper. While this activity may form a key element in the process of data validation immediately after the data have been entered into the computer, listing numbers fulfills very little purpose in terms of the presentation or description of data unless the lists are quite short. The simplest quantitative method that can be employed usefully to describe data is to construct tables, and we shall see that tables can be used effectively with all data types.

At their simplest, tables contain simple counts or frequencies of nominal variables. As such they would comprise a single column with several rows of frequencies. For example, the frequencies of blood groups for a sample of 1,687 individuals in the UK is tabulated below (Table 2.1.). Where the number of categories is small, the data can be summarised very succinctly in a table; difficulties do arise when the reader is required to interpret very large tables with many categories. The general rule is to ensure that you are quite clear of the function you

Blood Group	Frequency
O	727
A	755
B	147
AB	58
Total	1687

Table 2.1. Frequencies of blood groups in a sample of UK citizens.

Blood Group	Male	Female	Total
O	·342	385	727
A	380	375	755
B	65	82	147
AB	31	27	58
Total	818	869	1687

Table 2.2. Cross-tabulation of blood groups by gender.

wish the table to perform. Make it as simple as possible, and, if your table or other method of data summary does not portray the point clearly, you should question whether you are employing the best method.

You may encounter the term *cross-tabulation* in various texts. This term refers to the process of relating two or more nominal variables by means of a table, the frequencies of the various combinations of the cross-tabulated variables forming the main body of the table. For example, the survey of blood groups also included the gender of each individual and so we can cross-tabulate blood group with sex (Table 2.2.). What we should remember at all times, though, is that the construction of the table is still a *descriptive* operation. It does not tell us whether smoking is particularly associated with one sex or not (the method by which this can be achieved will be described in the chapter on problems of association).

Two examples of how Minitab enables us to construct tables are shown in Exhibit 2.12. The examples are based upon the pulse data set that was used earlier, and the tables illustrate the frequencies of males and females in the sample as well as the smoking habits cross-tabulated with gender. The menu-based route to constructing tables using Minitab is:

STAT > Tables > Cross Tabulation

Tables are not confined to containing only frequencies, nor is their use restricted to nominal variables. We will see later in this chapter that they are often used to list summary statistics, where they provide a very important medium for presenting numerical data.

Measures of central tendency

In everyday conversations, we have all encountered phrases of a kind similar to: *the normal resting pulse rate of adult humans is 72*

```
MTB > Table 'Sex'.

Tabulated Statistics


 Rows: Sex


         Count

 1         57
 2         35
 All       92

MTB > Table 'Sex' 'Smokes'.

Tabulated Statistics




 Rows: Sex    Columns: Smokes

           1        2      All

 1        20       37       57
 2         8       27       35
 All      28       64       92

  Cell Contents --
                 Count

MTB >
```

Exhibit 2.12. Examples of tables constructed using Minitab. The coding of the categories is as follows: Sex (1 = male, 2 = female) and Smokes (1 = smokes regularly, 2 = does not smoke regularly). The lower table summarises the relationship between smoking habits and gender: of the 57 men in the sample, 20 are smokers and 37 do not smoke.

beats.min⁻¹ (I used this phrase in Chapter 1). Similarly, statements of this kind can be used in a comparative sense: "… on average group *a* is larger than group *b*." In each statement, the meaning that is being expressed is that a group of measurements tend to be centred or located around a particular value. This gives rise to the general heading of measures of *central tendency,* which are also called *averages.* There are three main measures that fall under this heading: the *mean,* the *median,* and the *mode.*

The Mean: This measure is what most non-scientists understand by the term *average.* If we have a sample of measurements, we calculate the mean by adding all the values and then dividing the total by the number of measurements comprising the sample. So, if we have eight values (8, 9, 10, 11, 12, 12, 13, and 15) the mean would be calculated thus:

$$8 + 9 + 10 + 11 + 12 + 12 + 13 + 15 = 90$$

$$90 \div 8 = 11.25$$

In statistics, we represent the mean by the following formula:

$$Mean = \frac{\sum_{i=1}^{i=n} x_i}{n} \tag{2.1}$$

The notation in this equation may be unfamiliar (even though the calculation may not be new) and so it needs to be explained. We can see that the equation is in the form of a fraction with the letter n as the denominator, representing the number of measurements in the sample.

The numerator appears to be more elaborate, but remember that it is merely a mathematical way of saying the sum of all the measurements. First of all, let us deal with x_i, which represents each of the individual measurements. Imagine that each of the values in our sample has been given an identifying number (1, 2, 3, 4, 5, 6, 7, and 8) each of which, in the equation, is represented by the subscript i:

8	9	10	11	12	12	13	15	*measurement*
1	2	3	4	5	6	7	8	*identifying number(i)*
x_1	x_2	x_3	x_4	x_5	x_6	x_7	x_8	

Returning to the calculation of the mean, we have to add all the values. The symbol Σ (the Greek upper case letter *sigma*, equivalent to the letter *S*) is a shorthand way of saying *summation*, so Σx_i indicates that values of x_i have to be added together. The $i = 1$ and $i = n$ below and above the sigma indicate the range of values of i over which this summation should be carried out for every value of x_i where i lies in the range *1* to n inclusive (i.e., 1 to 8 in this example). In this case, it shows that all values of x_i should be included in the summation.

The equation that has been introduced here is shown in its most detailed form, but, because the calculation of the mean is generally carried out on all values, you will very often encounter the formula in a simplified form without the subscripts and superscripts ($i = 1$ and $i = n$). Thus, if you see a sigma in an equation with neither subscript nor superscript, you know that summation should use all possible values (Equation 2.2).

$$mean = \frac{\Sigma x}{n} \qquad (2.2)$$

Generally, we have to take a sample from the population under investigation, and, conventionally, the sample mean is denoted by an x with a bar above it, $\bar{x}$. In order to distinguish the sample mean from that of the population, the latter is generally denoted by the lower case Greek letter mu, μ (equivalent to the letter m). If we had taken two sets of measurements and denoted them by the letters x and y, then the population and sample means would, respectively, be denoted thus: μ_x, μ_y, $\bar{x}$ and $\bar{y}$.

Although statistics involves a lot of calculations, which, thankfully, the computer performs for us, the majority are not very taxing in terms of complicated mathematical techniques. Most involve only addition, subtraction, division, and multiplication. Therefore, the symbol, sigma (Σ), will be encountered very frequently and you should become familiar with it.

Since the mean is used so much in common parlance, we should not be surprised that it is extremely important as a data descriptor both in the sense of data exploration and data presentation. However, there are limitations to its use. For example, if one or a small number of measurements in the sample are considerably different from the rest, they will exert an undue influence on the value of the mean. For example, taking the set of eight values used above to illustrate the calculation of the mean and replacing the value of 11 with 26, would change the mean from 11.25 to 13.125, a 16% increase in the original value. Also, now only two of the measurements in the modified sample lie above the mean (26 and 15), and all the rest fall below this measure of central tendency. The mean is not really performing its required function; intuitively, it would appear to be rather biased as an estimate of central tendency.

On these occasions when we have outliers or skewed distributions, we generally have to utilise other measures of central tendency. One modified version of the arithmetic mean is the *trimmed mean*. This is calculated in exactly the same way as the method described above with the exception that portions of the data at the top and bottom of the range are removed prior to the calculation. In Minitab, the top and bottom 5% (rounded to the nearest integer) of the values are removed so, in this example, no values are removed at all (5% of 8 is 0.4, which, when rounded to the nearest integer, is equal to zero). This measure can sometimes overcome the problem of outliers and enable us to calculate a more meaningful value for the mean. However, in cases where outliers are not present, or only occur at one end of the range of data, we would discard useful information, so use of the trimmed mean is rather limited.

The Median: This measure is used when data are not distributed symmetrically (i.e., the measurements are clustered more at one end of the range than the other) because it is not overinfluenced by outliers. It is defined as the middle datum point of a data set that has been sorted into arithmetic order (i.e., ascending from smallest to largest, or vice versa) and, if there is an even number of values in the data set, then the median is taken as halfway between the two middle values. Again using the small sample of eight values the median value is 11.5, midway between 11 and 12:

$$8 \quad 9 \quad 10 \quad 11 \quad 12 \quad 12 \quad 13 \quad 15$$

In the modified sample, the median becomes 12, the two central values being identical:

$$8 \quad 9 \quad 10 \quad 12 \quad 12 \quad 13 \quad 15 \quad 26$$

The value of the median has moved very little in response to the outlying value of 26, whereas the mean changed to 13.125. Furthermore, if the number of outliers at each end of the range of data is similar, then the median will not move at all while the magnitude of the response of the mean will depend upon the magnitude of the outliers. The median is more robust to the effects of extreme values than the mean. Therefore, it is a very useful measure of central tendency when the data are not symmetrically dispersed around a central value.

The Mode: The third measure is the mode, which, although it provides a quick and easy method of gaining an impression of the central tendency, it is the least useful of the three measures. It is defined as the most frequent value in a sample. So, in the example data sets used to describe the other measures, the mode is equal to 12, that value being represented twice in both the unmodified and modified sample. Clearly, in the context of small samples (in this instance, $n = 8$), the usefulness of the mode is questionable because the modal value occurs twice, with the others occurring only once. Moreover, some data sets contain two or more modal values, each of which should be cited when presenting modes. If a data set is multi-modal, it should be examined to discover the origins of the separation into groups around each mode. Multi-modal data sets are generally indicative of several subsets of data caused by, for example, differences between sexes or between different life stages (juveniles and adults). In this particular context, the existence of modes provides useful prompts or questions about the underlying composition of the sample or population under examination.

An additional area where modal values are of use is when we deal with frequencies of nominal variables, such as in the blood group example shown in Table 2.1. Looking at such data, we can use the modal value to give an approximate guide to the predominant blood group in the population. In this particular example the modal group is Group A, but it is not much more numerous than Group O, thus showing that modes can only give approximate estimates, and nothing more than that!

Of all these measures of central tendency, the mean is the most important, and it is worth noting that, as sample size increases, the influence of a few extreme values diminishes. Also, as we shall see in the next section, the mean is closely related to some important measures of the variability or dispersion of data.

Measures of spread

The mean, median, and mode are three ways of expressing where the bulk of data in a sample or population occur on a scale of measurement. However, such measures do not provide a complete description of a data set. Consider the two histograms Figures 2.1.(a) and (b), which represent two distributions which share the same means, medians, and modes. They differ from each quite markedly in the degree to which the data vary around the central peaks. The importance of variation was introduced during the discussion of hypothesis testing in Chapter 1 and we have also just seen that outliers can influence measures of central tendency. Various methods of measuring the variability or dispersion of data will be introduced in this section and their utility will be illustrated.

Range: This is the most straightforward measure of variability and is simply the difference between the maximum and minimum values in a data set. It is a very straightforward measure of spread. Its calculation can be illustrated using the small sample from the calculations above:

8   9   10   11   12   12   13   15

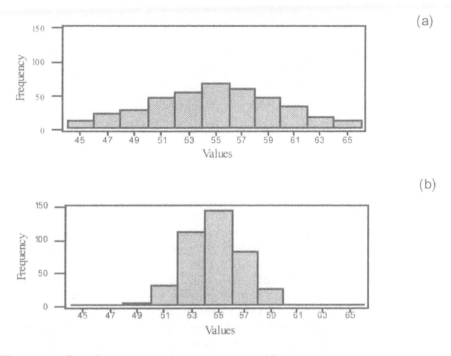

Figure 2.1. Two distributions of measurements with similar means, medians, and modes. Clearly, they differ considerably with respect to the degree to which the measurements spread around the mean.

The maximum and minimum values are 15 and 8, respectively, and the range is the difference, i.e, 7.

Although the range is influenced by extreme values, it is particularly useful where the frequency distribution is asymmetrical, hence it is used very often in conjunction with the median.

Interquartile Range: This measure was developed in order to overcome some of the deficiencies of the range. It is calculated as the difference between the maximum and minimum values in the data set after the top and bottom quartiles (or quarters) have been excluded, thereby mimicking the principle underlying the trimmed mean by removing the most extreme values of the data set. The truncation reduces the influence of extreme values, but it does make an arbitrary modification to the data set irrespective of whether there are outliers or not. There is a loss of information which may or may not be important. With the data set above, the interquartile range would be equal to 2, the 8, 9, 13, and 15 having been removed as the lower and upper quartiles, respectively (i.e., it is the central 50%).

Another, more general point relating to both the range and the interquartile range is that these two measures are calculated by reference to only the maximum and minimum values of either the full sample or the truncated group. No reference is made to the degree to which each of the values clusters around the midpoint of the distribution. In contrast, an ideal measure of dispersion should be based upon every value in the data set because, by so doing, a more complete or faithful representation of the data set can be achieved.

Variance: This is the basis of the most important measures of dispersion that can be applied to a normal distribution. It satisfies the constraint outlined above in that its calculation utilises all the data, and, therefore, minimises the influence of outliers.

Given that the variance is the source of several measures of dispersion, and is, itself, the basis for an extensive family of analytical techniques, I shall develop the method of its calculation in some detail. To grasp the meaning and method of calculating the variance now will facilitate an understanding of many other important techniques later on.

A measure of dispersion should possess several properties, notably it should be based upon all values and it should somehow summarise the degree to which the values differ from the mean. For an entire set of measurements the variance, in essence, calculates the average deviation from the mean. We might simply calculate the difference between each value and the mean, but, since the values above the mean will be balanced out by those values below, such a calculation would produce a value of zero. We can overcome this difficulty by squaring all of the differences as shown in Equation 2.3. The squared differences are then summed and divided by the number of values to give an *average squared deviation or variance*. Sometimes this is referred to as the *mean square error* (a term that we will encounter in Chapter 7 when we look at methods to compare several groups) and it is generally denoted using the lower case form of the Greek letter sigma (σ^2):

$$Variance \ (\sigma^2) = \frac{\sum_{i=1}^{i=N}(x_i - \mu)^2}{N} \tag{2.3}$$

Equation 2.3 represents the variance of a *population* (with μ and N representing the population mean and size, respectively) but, as we noted in the discussion of the mean, we are rarely in a position to measure or census every individual in a population. We generally take a *sample* from which we draw conclusions and extrapolate to make general statements about the wider population from which the sample was extracted. The function of a sample is to provide an unbiased representation of the population. Although we may be extremely careful when designing our sampling strategy, whenever a population parameter (e.g., population mean or variance) is estimated by a sample statistic (e.g., sample mean or variance), there is generally some error. That is one of the reasons why we have to study statistics. The significance of this error decreases as the size of the sample increases. The logic of this relationship between sample size and error is reasonably straightforward. With very large sample sizes, the sample will approximate more closely to the population, and with small sample sizes, this error will become more important. Fortunately, for the calculation of sample statistics, a modified version of the variance has been developed, which takes into account the error due to sample size. The formula for the sample variance is given by Equation 2.4, where the denominator is $n - 1$ and not n ($n - 1$ is also referred to as the *degrees of freedom,* a term that will be encountered frequently in subsequent chapters). Notice that, for the sample variance, sigma is replaced by the letter s thereby distinguishing the sample variance from that of a population (in the same way as x was used to represent the sample mean and μ denoted the population mean).

$$Sample\ \ Variance\ (s^2) = \frac{\sum_{i=1}^{i=n}(x_i - \bar{x})^2}{n\ \ 1} \tag{2.4}$$

The impact of this subtraction $(n - 1)$ will be greatest when the sample size (n) is small. This can be illustrated by contrasting the relative effects on $n - 1$ of this subtraction when $n = 10$ and when $n = 100$. In the former case, the modification alters the denominator by 10%, while in the latter the change is only 1%.

For computational purposes, these two formulae are not particularly convenient because they require the mean to be calculated prior to calculating the variance. The following two versions of these formulae are much easier to use:

$$Population\ \ Variance\ (\sigma^2) = \frac{\sum_{x=1}^{x=N}x_i^2 - \frac{\left(\sum_{x=i}^{x=N}x_i\right)^2}{N}}{N} \tag{2.5}$$

$$\textit{Sample Variance } (s^2) = \frac{\sum\limits_{i=1}^{i=n} x_i^2 - \dfrac{\left(\sum\limits_{i=1}^{i=n} x_i\right)^2}{n}}{n-1} \qquad (2.6)$$

Examples of the calculation of both population and sample variances are shown in Boxes 2.1. and 2.2. These example calculations also demonstrate the effect on the magnitude of the variance of using $n - 1$ rather than n. They are based upon the same data that were used for the calculation of the mean and median.

While the variance possesses the most essential property of an ideal measure of variability, namely that it is based upon all data in a sample or population, it has one major disadvantage. You will recall from the development of the method of its calculation (Equation 2.3.) that we had to square the difference between each measurement and the mean

x	x^2
8	64
10	100
12	144
12	144
13	169
9	81
11	121
<u>15</u>	<u>225</u>

$\Sigma x = 90 \qquad \Sigma x^2 = 1048 \qquad n = 8$

$(\Sigma x)^2 = 8100$

$(\Sigma x)^2 \div n = 8100 \div 8 = 1012.5$

$\Sigma x^2 - ((\Sigma x)^2 \div n) = 1048 - 1012.5 = 35.5$

Population Variance $(\sigma^2) = 35.5 \div n = 35.5 \div 8 = \underline{4.44}$

Box 2.1. Calculation of the population variance. The formula used is Equation 2.5.

x	x^2
8	64
10	100
12	144
12	144
13	169
9	81
11	121
<u>15</u>	<u>225</u>

$\Sigma x = 90$ $\qquad$ $\Sigma x^2 = 1048$ $\qquad$ $n = 8$

$(\Sigma x)^2 = 8100$

$(\Sigma x)^2 \div n = 8100 \div 8 = 1012.5$

$\Sigma x^2 - ((\Sigma x)^2 \div n) = 1048 - 1012.5 = 35.5$

Sample Variance $(s^2) = 35.5 \div (n - 1) = 35.5 \div 7 = \underline{5.07}$

Box 2.2. Calculation of the sample variance; the same data are used as in Box 2.1., but treated here as a sample. The formulae used to calculate the sample variance is Equation 2.6.

in order to make all differences positive. As a consequence, the units in which the variance is expressed are the *square* of the units of the mean so, for example, mean height would be expressed in *metres* while the variance would be measured in *square metres*. Such a disparity in the units of two descriptors of the same data is confusing and so the variance is not used for presenting or summarising data; other measures are used for that purpose. However, the variance is important as an intermediate entity in calculations and, as we shall see later, forms the basis of a large family of important analytical techniques.

Standard Deviation: This measure of variability was developed to overcome the problem of the units of the variance. The standard deviation is simply the square root of the variance, and so is expressed in the same units as the mean (Equation 2.7).

$$Standard \ \ Deviation = \sqrt{VARIANCE} \qquad (2.7)$$

The standard deviation, being in the same units as the mean is particularly useful for descriptive and presentation purposes. However, for computational work, the variance is far more useful. In scientific journals, you will often see data summarised in the form of tables of results including means and standard deviations, thus illustrating the central tendency and degree of spread of the measurements. Following the same conventions for notation as the mean and variance, the population standard deviation is represented by σ and the sample standard deviation is denoted by *s*.

Coefficient of Variation: The standard deviation expresses the spread of a frequency distribution or data set in the same units as the mean. However, by itself, a stark statement of the standard deviation is not particularly meaningful. If we wished to compare the variability of two closely-related species that differed greatly in size (such as blue whales and their smaller relatives the minke whales), their respective standard deviations would differ tremendously. As a general rule, standard deviations should be stated in the company of the mean and the sample size upon which these two statistics were based.

Often it is useful to express the dispersion relative to the mean, and this is the function of the *coefficient of variation.* For example, a spread of 10 in 100 is relatively greater than a spread of 10 in 10,000. The coefficient of variation (often abbreviated to *CV*) is simply the ratio of standard deviation to mean and it is often expressed as a percentage (Equation 2.8).

$$CV = \left(\frac{Standard\ \ Deviation}{mean} \right) 100 \qquad (2.8)$$

Minitab methods of numerical data description

Minitab includes several commands that calculate the various descriptive statistics that I have introduced. The majority of the numerical descriptors of central tendency and spread can be obtained using the **DESCRIBE** command, which is illustrated in Exhibit 2.13. The example data set is taken from a study of the effect of transport on stress in pigs, and the measurements record blood glucose concentrations (mmol.l^{-1}) of two batches of pigs, the samples having been taken immediately after slaughter.

The interpretation of the output of the **DESCRIBE** command is fairly straightforward, but a little explanation is required. The sample

```
MTB > PRINT C1 C2
```

Data Display

```
Row   batch 4   batch 2

  1     5.47      6.83
  2     6.71      6.42
  3     5.79      5.97
  4     6.17      6.47
  5     3.82      6.77
  6     8.56      7.50
  7     6.51      8.53
  8     4.39      9.89
  9     7.55      6.33
 10     6.92      7.50
 11     8.36      6.80
 12     5.22      5.97
 13     6.20      5.20
 14     5.44      7.91
 15     6.43      5.35
 16     7.07     11.59
 17     5.88      7.73
 18     5.81      6.81
 19     7.67      8.18
 20     6.64
```

```
MTB > DESCRIBE 'BATCH 4' 'BATCH 2'.
```

Descriptive Statistics

Variable	N	Mean	Median	TrMean	StDev	SEMean
batch 4	20	6.331	6.315	6.346	1.196	0.267
batch 2	19	7.250	6.810	7.115	1.548	0.355

Variable	Min	Max	Q1	Q3
batch 4	3.820	8.560	5.550	7.033
batch 2	5.200	11.590	6.330	7.910

```
MTB >
```

Exhibit 2.13. Minitab output of the **DESCRIBE** command, showing basic descriptive statistics of blood glucose measurements taken from two batches of post-slaughter pigs.

size (**N**) is given along with the mean, median, trimmed mean (**TrMean**), *sample* standard deviation, minimum, maximum, and the first and third quartiles (**Q1** and **Q3**) from which the interquartile range can be calculated. The other statistic that is provided by **DESCRIBE** is the standard error of the mean (SEMean) which will be introduced in Chapter 3.

```
MTB > NAME K1 'Coeffvar'
MTB > NAME K2 'Range'
MTB > NAME K3 'PopVar'
MTB > NAME K4 'PopStdev'
MTB > Note: Calculate Population variance
MTB > LET K1 = STDEV(C1)/MEAN(C1) * 100
MTB > LET K2 = MAX(C1) - MIN(C1)
MTB > LET K3 = ((STDEV(C1)**2)* (COUNT(C1)-1))/COUNT(C1)
MTB > LET K4 = SQRT(K4)
MTB > PRINT K1-K4

Data Display

Coeffvar 18.8972
Range    4.74000
Popvar   1.35890
PopStdev 1.1657

MTB >
```

Exhibit 2.14. Examples of the use of Minitab's arithmetic and statistical functions with the **LET** command to calculate coefficient of variation and range as well as the variance and standard deviation of a population. The calculations have been performed on column 1 of the blood glucose concentrations shown in Exhibit 2.13.; for the purposes of this Exhibit, those data have been regarded as a population rather than a sample.

The **DESCRIBE** command can be accessed via the menus thus:

STAT > Basic Statistics > Display Descriptive Statistics

Exhibit 2.14. illustrates how Minitab's built-in arithmetical and statistical functions can be combined to calculate other descriptive statistics. In particular, the coefficient of variation and range are calculated and ascribed to the constants, **K1** and **K2**, respectively. The functions applied to the columns of data in these calculations are readily identifiable: **MEAN**, **STDEV**, **MAX**, and **MIN** and they can used via menus as follows:

CALC > Columns Statistics

Since the **DESCRIBE** command and the **STDEV** function both calculate the *sample* standard deviation, Exhibit 2.14. also includes a method for computing the variance and standard deviation for a *population* (**K3** and **K4**, respectively). These computations utilise the functions to calculate the number of data points in a column (**COUNT**) and the square root (**SQRT**). Compare the Minitab commands used to calculate the population and sample variances with Equations 2.5 and 2.6 and the example calculation in Boxes 2.1. and 2.2. to confirm the way in which these Minitab commands perform the calculation.

These two exhibits illustrate how Minitab can be used to describe data numerically. The arithmetic and statistical functions used in these examples are a small selection of those that are available. You should take the opportunity to explore the other functions that are available. One of the problems at the end of the chapter challenges you to calculate the population variance and standard deviation using some of the other functions.

We shall see in later chapters that measurements from many treatment groups or batches are often stored in a single column, with a second column containing a treatment code. In fact, we encountered such an arrangement of data in Exhibit 2.10., where glucose concentrations of 10 subjects were stored in Column 1 and their sex was stored in Column 2 (1 = male and 2 = female). This method of storage is particularly efficient when there are more than two treatment groups involved in the experiment. By storing data this way, we would always have two columns (one holding data, the other storing the other group codes) irrespective of the number of subgroups into which the data were divided. In such cases, we can use the **DESCRIBE** command along with its subcommand, **BY,** which instructs Minitab to produce descriptive statistics on each subgroup contained within a single column of data. The dialog box that enables us to obtain these descriptive statistics is shown in Exhibit 2.15., and Exhibit 2.16. shows the results of this operation.

The Minitab output, though very neat, is only fit for describing results. Like all statistics packages, it is not in a form that would be acceptable for a report or a paper to be published in a journal of a learned society. All publishers specify the way in which results should be presented; each has its own *house style* to which authors have to conform. For example, the pig blood glucose results should be summarised in a table and many publishers would require that the means, standard deviations, and sample sizes should be included in a table of results similar to that shown in Table 2.3.

Notice that the explanatory legend and the units of measurement are specified. Furthermore, in a real report some details of the two batches would also be included so that the reader could be informed of the different treatments to which the various batches were subjected.

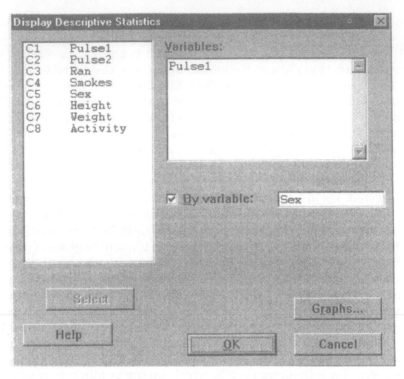

Exhibit 2.15. The Minitab dialog box to produce descriptive statistics of the resting pulse rate data, split by sex.

```
MTB > DESCRIBE 'PULSE1';
SUBC>    BY 'SEX'.

Descriptive Statistics


Variable     SEX         N      Mean    Median    TrMean    StDev    SEMean
PULSE1        1         57     70.42     70.00     70.27     9.95      1.32
              2         35     76.86     78.00     76.65    11.62      1.96

Variable     SEX       Min       Max        Q1        Q3
PULSE1        1      48.00     92.00     63.00     75.00
              2      58.00    100.00     66.00     86.00

MTB >
```

Exhibit 2.16. Descriptive statistics of the resting pulse rates of 92 individuals grouped by gender: 1 = male, 2 = female. Note that the **BY** subcommand specifies the column (**C5**) holding the subgroup codes.

Batch No.	Mean glucose concentration (mmol.l^{-1})	Std. Dev.	n
2	7.25	1.55	1 9
4	6.33	1.20	2 0

Table 2.3. Mean blood glucose concentration (mmol.l^{-1}) of two batches of pigs. Samples were taken immediately post-slaughter.

2.4 Graphical methods of data description

While numerical methods of data description provide an unambiguous summary of a set of measurements, pictorial representations can very often provide much clearer images, the significance of which can be grasped more immediately than through the medium of bare numbers. For example, summarising a relationship between two variables in the form of a graph is an extremely effective method of communication (as demonstrated in Figure 2.2. which shows the levels of injuries sustained in agonistic encounters between sows as a function of reproductive experience). Likewise, a histogram depicts very clearly indeed

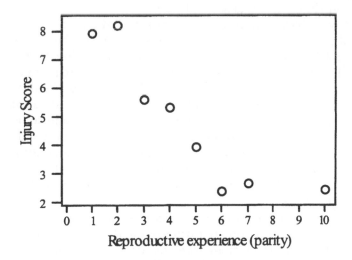

Figure 2.2. The relationship between injury score of sows and reproductive experience. There is a clear decrease in injuries sustained in agonistic encounters as the sows become more experienced.

the variability in a series of measurements. These two important types of diagrams along with other traditional methods of displaying data such as pie charts have now been joined by relative newcomers such as *boxplots* and *stem-and-leaf diagrams*.

Each method has its own role to play in representing data and is associated both with specific data types and also with particular categories of problems. Furthermore, the choice of diagram also depends to a certain extent on whether the primary objective is exploration or presentation of the data. This section will illustrate various methods of graphical description and the circumstances where each method can be used most profitably. Included in these examples will be an introduction to the methods through which such diagrams can be produced using Minitab.

Histograms, bar charts, and stem-and-leaf diagrams

These types of diagram are found in many contexts and traditionally are used to display the distribution of a set of measurements or frequencies by means of bars (examples are shown in Figures 2.3., 2.4., 2.5., and 2.6. and Exhibit 2.17.). The area of the bars in a histogram or bar chart are proportional to the abundance or frequency that they represent. However, there are a number of variations on this theme, and they will be described here.

We have already encountered several data sets that might be displayed as histograms. Two examples are shown below in Figures 2.3. and 2.4., the first of which shows the distribution of the number of red cells present in the squares of a haematocrit (e.g., four squares had no red cells present). In this particular example, the data are not distributed symmetrically around a central peak and are spread more to the right of the peak. This sort of distribution is said to be skewed to the right. We will see in later chapters that some analytical methods are based upon an assumption of a symmetrical distribution; therefore, histograms provide a visual check of the validity of such an assumption as well as giving an immediate impression of the distribution of the data.

Clearly this single diagram has conveyed a great deal of useful information. A general presentation point that also should be noted here is that the units of measurement are included on the axis label (in this case, *number of red blood cells present*). The essential point being made by any figure in a report should be evident from the figure alone; the reader should not have to refer to the main text in order to

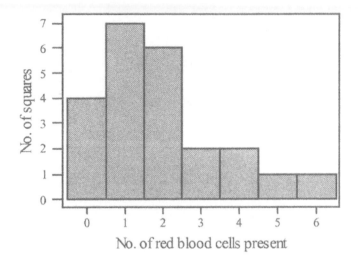

Figure 2.3. Frequency histogram of red blood cell counts in the squares of a haematocrit. The distribution is skewed to the right.

understand the figure. It is also legitimate, and necessary on occasions, to use the legend to remind the reader of the main point that the figure is conveying. The golden rule is to keep each figure as simple as possible so that it has maximum impact.

A second example of a histogram (Figure 2.4.) uses column one of the pulse data that contains the resting pulse rates of 92 individuals. The figure shows that the data range from approximately 50 to 100 beats.min^{-1}, and that the pulse rates seem to be distributed fairly symmetrically around 70 beats.min^{-1}. The difficulty for the investigator in constructing a histogram from these data is that each bar has to span a range of pulse rates (45–54, 55–64, etc.). Contrast this with the red blood cell example in Figure 2.3. where each bar represented a single integer value (0, 1, 2, 3, 4, 5, or 6 blood cells per square). The pulse rate data had to be split into arbitrary groups or categories before plotting.

This problem will always present itself when we have to draw histograms based upon variables measured as real or decimal numbers (such as length, pressure, thickness, weight); the specification of appropriate axis scales will always require careful thought.

There is no fixed rule about the sizes of the class divisions, but one should try to divide the data into a sufficient number of groups so that the general distribution pattern is apparent without giving too much detail. Remember, pictorial data descriptions are meant to provide

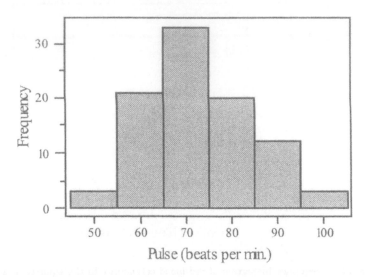

Figure 2.4. A histogram showing the distribution of resting pulse rates.

immediate, simple, and general impressions; detailed, more unambiguous summaries are provided by quantitative descriptors such as the mean, median, standard deviation, and range.

Frequency histograms are useful in showing the symmetry of a sample and so help us to decide whether we should use either the mean and standard deviation or median and range to describe the data. However, we shall see in Chapter 3 that there are some well-known statistical distributions that, even though they are not symmetrically dispersed around the mean, can be described using their means.

The sequence of Minitab commands used to construct Figure 2.4. are listed in the first half of Exhibit 2.17.

Notice also the subcommand **MIDPOINT** that controls the division of the bars. The various subcommands add refinements to the histograms, but simpler forms of these diagrams could be produced using only the command:

MTB > HIST C1

These macros (combinations of commands) were developed using the dialog boxes available via the menu system within Minitab which can be accessed via the menus thus:

GRAPH > Histogram

```
MTB > Histogram 'PULSE1';
SUBC>   MidPoint;
SUBC>   Bar;
SUBC>     Type 2;
SUBC>     Color 0;
SUBC>   Axis 1;
SUBC>     Label "Pulse (beats per min.)";
SUBC>     TFont 1;
SUBC>   Axis 2;
SUBC>     TFont 1;
SUBC>   Tick 1;
SUBC>     TFont 1;
SUBC>   Tick 2;
SUBC>     Number 4 1;
SUBC>     TFont 1.
MTB >
MTB > Note: Commands to construct Figure 2-5
MTB >
MTB > RETRIEVE 'A:\BOOK\DATA\FIG2-5'
Retrieving worksheet from file A:FIG2-5.MTW
Worksheet was saved on 11/1/1994
MTB >   Chart Count( C1 ) * C1;
SUBC>     Cluster C2;
SUBC>     Bar C2;
SUBC>     Axis 1;
SUBC>       Label "Species";
SUBC>       TFont 1;
SUBC>     Axis 2;
SUBC>       Label "Abundance (no. of quadrats)";
SUBC>       TFont 1;
SUBC>     Tick 1;
SUBC>       TFont 1;
SUBC>     Tick 2;
SUBC>       TFont 1.
MTB >
```

Exhibit 2.17. Minitab commands to construct Figures 2.4. and 2.5.

The data that were plotted in the previous two figures (number of red cells in a grid square, pulse rate) were measured on ratio scales, and adjacent bars of the histograms touched each other (this is referred to as being *contiguous*). We can also describe frequency data associated with nominal or categorical data using bar charts that are very similar to the two previous examples with the exception that adjacent bars do not touch and are said to be *non-contiguous*. An example of this type of chart is shown in Figure 2.5., where the relative abundances of six species of flowering plant found in two locations are plotted. The abundance was measured in terms of the number of quadrats in which each species was observed, and there were 100 quadrats of 0.25 m² in each location. Discrete, nominal data are found in all areas of biology

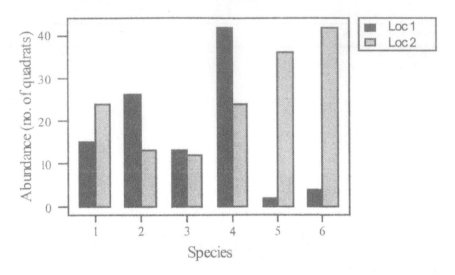

Figure 2.5. Bar chart showing the relative abundance of six species of flowering plant in two different locations.

(e.g., sex, life-history stage, or phenotypic characteristic) and so this type of pictorial representation of frequency data is very common.

In general, the term *bar chart* is used to describe these diagrams of nominal or categorical data (i.e., where the bars are non-contiguous); and *histogram* is applied to figures based on data where the bars are contiguous. Bar charts are produced by Minitab using the **CHART** command (not **HISTOGRAM**) and a listing of the commands used to produce Figure 2.5. is shown in the second half of Exhibit 2.17.

Although the frequency of the species has been used in this example, we are not restricted to using only frequencies in bar charts. We could, for example, use the mean or median values. An example, using the pulse data, shows the mean resting pulse rate for smokers and non-smokers (Figure 2.6.). Clearly, this form of bar chart has the nominal variable (smoking habit) on the horizontal (x) axis and the continuous variable (mean pulse rate) on the vertical (y) axis. Exhibit 2.18. shows the Minitab commands that created this figure. As with the relative abundance bar chart, the mean pulse rate figure provides us with some help in making comparisons between groups. We shall see this a little more clearly when we look at box and whisker plots later in this chapter.

Histograms and bar charts provide an immediately accessible image of the distribution of the variability in a data set. They can also be used as a first stage in making comparisons (as in the pulse

```
MTB > Note: Minitab command to create Figure 2-6.
MTB > Chart Mean( PULSE1 ) * 'SMOKES';
SUBC>    Bar;
SUBC>    Axis 1;
SUBC>      Label "Smoking habit";
SUBC>       TFont 1;
SUBC>    Axis 2;
SUBC> Label "Mean resting pulse rate (beats per minute)";
SUBC>       TFont 1;
SUBC>    Tick 1;
SUBC>       TFont 1;
SUBC>    Tick 2;
SUBC>       TFont 1.
MTB >
```

Exhibit 2.18. Minitab commands to construct a bar chart (Figure 2.6.) of the mean resting pulse rates of smokers and non-smokers.

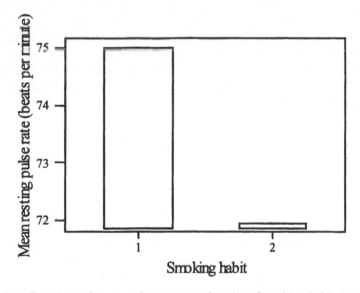

Figure 2.6. Bar chart of mean pulse rate as a function of smoking habit (1 = smoker, 2 = non-smoker). This figure was constructed in Minitab using the macro shown in Exihibit 2.18.

rate/smoking habit example). The chart command can be obtained via the menu system using the graph menu:

GRAPH > Chart

A full exploration of a data set prior to analysis should provide not only a feel for the data set, such as an impression of its symmetry and variability (which can be gauged subjectively from examining histograms and bar charts), but also a more detailed, quantitative knowledge of the identity of outliers, the location of gaps and concentrations in the data. While traditional histograms can provide this information to a certain extent, when one is confronted by large bodies of data, the detail that they can provide is rather limited. *Stem-and-leaf* diagrams display the data values themselves in a graphical format. As such, they provide a better vehicle for data exploration than histograms and bar charts, whereas the latter two display types are better suited for presentation purposes.

In stem-and-leaf diagrams, each datum is split into two sections, *the stem* and *the leaf*, and these are used to construct the diagram. The whole process is best illustrated by reference to an example using the pulse data set that we have already encountered.

First of all, we must sort the data into ascending order of magnitude, and then stems and leaves are created by splitting each value at a specified position between two digits. If we examine the data stored in column 1 of the pulse data, we can see that the values range between 48 and 100 beats.min^{-1}. A suitable position for the split would, therefore, be between the tens and the units as shown below for the four lowest pulse rates:

Pulse Rate	Stem	Leaf
48	4	8
54	5	4
54	5	4
58	5	8

The actual diagram is constructed by listing the stems vertically with the associated leaves stretching horizontally from each stem (as shown in Exhibit 2.19.) and by using the menu:

GRAPH > Stem-and-Leaf

```
MTB > Stem-and-Leaf 'PULSE1'.

Character Stem-and-Leaf Display


Stem-and-leaf of PULSE1     N  = 92
Leaf Unit = 1.0


     1      4 8
     3      5 44
     6      5 888
    24      6 000012222222224444
    40      6 6666688888888888
   (17)     7 00000022222244444
    35      7 6666688888
    25      8 0002224444
    15      8 67888
    10      9 0000224
     3      9 66
     1     10 0

MTB >
```

Exhibit 2.19. A stem-and-leaf diagram of the resting pulse rate data.

The stems are equivalent to class divisions in a histogram, and, in this example, they are five units wide: the leaves including either units 0-4 or 5-9.

The visual impression given by this diagram is similar to a histogram, even though the two diagrams differ in their orientation by 90°. However, since the stem-and-leaf diagram is composed of the actual numbers, it provides a more detailed description of the data set. Furthermore, on the left-hand margin of the diagram there is a list of the cumulative frequencies of measurements as counted from the two extremes towards the centre of the distribution (the number in parenthesis, 17, is the frequency represented by the central class division). This element of the diagram illustrates, more quantitatively than the histogram, whether or not the data are distributed symmetrically. The stems in this form of diagram are the equivalent to the class intervals in a conventional histogram.

This type of diagram will not only identify outliers or groups of outliers, but it will also highlight other properties of the data set. By including the actual numbers, stem-and-leaf diagrams can identify gaps or concentrations in the data that would not be shown by a traditional histogram in such clear detail. Minitab will omit from the diagram any extreme outliers,* but the actual values will be listed on the output so the researcher will immediately be made aware of such outliers.

In this particular example, we can see that only two values (61 and 87) are odd numbers. One might imagine that the pulse rates were measured over a thirty-second period and then doubled, in which case these two values are perhaps mistakes of some sort. In this particular case, we cannot determine whether these data were errors or not, but this example illustrates the way in which this type of diagram, because it includes the actual data, can identify possible data transcription errors. As mentioned in the introduction to this chapter, one of the functions of data exploration is to detect any possible errors; therefore, stem-and-leaf diagrams provide a very useful tool for that purpose.

In this specific example, the range of data was quite limited. However, wide ranges can require the investigator to make further decisions about data display. For example, what should be done when the data are very large and not rounded to convenient values? How would we cope with a data set that included the following measurements: 1005, 1006, 1011, 1103, 1111, 1672, and 2314? The person constructing the diagram has to perform the rounding operation, which simply involves discarding the last digit altogether and then splitting the remaining digits into stems and leaves. Remember, stem-and-leaf displays have an *exploratory* function, so the units in values measured in thousands are not really essential in an exploratory context. So, we would end up with:

Data	Stem	Leaf	Discard
1005	10	0	5
1006	10	0	6
1011	10	1	1
1103	11	0	3
1111	11	1	1
1672	16	7	2
2314	23	1	4

Stem-and-leaf displays are not as pretty as traditional histograms, but as exploratory tools they are far more informative. Like histo-

* Outliers are defined by Minitab as those values that lie either above or below the *inner fences*. The calculations of inner and outer fences are shown in the section entitled *Box-and-whisker plots* (in Section 2.4) in relation to box-and-whisker plots.

grams, they can be used with integer and real data and are best utilised when we wish to examine frequency distributions or relative abundance.

Box-and-whisker plots

Boxplots (the abbreviated name by which this type of diagram is generally known) provide a useful graphical description of a set of measurements. They are used to illustrate the variability in a set of measurements, and, when several are plotted together, they can illustrate comparisons very effectively since they show the median, range, and interquartile ranges of the groups being compared. Consequently, if we need to compare groups of measurements (two or more), we can contrast not only central tendencies, but also the variability of the groups. We shall see the usefulness of the boxplot in Chapter 6 (Comparing Two Samples). An additional quality of the boxplot is that it can illustrate the skewness of a distribution because it shows the range and interquartile range (means and standard deviations would only be suitable with symmetrically-distributed data sets). The rules concerning data types suitable for use with boxplots are the same as those applying to histograms. The data plotted on the y-axis must be measured on an ordered scale of some sort (ordinal, interval, or ratio). If several boxplots are to be displayed together, the different plots may relate to different nominal variables. For example, Figure 2.7. shows the pulse rates of smokers and non-smokers (smoking habit being a nominal variable).

The structure of the boxplot is very simple and it is illustrated in Figure 2.7. Accessing the commands via the Minitab menu system is as follows:

GRAPH > Boxplot

The box encloses the interquartile range and the vertical lines or *whiskers* terminate at the minimum and maximum value in the set of measurements. The horizontal line through the box is at the level of the median value. Any substantial differences in these three characteristics of the two data sets will be revealed very clearly by this diagram. In this particular example, we can see that the median pulse rates of smokers and non-smokers are very similar, but there is an obvious difference between the two groups with respect to both their

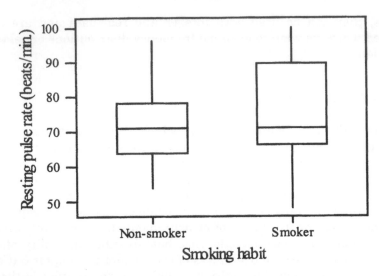

Figure 2.7. Boxplots of the resting pulse rates of smokers and non-smokers. Notice that the medians of the two groups seem to be fairly similar, but the variability of the smokers is much greater (both in terms of the range and inter-quartile range).

interquartile ranges and their overall ranges. Such a diagram should prompt a physiologist to explore the data further to discover the source of the difference in variability.

When Minitab constructs a boxplot, it will identify extreme values as possible outliers. In fact it categorises these measurements into two groups dependent upon the magnitude of distance that separates them from the main body of data. The less extreme values are denoted by an asterisk (*) beyond the ends of the whiskers, and the more extreme outliers by a zero (0). The quantitative criteria by which Minitab judges measurements to be outliers are described in Box 2.3. and are the same as those used to identify data as outliers in stem-and-leaf diagrams.

In microbiology and ecology, we often encounter population counts within a sample that may differ by several orders of magnitude and this presents us with particular difficulties both for data analysis and presentation. An example of this type of difficulty is illustrated in Figure 2.8. using boxplots, where a comparison is made between five different methods of estimating bacterial populations. The example also serves to illustrate the general principle of how we can cope with such variations in data.

The data comprise bacterial counts using the five different methods and are very variable (range: 200–910,000). Figure 2.8. shows a boxplot

Minitab calculates two sets of limits, *inner* and *outer fences* which are the criteria by which measurements are judged to be outliers. The calculations are as follows:

Inner Fence Upper = Q3 + 1.5 × Interquartile range
             Lower = Q1 − 1.5 × Interquartile range

Outer Fence Upper = Q3 + 3.0 × Interquartile range
             Lower = Q1 − 3.0 × Interquartile range

Where Q1 and Q3 refer to the first and third quartiles, respectively.

In terms of the pulse rate and smoking habit example, the fences are:

Non-smokers:
Inner Fence Upper = 78 + 1.5 × 14 = 99
             Lower = 64 − 1.5 × 14 = 43

Outer Fence Upper = 78 + 3.0 × 14 = 120
             Lower = 64 − 3.0 × 14 = 22

Smokers:
Inner Fence Upper = 89.5 + 1.5 × 23.5 = 124.75
             Lower = 66 − 1.5 × 23.5 = 30.75

Outer Fence Upper 89.5 + 3.0 × 23.5 = 160.0
             Lower = 66 − 3.0 × 23.5 = −4.5

Box 2.3. Quantitative criteria by which Minitab identifies possible outliers in stem-and-leaf diagrams and boxplots. If a measurement falls between the inner and outer fences, then it will be signified by an asterisk; if it falls outside the outer fence, then it will appear as a zero on the boxplot.

of the count data for each method, and, very obviously, this figure is quite worthless because the variability in counts from method 3 is so vast that the boxes for the other methods are virtually non-existent.

When we investigate population variation, we are often interested in relative abundance or percentage change over time rather than absolute numbers and so, if we take logarithms of the population

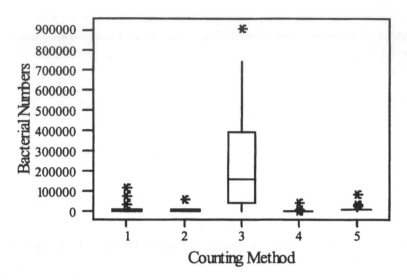

Figure 2.8. Bacterial counts determined using five different methods. The raw data have been plotted and, given the large range, the boxplots are not very useful. Contrast this plot with that shown in Figure 2.9.

counts, we can satisfy those requirements and, at the same time, transform the data to a more manageable scale. The resulting boxplot is shown in Figure 2.9. and it clearly provides a much more informative comparison of the different methods of determining bacterial counts. Whenever we are confronted by a data set from any source that varies to such a great degree, plotting the data on a logarithmic scale enables us to see the relative effects much more easily. The use of logarithmic transformations is not confined to boxplots; it can be used with any type of graph and is seen very often, for example, in plots of population numbers over time. We shall see in Chapters 7 and 8 why the logarithmic transformation is so useful.

Transforming data is very useful, but can also be potentially dangerous, after all, we have just modified the data! Since transformations are so important, they will be discussed in some detail in later chapters (cf Chapter 7, in particular).

As far as this chapter is concerned, logarithmic transformations should be regarded as a means of overcoming a presentational problem with data sets that have large ranges. When we report results based upon transformed data, we have to specify the transformation used (there are several commonly-encountered transformations), and when we summarise data in tables, we must always state what transforma-

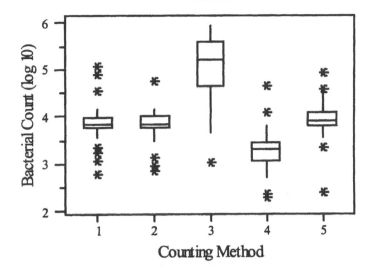

Figure 2.9. Bacterial counts plotted on a logarithmic scale showing the relative variability of the five different methods of determining bacterial populations.

tion has been applied for the purposes of analysis even if we use the raw measurements in the table. As we have seen in the various examples in this section, if a data transformation is required, then we should plot that rather than raw data on graphs.

Graphs and scatter diagrams

In the introductory chapter, several different groups of problem types were identified and one of those groups included sequential relationships. One cannot pick up a textbook or journal relating to any scientific discipline without finding examples of investigations into such relationships. For example, we frequently need to record quantitatively the responses elicited in organisms (plant or animal) to various stimuli whether the stimuli are added nutrients, environmental modifications, or changes in social groups. Also, in many disciplines, the change in a variable over time has to be recorded, such as population growth in microbiology and ecology.

Visually, we describe such relationships in the form of graphs. While all biologists have encountered graphs and are, therefore, in no need of a substantial introduction to their function, I shall include here a few examples of how graphs of various sorts can be constructed

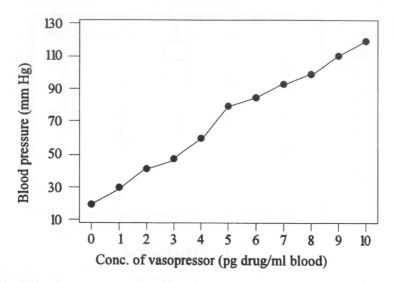

Figure 2.10. Graph showing the response in blood pressures of mice to the concentration of an injected dose of vasopressive drug.

using Minitab and also take the opportunity of mentioning a few conventions that need to be followed when presenting data in the form of graphs.

Conventionally, the horizontal and vertical axes of the graph are known as the *x-axis* and *y-axis,* respectively. Figure 2.10. illustrates how the blood pressure of mice increases with increasing concentration of a vasopressive drug. The blood pressure is known as the *dependent variable* because we are trying to find out whether it is dependent upon drug concentration, which is referred to as the *independent variable*. The dependent variable is always plotted on the *y*-axis and the independent on the *x*-axis. It is important to make this distinction between the two variables as we shall see in later chapters. Minitab includes many options within the **PLOT** command that are shared with the **CHART** and **HISTOGRAM** commands. The menu-based route to the **PLOT** command is:

GRAPH > Plot

We saw that histograms can be used with particular data types (e.g., frequencies) and so it is with graphs. The data that can be plotted on the *y*-axis of a graph must be measured on an ordinal, interval, or ratio scale, i.e., nominal data are excluded; the *x*-axis can be represented by variables of any data type including nominal data. There are no other restrictions governing data that can be plotted on a graph, so they can be continuous (e.g., length, pressure, or concentration) or they can be measured on a discrete scale (such as parity, counts, or ranks). Furthermore, the variables plotted against each other do not have to be of the same data type (look back at Figure 2.2., where the dependent variable, injury score, is ordinal and the independent variable, parity is measured on a ratio scale). In this particular example, successive points on the x-axis are joined together by a line; this is permissible because the independent variable, drug concentration, is measured on a ratio scale. However, if the independent variable is a nominal variable, we should not be permitted to join adjacent points because the values are not on a truly ordered scale. An example of this is shown in Figures 2.11. (a) and (b), where the mean yields of six varieties of alfalfa are plotted. The varieties are coded by numbers,

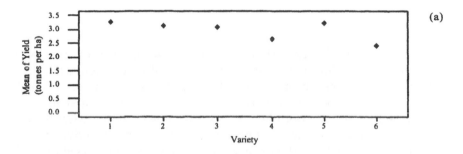

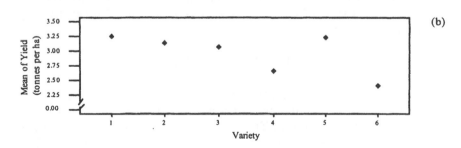

Figure 2.11. (a) and (b): Mean yield (tonnes.ha⁻¹) of six varieties of alfalfa. The varieties are numbered 1–6. As such they are nominal variables and, therefore, the points on the graph should not be joined by a line. Notice also that in (a) there is a lot of white space below the points and so the y-axis should be broken to use the space more effectively as in (b).

but the numbers are merely numeric codes for six names and do not imply any ordered sequence.

Figures 2.11. (a) and (b) also demonstrate very clearly the need to pay particular attention to the axis scales. We can see in Figure 2.11. (a) that there is a lot of clear space below the data points because the y-axis starts at the origin and continues uninterrupted to 3.5 t.ha^{-1}. By interrupting the axis as shown in 2.11. (b), the differences between the points are more clearly seen. However, one must be careful in this procedure both in presenting data and also in examining and drawing conclusions from graphs because, by expanding the y-axis, one can give a false impression of the difference between data points. Always be careful to look at both the points and the scale of the axes in order to gauge the real meaning of a graph. As with all other graphical diagrams, the axes of a graph must be labelled with the name of the variable and the units of measurement.

Pie charts

The final form of diagram that I shall review is, perhaps, the least used in biology. However, it does have a useful function where we consider percentage or proportional composition. Examples of such problems include behavioural time budgets and diet composition as established from, for example, regurgitated owl pellets, oesophageal fistulas, or stomach contents of grazing animals. The objective of such diagrams is to illustrate the proportion of the whole comprised by each of the constituents. The example below (Figure 2.12.) shows the dietary composition of barn owls as established from pellets found at nest sites. This type of diagram is accessed within Minitab through the menu system thus:

GRAPH > Pie Chart

The examples in this chapter have illustrated the variety of numerical and graphical techniques through which data can be explored, summarised, and presented. The methods by which these three tasks can be performed share many common elements, but each technique possesses its own strengths and weaknesses which I have highlighted. For example, specific data and problem types are best represented by particular descriptive methods, and some techniques are more appropriate for exploring data rather than for presentation purposes.

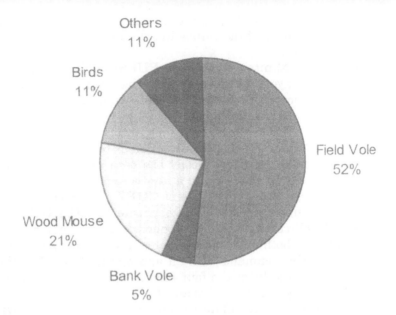

Figure 2.12. Diet composition of barn owls as estimated by percentage composition of pellets.

I cannot stress too strongly the need to explore data prior to analysis and, with the computing power readily available to us, we can plot data in a variety of ways very quickly and easily indeed. Remember that null hypotheses are framed in terms of populations. In later chapters, we shall see the importance of data exploration in helping us to establish that the samples that we analyse provide a good estimate of the populations about which we wish to generalise. Therefore, we must take every opportunity to ensure that our samples are reliable images of the populations, and, in Chapter 3, we shall encounter measures of reliability that are based upon some of the numerical methods of data description introduced here.

2.5   Exercises

1.   On your own computer system, enter the data shown in columns 1 and 2 of Exhibit 2.10. and copy Column 1 into a new column. Delete rows 2-4 inclusive of column 1 and print all three columns on a printer. Now try to recreate

column 1 by inserting the data that you removed. Now find the square of the values in column 1. Which function did you use?

2. Open the Minitab worksheet **PULSE.MTW.** (You may have to ask your system manager where this file is located on your computer system.) Separate column 2 (**PULSE2**), which holds the pulse rates of 92 people some of whom have run while others have not, into two separate columns. Column 3 holds the selection criteria (1 = ran, 2 = did not run). Now try to copy the data in column 1 that relates only to females into a new column, but do not use **COPY** and **USE,** make use of **COPY** and **OMIT.**

3. Again, using the pulse data, separate the **HEIGHT** and **WEIGHT** columns into four new columns: height and weight for both males and females. Do this using a single **UNSTACK** command with an appropriate **SUBSCRIPT** subcommand. Print the four new columns on the printer.

4. Using the pulse data, consider Column 1 (**PULSE1**) as a *population* and calculate its variance and standard deviation using the following Minitab functions: **SSQ, SUM, COUNT,** and **SQRT** (do not use any other functions, but you can use the arithmetic operators + – * / and **). Check whether you have obtained the correct answers by using the method given in Exhibit 2.14.
    What percentage of the observations lie within one standard deviation of the mean? What percentage of the observations lie within two standard deviations of the mean? What percentage of the observations lie within the inter-quartile range?

5. Using the pulse data again, cross-tabulate the levels of activity (1 = slight, 2 = moderate, 3 = a lot) with smoking habit (1 = smoker, 2 = non-smoker), calculating descriptive statistics (mean, median, standard deviation, and range) for pulse rate *after* exercise for each combination of activity and smoking habit. Be careful, remember that not all the data relate to individuals who exercised, so you will have to select the appropriate data (**C3** holds codes for running: 1 = ran, 2 = did not run).

6. Produce a bar chart of mean resting pulse rates for smokers and non-smokers for each level of activity.

7. The data tabulated below (Jerison, 1973), show the brain and body weights of 28 animal species. The data were used

to assess whether there was any relationship between
body and brain size. Present these data in the most appropriate way for a report that demonstrates such a relationship exists. With what particular presentational problems
do you have to cope? Can you identify any obvious outliers?
If so, on what grounds might you explain their dissimilar-
ity to the rest of the data?

Species	Body weight (kg)	Brain weight (g)
Brachiosaurus	87000.00	154.50
Rat	0.28	1.90
Beaver	1.35	8.10
Cow	465.00	423.00
Grey Wolf	36.33	119.50
Goat	27.66	115.00
Guinea Pig	1.04	5.50
Diplodocus	11700.00	50.00
Asian elephant	2547.00	4603.00
Donkey	187.10	419.00
Horse	521.00	655.00
Potar monkey	10.00	115.00
Cat	3.30	25.60
Giraffe	529.00	680.00
Gorilla	207.00	406.00
Human	62.00	1320.00
African elephant	6654.00	5712.00
Triceratops	9400.00	70.00
Rhesus monkey	6.80	179.00
Kangaroo	35.00	56.00
Hamster	0.12	1.00
Mouse	0.02	0.40
Rabbit	2.50	12.10
Sheep	55.50	175.00
Jaguar	100.00	157.00
Chimpanzee	52.16	440.00
Mole	0.12	3.00
Pig	192.00	180.00

Courtesy of Jerison, H.J. (1973) *Evolution of the Brain and Intelligence*, Academic
Press, New York. With permission.

8.   Four varieties of soya bean (A, B, C, and D) were har-
     vested after 12 days of growth. The total leaf area (cm^2)
     was measured and the data are tabulated below. Calculate
     the mean, median, standard deviation, and range of the
     leaf area for the four varieties as well as producing his-
     tograms, stem-and-leaf diagrams, and boxplots for each.

Which numerical and pictorial descriptions are most appropriate for these data? How would you summarise these results in a report in which you wished to compare the four varieties?

A	B	C	D
230	278	259	307
184	306	196	313
191	285	220	304
201	254	262	333
208	212	211	293
210	197	236	263
179	261	227	338
211	319	251	266
266	237	225	279
160	286	283	303
225	263	249	330
250	276	282	276
198	252	226	268

9.   The red blood cell data used to construct Figure 2.3. are as follows:

No. of cells per grid square	Frequency
0	4
1	7
2	6
3	2
4	2
5	1
6	1

Calculate the mean, mode, median, range, interquartile range, variance, standard deviation, and coefficient of variation for the number of cells per grid square. You should perform the calculation both manually (use the arithmetic operators on a calculator, but not function keys) and using Minitab.

Which measures of central tendency and dispersion would you suggest would be the most appropriate for describing these data in a report?

10. A group of rats were weighed at weekly intervals for a period of nine weeks. The weights for Weeks 1, 5, and 9 (in grams) are tabulated below. Calculate the coefficient of variation for each of the weekly weights. Comment on the results that you obtain. If the objective of a report was to show weight gain over the experimental period and the entire set of weekly weights were available (i.e. weights on days 1, 8, 15, 22, 29, 36, 43, 50, 57, and 64), how would you present the data? In fact, in this experiment, there were three sets of rats, each group having been subjected to a different diet. How would you present those data if the objective was to compare the effects of the three diets?

Weight (g)		
Day 1	Day 36	Day 64
240	258	278
225	245	260
245	265	269
260	268	275
255	273	280
260	277	281
275	274	284
245	265	278

2.6   Summary of Minitab commands

General

COPY (with **USE** and **OMIT**)

DELETE and **INSERT**

HELP

LET

PRINT

RETRIEVE

SAVE

STACK and **UNSTACK** (with **SUBSCRIPTS**)

STOP (to terminate Minitab)

Calculation and manipulation

COUNT

DESCRIBE (with subcommand **BY**)

MEAN

MAX

MIN

SQRT

SSQ

STDEV

SUM

TABLE (with various subcommands)

Graphical

BOXPLOT

CHART

HISTOGRAM

PLOT

STEM-AND-LEAF

2.7   Summary

1. Various Minitab techniques to manipulate and organise data were introduced.
2. Listing and tabulating data were shown to be simple and informative methods of displaying data, but summary statistics of central tendency and spread provided succinct methods of numerical data description.
3. Various graphical methods of displaying data were introduced, and the usefulness of each was related to the different data types defined in Chapter 1.

CHAPTER 3

Reliability, Probability, and Confidence

3.1   Introduction

In Chapter 2, we examined various methods of describing data both numerically and pictorially. The descriptions of the various graphical methods included several examples of the variability that are present in biological data. We saw the relatively symmetrical variation around the mean resting pulse rate (Figure 2.4.) and the rather skewed frequency distribution of red blood cell counts on a haematocrit (Figure 2.3.). The various quantitative descriptors showed how samples and populations that were distributed either symmetrically or in a skewed fashion could be described in an unambiguous manner. Those examples served to illustrate not only methods of data description, but also the very important fact that the variability implicit in all data sets can take many forms.

Several measures of dispersion were introduced (e.g., standard deviation and range), which help us to answer a few questions. How variable are the numbers of the population that we are studying? How variable are the numbers of the sample that we have collected? What we have to remember at all times is that we take samples only because we are generally unable to examine every member of the population, and that the properties of the samples that we collect provide our best estimates of those of the larger populations in which we are interested. The sample mean, for example, is an estimator (or *statistic*) of the population mean, which, in turn, is referred to as a *parameter*. (Statistics are attributes of samples, while the equivalent characteristics of populations are called parameters.)

Estimators may be good or bad, and, therefore, it is not only very important to be able to take samples that mimic the population as closely as possible, but also to be able to measure the *reliability* of any sample statistics. For instance, think how useful it would be if we could attach a measure of reliability to a sample mean that expressed its precision as an estimator of the population mean. Such measures exist, and, not surprisingly, are closely related to some of the measures of variability that we encountered in Chapter 2. The objective of this chapter is to introduce measures of reliability and methods by which they can be used effectively in an investigation. Clearly, the actual methods by which samples are selected will be critical to the accuracy with which a sample mimics a population; a biased sample will lead to biased conclusions. The following chapter (Chapter 4) will be devoted to a discussion of various sampling techniques that reduce the potential for bias. The contents of Chapters 3 and 4 are, therefore, very closely related to each other.

This chapter is divided into three main sections. Firstly, I shall introduce a measure that shows the reliability of the sample mean as an estimator of the population mean. Although measures of reliability can be derived for any sample statistic (e.g., for means, medians, or variances), the most important and frequently encountered is that relating to the mean. Secondly, I shall outline some basic principles of probability that underpin many of the methods of hypothesis testing in later chapters. This section will demonstrate how we can calculate the probability of particular events occurring, and it will also show how the shapes of frequency histograms are related to probability. Thirdly, I shall illustrate how the properties of one particular distribution, the *normal distribution,* are utilised more specifically along with measures of reliability in order to quantify the level of confidence that we can attach to our estimates of population parameters. Once we are able to quantify confidence, by producing *confidence intervals,* we will be in possession of the necessary analytical tools to commence testing hypotheses.

3.2   Reliability

How do we measure reliability? How is such a measure related to the measures of variability that were introduced in Chapter 2? We can find answers to these questions by deriving a measure of reliability in a worked example.

Let us first assume that we have a large population of measurements from which we can randomly select many samples, and, after taking each sample, we calculate its mean and then return the sample back to the population. If we performed this operation repeatedly, taking many samples and then constructing a frequency distribution of the sample means, we would obtain a bell-shaped curve that had a

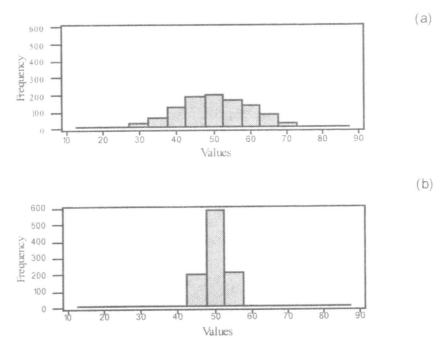

Figure 3.1. Frequency distributions of (a) a population ($n = 1000$) with a mean of 50.183 and standard deviation of 9.907; and (b) the means of 1000 samples ($n = 10$) drawn from the same population. The mean of the sample means (50.133) is very similar to that of the population but the standard deviation of the sample means is much smaller, 3.119.

mean that was very similar to the true population mean. An example is displayed in Figure 3.1., where a population of 1000 measurements are plotted as a frequency histogram (Figure 3.1a.). From this population, 1000 samples, each comprising 10 individuals, have been taken, and the means of these samples are plotted in Figure 3.1b. Notice that the means of the two distributions are very similar (population mean = 50.183; mean of the distribution of sample means = 50.133), but the standard deviation of the sample means is considerably smaller than that of the population (population standard deviation = 9.907; standard deviation of distribution of the sample means = 3.119). This is vividly demonstrated in Figures 3.1a. and b.

In fact, the mathematical theorem, *The Central Limit Theorem*, states that if an infinite number of samples is drawn from a population, the mean of those sample means would be the same as the population mean. The standard deviation of the distribution of sample means will be smaller than the actual population standard deviation, and it will decrease with increasing sample size. This is not surprising since, by plotting measures of central tendency, rather than the raw measurements themselves, we remove some variability.

One other feature of the distribution of sample means that is perhaps more remarkable is that its shape will always be approximately symmetrical and bell-shaped *irrespective* of the form of the frequency distribution of the original population. An example of this phenomenon is illustrated in Figure 3.2., where a skewed frequency distribution is shown (Figure 3.2a.) along with an approximately symmetrical frequency distribution of many means of samples drawn from that original skewed distribution (Figure 3.2b.). The means of the two distributions are fairly similar, but the standard deviations are very different.

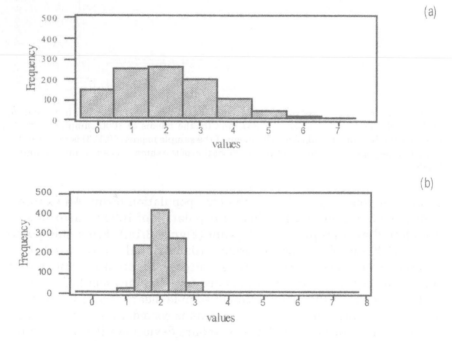

Figure 3.2. Frequency distributions of (a) a skewed population ($n = 1000$) with a mean of 2.049 and standard deviation of 1.429; and (b) 1000 sample means ($n = 10$) drawn from that population. The distribution of sample means is symmetrical around a mean of 2.044 with a standard deviation of 0.441. The sample means are distributed according to a symmetrical, bell-shaped distribution even though they have been drawn from an asymmetrical distribution.

Frequency distributions come in all shapes and sizes, but some shapes occur so frequently that they have been given specific names. We shall encounter several throughout this and subsequent chapters. The symmetrical, bell-shaped distribution seen in Figures 3.1. and 3.2. is known as the *normal distribution*. It is one of the most widely known, and is intimately involved with a great deal of quantitative analysis in biology. In the next section, where the principles of probability are introduced, we shall encounter two other important distributions, the *binomial* and *Poisson*. The properties of all three distributions will be discussed in some detail. I mentioned in Chapter 2 that many statistical tests are based upon assumptions about the data to which they are applied. Some of those assumptions concern the nature of the frequency distribution to which the data conform.

Returning now to the development of a measure of reliability, consider the distribution of the sample means (e.g., Figures 3.1b. and 3.2b.) and, in particular, the importance of the size of the variance and standard deviation. If the variability of this distribution is very small, then we can conclude that the means producing the distribution are quite close together, or that the samples are consistently fairly similar. Furthermore, since the central limit theorem states that the mean of an infinite number of sample means is the same as the population mean, then low variability indicates that our samples are reliably estimating the population mean. Therefore, *the measure of variability of the distribution of sample means is also a measure of reliability.*

The task now is to define the measure of reliability. It can be shown mathematically that, with an infinite number of samples, the variance of the distribution of sample means is given by Equation 3.1 in which σ^2 is the population variance and n is the sample size. Note that the variance of the distribution of sample means will decrease with increasing sample size because the denominator, n, is the sample size.

$$\sigma_{\bar{x}}^2 = \frac{\sigma^2}{n} \tag{3.1}$$

The symbol, $\sigma_{\bar{x}}^2$, denotes the variance of the sample means. The best estimate of $\sigma_{\bar{x}}^2$ that will generally be available to us, is based upon the sample variance (s^2), and this estimate is given by Equation 3.2.

$$s_{\bar{x}}^2 = \frac{s^2}{n} \tag{3.2}$$

We saw earlier that the units of variances are the square of the units of the original measurements and, as such, are inconvenient as measures of dispersion or spread, giving rise to the use of the standard deviation (the square root of the variance). Likewise, here we can calculate the standard deviation of the frequency distribution of the sample means, which is called the standard error of the mean (S.E.M.), but often referred to simply as the *standard error.* Its general formula is shown in Equation 3.3. Therefore, the units of the standard error of the mean are the same as those of the mean.

$$\textit{Standard Error of the Mean } (s_{\bar{x}}) = \sqrt{\frac{s^2}{n}} = \frac{s}{\sqrt{n}} \qquad (3.3)$$

This is our measure of the reliability of the sample mean. However, before proceeding further, let us consider this expression in more detail. Whenever we are presented with a new equation, we should examine it and ask some questions. What do each of the components in the equation represent? More specifically, can we visualise the meaning of each term in the equation? In what type of units are the different variables measured? What happens to the value of the variable that is defined by the equation when each of the other variables increases or decreases? Let us examine this particular equation and hopefully, in doing so, understand why this particular statistic can truly be regarded as a measure of reliability (or *precision*) of a sample statistic.

The standard error of the mean is the standard deviation of the distribution of the sample means and thus it measures the spread or variability of the sample means. A small spread indicates that samples are consistently being produced that have means which are very similar in value. Thus, it follows that a small standard deviation of sample means indicates that the population mean is being estimated reliably.

Two examples of the calculation of standard errors of the mean are illustrated in Box 3.1. These calculations are based upon the pulse rate data that we have already encountered. The **DESCRIBE** command within Minitab, which was introduced in Chapter 2 (Exhibit 2.13.), calculates several descriptive statistics including the standard error of the mean. Exhibit 2.15. illustrated this command using these same pulse rate data.

Equation 3.3 shows that as sample size increases, the standard error will decrease, and so the reliability of the sample mean as an estimator will increase. Therefore, it is within our power to increase

---

Male Resting Pulse Rate:
Standard deviation = 9.95 beats.min^{-1}
Sample size = 57
Standard error = 9.95 ÷ $\sqrt{57}$ = 1.318 beats.min^{-1}

Female Resting Pulse Rate:
Standard deviation = 11.62 beats.min^{-1}
Sample size = 35
Standard error = 11.62 ÷ $\sqrt{35}$ = 1.964 beats.min^{-1}

Box 3.1. Example calculations of standard error of the mean for resting pulse rates of males and females. The data used in these calculations are stored in the file **"PULSE.MTW."** The Minitab command, **DESCRIBE,** calculates the standard errors, and this is illustrated for this data set in Exhibit 2.15.

the reliability of the sample mean as an estimator of the population mean by increasing the sample size. This is not surprising since, as sample size increases, the sample will more closely approximate to the population, and thus we would expect it to be more reliable. This observation is particularly important to note as it will appear again in our discussion of confidence intervals, which follows later in this chapter.

Increasing the sample size will reduce the standard error, thereby increasing reliability. However, be warned that the reduction in standard error resulting from an increase in sample size may not justify the extra effort required to gather or process the extra measurements. This problem results from the denominator being the square root of the sample size. To illustrate this, if a standard deviation is equal to 10, work out the effect on the value of a standard error of the mean by increasing a sample size from 50 to 51, and from 5 to 6.* This emphasises the need for thorough planning *prior* to data collection. We must make as much of an effort as possible to ensure that we have collected sufficient data to answer the key questions of our investigation, but we must also be very conscious not to exhaust our resources by gathering excessive quantities of data. A good rule of thumb to remember in this respect is that we have to increase the sample size by a factor of 4 to obtain a doubling of precision.

Finally, the reliability of all sample statistics (not just the mean) can be estimated by standard errors, each of which is calculated by

---

* The standard errors will be equal to 1.414, 1.400, 4.472, and 4.082 when the sample sizes are 50, 51, 5, and 6, respectively.

its own specific formula, and we shall meet a few of them in later chapters. We will also see later that the standard error is extremely important in hypothesis testing and in a variety of statistical tests.

3.3   Probability

Frequency histograms of various shapes have been plotted in this and the previous chapter. Many biometrical measurements are described by the normal distribution that has a characteristic bell shape. Two other distributions of which you should be aware are the binomial and Poisson distributions, which are also found extensively in biological investigations. As biologists, there are two very important reasons for becoming familiar with the properties of those distributions that occur most frequently. Firstly, we shall be able to describe the data and test hypotheses more rigorously if we can utilise the properties of a specific and appropriate statistical distribution. Many methods for testing hypotheses are based around the normal distribution in particular and, as a consequence, many of the analytical techniques used by biologists assume that the data being analysed do actually conform to a normal distribution. We shall also see that the binomial and Poisson distributions are closely related to the normal distribution. Secondly, we need to be familiar with these distributions because some data do not conform to a specific pattern, and we have to adopt different methods to analyse these data. If we can recognise particular distributions, then we should also be able to recognise more easily those data that do not conform to specific statistical distributions. As a consequence, we must select the most appropriate techniques to solve the problems. As biologists, we use statistical techniques as tools. If we can recognise that our data conform to a particular distribution, then we are better placed to choose the appropriate statistical tool for analysis. One of the functions of exploratory data analysis is to aid us in this identification process.

Particular statistical tests are linked to statistical distributions through the relationship between the measurement of probability and the form or shape of the distribution; this section will describe that relationship. The general principles outlined here apply to all statistical distributions. A sound understanding of these principles will provide a firm basis upon which to develop the ideas of hypothesis testing and statistical testing in general. While the prospect of working through what may appear to be statistical theory may be unwelcome,

I can assure you that, as a biologist who felt that way some years ago, I have found that by grasping the basic principles of probability, the use of statistical tools has become much simpler and more readily understandable.

First of all, we need to understand what is actually meant by the term "probability," and how it is measured. Put quite simply, probability is the chance of a particular event occurring, and this can be illustrated by a simple example.

Imagine that we had a litter of eight piglets of which six were male and two were female, and we selected one piglet at random. Intuitively, we would quite rightly say that the probability of randomly selecting a male is 6 in 8 or 0.75 (the number of males divided by the total number of piglets).

The probability of selecting a male from another litter of eight piglets that were all females would be 0 in 8 or 0 (an impossibility). Likewise, the probability of selecting a person of height 2.05 m from a group whose heights range between 1.5 m–2.0 m is also 0. Conversely, a certainty would be to select a female piglet from that same all-female litter of eight piglets: 8 out of 8, i.e., the probability of a certainty is 1. Similarly, using that same group of people, selecting an individual whose height was between 1.5 m and 2.0 m would be a certainty.

We can summarise the measurement of probability by saying that it ranges from a minimum of 0, which represents an impossibility, to a maximum of 1, which denotes a certainty. Probabilities are conventionally expressed as proportions (although sometimes you may see them shown as percentages) and they are conventionally denoted by P, as shown in the following examples.

If a litter of 8 piglets includes four males and four females, then the probability of selecting a female will be:

$$P(\text{female piglet}) = 0.5$$

If the heights of a group of individuals range from 1.5 m to 2.0 m, then the probability of selecting someone with a height of 2.05 m is given by:

$$P(\text{height} = 2.05\text{m}) = 0$$

Before moving on to discuss specific statistical distributions, there are some general principles of probability that we should consider.

There are three fundamental laws of probability of which we should be aware.

1. The law of addition: *The probability of either of two events occurring is equal to the sum of the probabilities of each event.*

This law of probability can be illustrated very clearly by considering the probability of the first piglet to be born in a litter being *either* a male *or* a female.

$$P(\text{male or female}) = P(\text{male}) + P(\text{female})$$
$$= 0.5 + 0.5 = 1$$

This is a certainty: a piglet must be either a male or a female, and we can represent these probabilities graphically using *Venn diagrams*. These diagrams essentially comprise circles that represent each possible outcome (in Figure 3.3., they symbolise the probabilities of the first piglet being male or female). Notice that the two circles are similar in size, denoting equal probabilities, and that there is no overlap, illustrating that the two events are mutually exclusive. Thus, the piglet can be male *or* female, but not both.

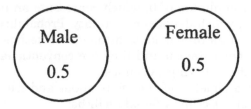

Figure 3.3. A Venn diagram representing the probability of a piglet being either male or female. The two circles do not overlap denoting that the piglet may be either male or female, but not both.

We can continue with this example by considering the probabilities of both the sex of the piglet and perinatal death. If we assume that the probability of a piglet dying within 21 days of birth is 0.05 and that the probability of a male piglet dying within 21 days of birth is 0.025, what is the probability of the piglet selected at random being male or dying within 21 days or both? We can denote the probabilities of each individual event as:

P(male) = 0.5

P(dies within 21 days) = 0.05

P(male and dies within 21 days) = 0.025

The corresponding Venn diagram is shown in Figure 3.4., where two overlapping circles represent the probability of the piglet being male (P = 0.5), the probability of the piglet dying within 21 days of birth (P = 0.05), and the overlapping region signifies the probability of the piglet being male and dying (P = 0.025). The probability of any of these three outcomes will be equal to the sum of the probabilities of the piglet being male and the piglet dying *minus* the probability of both being true. We subtract because the area of overlap would otherwise be counted twice (once within each circle), as can be seen clearly in the Venn diagram (Figure 3.4.).

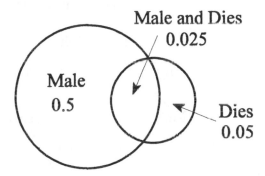

Figure 3.4. A Venn diagram representing the probabilities of a piglet being a male (P = 0.5), of dying within 21 days of birth (P = 0.05) or both (P = 0.025).

In mathematical notation, we can state this general law of addition as follows:

P(either event) = P(event 1) + P(event 2) − P(both events)

Applying it to the two examples:

P(male or female) = P(male) + P(female) − P(male and female)

P(male or female) = 0.5 + 0.5 − 0 = 1

and,

P(male or dies) = P(male) + P(dies) − P(male and dies)

P(male or dies) = 0.5 + 0.05 − 0.025 = 0.525

2.   The Law of Multiplication: *The probability of two independent events occurring is equal to the product of the probabilities of the two separate events.*

In mathematical notation, the probability of a male piglet being born and then surviving beyond 21 days is given below. We can visualise this result (and all the other possible outcomes) in the form of a tree diagram as shown in Figure 3.5.

$$P(\text{male then lives}) = P(\text{male}) \times P(\text{lives})$$
$$= 0.5 \times 0.95 = 0.475$$

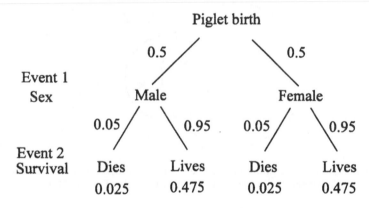

Figure 3.5. A tree diagram illustrating the law of multiplication with the four possible outcomes of the birth of two piglets with respect to their sex and survival. The final probabilities are obtained by multiplying the probabilities along each branch of the tree. The sum of the probabilities at the bottom of the diagram is equal to 1: it must include all possible outcomes.

In this type of diagram, the top branches lead to the possible outcomes of the first event (in this example a male or female). The branches are labelled with the respective probabilities of those outcomes. The possible outcomes of the second event are detailed at the next level down, again with their respective probabilities. Finally, the probabilities of the various combinations of outcomes, which are obtained by multiplying the probabilities along each branch of the tree, are noted at the bottom of the diagram. Notice that the sum of the final probabilities is equal to 1; they define the sum of the probabilities of all possible outcomes.

The probability that a piglet survives beyond 21 days given that the piglet is male is a *conditional probability*. In other words, the probability of a result being observed given that another, specified result has occurred previously. The conventional notation for expressing conditional probabilities is shown below. The vertical bar divides the outcome of the event that occurs second (lives) from that of the event that occurs first (male).

P(lives | male) = the probability that a piglet lives
given that it is a male = 0.95

We would denote the probability of a male piglet being born and surviving 21 days by:

P(lives and male) = P(male) × P(lives | male)

There is a subtle distinction between these two examples that needs to be highlighted. If we are confronted with a male piglet that has just been born, the probability that it will survive beyond 21 days is 0.95. The second mathematical statement denotes the probability that a piglet that has yet to be born will be male and will also survive beyond 21 days. Referring to the tree diagram (Figure 3.5.), the first statement is made halfway down with the sex of the piglet already known, whereas the second statement is made if we were right at the top of the diagram awaiting the birth of a piglet.

There are two key words that identify which of these two laws applies to a particular problem. If the word *or* is present linking the two events, then the addition law applies. The other keyword is *and*, the presence of which indicates that the multiplication law applies.

The final law of probability is one that we have already mentioned in passing. It is called the *law of complementarity*, which is based upon the fact that the probabilities of all possible outcomes of an event must add up to 1.

3. The Law of Complementarity: *The probability of an event occurring is equal to 1 minus the probability of that event not occurring.*

P(one piglet of each sex)

P(female then male) $= P(female) \times P(male \mid female) = 0.5 \times 0.5$
$= 0.25$

P(male then female) $= P(male) \times P(female \mid male) = 0.5 \times 0.5 =$
0.25

$\therefore$ P(one piglet of each sex)$= 0.25 + 0.25 = 0.5$

The two other possible outcomes are that the first two piglets born
are of the same sex, the probabilities of which would be:

P(both male) $= P(male \; AND \; male) = 0.5 \times 0.5 = 0.25$
P(both female) $= P(female \; AND \; female) = 0.5 \times 0.5 = 0.25$

P(two males) + P(both sexes) + P(two females) $= 0.25 + 0.5 + 0.25 = 1$

Box 3.2. Calculation of the probabilities of obtaining all possible combinations of two
piglets with respect to their sex. The calculations use both the law of addition and the
law of multiplication.

Stated formally in the context of the piglet example:

$$P(male) = 1 - P(female)$$

By using the first two laws in combination, we can calculate the
probability that the first two piglets born in a litter are a male and a
female, although not necessarily born in that order. The calculations
are shown in Box 3.2. and you could also represent the calculations
graphically in a tree diagram.

So far, we have discussed these laws of probability in general terms.
However, before moving on to an examination of their wider application, we should look at two specific conditions, *independent events* and
mutually exclusive events.

Independent events are those events where the outcome of one event
does not influence the outcome of another. For example, if two piglets
are born, then the fact that the first is female does not alter the
probability of the second also being female; there is still an equal
chance of the second being male or female. The example shown in
Figure 3.5. also illustrates independent events; the probability of survival of a newly born piglet does not depend upon its gender. In the
mathematical notation that describes probabilities, where two events

are independent in the context of these laws, we can reduce the general equation to:

$$P(A \text{ and } B) = P(A) \times P(B)$$

rather than

$$P(A \text{ and } B) = P(A) \times P(B \mid A)$$

because there is no conditional component to the relationship between the two events, A and B. In other words, when two events are independent,

$$P(B \mid A) = P(B)$$

or, in the context of the piglet example:

P(second piglet is male given that the first piglet is female)

$$P(\text{male} \mid \text{female}) = P(\text{male})$$

When two (or more) events are mutually exclusive, we mean that they cannot occur simultaneously. Therefore, in general terms we can say that for two mutually-exclusive events A and B:

$$P(A \text{ and } B) = 0$$

or, in the context of the piglet example, we can rephrase this mathematical statement as:

$$P(\text{a piglet is male AND female}) = 0$$

The discussion of these three fundamental laws of probability has introduced, in general terms, many of the important principles that will be employed either implicitly or explicitly in subsequent discussions of statistical analysis. However, before moving on to consider specific statistical distributions, I think it would be worthwhile to spend a little time consolidating your understanding by completing those exercises at the end of the chapter that relate specifically to these laws of probability (Exercises 1.3).

The binomial distribution and binomial probabilities

The piglet illustration of probability used two discrete variables, sex and survival, both of which could take only two possible values, *male* or *female* (for the sake of argument, we will ignore the possibility of hermaphroditism) and *lives* or *dies*. This type of nominal data is referred to as binary since there are only two possible categories. Throughout biology we find many examples of binary data, such as: male and female; alive and dead; success and failure in hatching or in learning or performing a task; left and right; positive and negative; presence and absence in botanical surveys. We can ask questions like, is the probability of obtaining a litter of six males and two females consistent with a 1:1 sex ratio? Or, in a trial of a new antibiotic, does the ratio of dead:alive bacterial colonies indicate that the new agent is effective in eliminating the bacteria? These examples serve to illustrate the types of questions that can be answered using binary data. In this section, I shall describe the way in which the probability of such events can be calculated, and, in later chapters, we will see in detail exactly how such questions can be answered. The binomial distribution is used to calculate the probability of particular outcomes that arise from binary data.

In the piglet example, we were clearly considering binomial probabilities; there were only two possible outcomes of each event. The binomial distribution only describes *discrete* events, i.e., outcomes that can only take whole numbers or integers, which are often called *binomial* or *Bernoulli trials* (named after Jakob Bernoulli, a Swiss mathematician of the late 17th century). A series of such independent trials is called a *Bernoulli* or *binomial experiment*.

Often we need to evaluate the probability of obtaining a specific combination of outcomes from a series of binomial trials. For example, when two varieties of peas are crossed, they produce plants that bear either round peas (probability = 0.65) or wrinkled peas (probability = 0.35). We might ask what would be the probability of obtaining 3 pea plants that produce round peas if we examined the progeny of 4 trials?

We could answer this question by examining the different ways that three round pea plants could arise from four trials, calculating the probability of that set of results (Box 3.3.). However, with only four trials in the experiment, this method already appears to be very time consuming. What is needed is a general formula that will enable us, and, more importantly, a computer, to calculate the probabilities.

If we examine the various possible outcomes of the experiment listed in Box 3.3., we can see that there are four ways of obtaining

The 16 possible combinations of round and wrinkled peas that could arise from four trials are as follows (R = round and W = wrinkled):

RRRR   RRRW*   RRWR*   RWRR*   WRRR*   RRWW
RWWR   WWRR   WRRW   RWRW   WRWR   RWWW
WRWW   WWRW   WWWR   WWWW

The probability of obtaining three plants producing round peas *and* 1 plant producing wrinkled peas is calculated by the law of multiplication (note the use of the word *and* in relation to this law of probability):

$$P(\text{round}) \times P(\text{round}) \times P(\text{round}) \times P(\text{wrinkled})$$
$$= (0.65)^3 \times (0.35) = 0.096$$

But we have 4 different ways (marked by *) or *combinations* of obtaining 3 plants that produce round peas, so the probability of getting three pea plants producing round peas from an experiment of four trials is:

$$\text{No. of combinations} \times P(\text{3 round pea plants})$$
$$= 4 \times 0.096 = 0.384$$

Box 3.3. Calculation of the probability of obtaining three plants that produce round peas from a Bernoulli experiment consisting of four trials. It is equal to the probability of obtaining the specified outcome from a single trial multiplied by the number of trials. This is a general rule, which applies in all calculations of probability.

three pea plants that produce round peas. The probability of each of the possible combinations is 0.096. This probability was obtained thus:

$$[P(\text{round})]^3 \times [P(\text{wrinkled})]^1$$

If we generalise this formula, the probability of getting X successes in an experiment of n trials can be described by Equation 3.4 in which the probabilities of a success (e.g., a round pea plant) and failure (e.g., a wrinkled pea plant) are denoted by p and q, respectively. These letters (n, p, and q) are conventionally used in the statistical literature to describe binomial probabilities.

$$p^X q^{n-X} \tag{3.4}$$

Equation 3.4 provides the method by which we can calculate the probability of obtaining any one of the possible combinations. Jakob Bernoulli derived a formula that established how to calculate the number of combinations and, in conjunction with Equation 3.4, produced a general formula for calculating the probabilities for the outcomes of binomial or Bernoulli experiments. In general, the number of combinations for an experiment comprising any number of trials is given by Equation 3.5.

$$\frac{(no \ of \ trials)!}{(no. \ of \ successes)! \ (no. \ of \ failures)!} \tag{3.5}$$

The exclamation mark indicates a *factorial*, which is a shorthand method of stating, in the case of 4!

$$4! = 4 \times 3 \times 2 \times 1$$

Factorials can be calculated for any positive integer very simply, and most modern calculators possess a factorial function. On some occasions, we may find ourselves in the special situation of needing to utilise the factorial of 0, in which case its value is 1.

According to his formula, the number of combinations of getting, say, 3 round pea plants out of 4 trials is given by:

$$\frac{4!}{3!(4-3)!}$$

Use the general formula (Equation 3.5) to calculate the number of combinations of getting 2 plants that produce round peas in a Bernoulli experiment that comprises four trials. Check your answer with the list of all the possible combinations that are listed in Box 3.3.

The formula to calculate probabilities for an experiment is a combination of Equations 3.4 and 3.5, and is given in Equation 3.6.

$$P(X \ successes) = \frac{n!}{X!(n-X)!} p^X q^{n-X} \tag{3.6}$$

Apply this formula to calculate the probabilities of obtaining 0, 1, or 2 plants producing wrinkled peas in an experiment of 5 trials.*

You will see that the probability of X successes is calculated by reference to n, p, and q, and the value of q is dependent on that of p. Therefore, we can calculate the probability of X successes in a binomial experiment if we know n and p, the number of trials and the probability of a success, respectively. These two quantities are known as the parameters of the binomial distribution since we can describe a binomial distribution uniquely if we know the magnitude of these two parameters.

Minitab can calculate probabilities based upon many different statistical distributions including the binomial. There are a number of generic commands concerned with the calculation of probabilities and related quantities (examples are shown in Exhibits 3.1. and 3.2.). Each of these generic commands takes a subcommand that specifies the particular statistical distribution of interest, and it requires us to supply the parameters of the specific distribution. Therefore, once you have gained an understanding of these commands in the context of one statistical distribution, you possess versatile tools that can be applied readily to other important distributions such as the *Poisson* and *normal* without any great extra effort of learning. The same statistical principles apply and the structure of the Minitab commands are consistent across all the distributions.

The two Minitab commands that are of particular interest in the context of calculating binomial probabilities are **PDF** (probability density function) and **CDF** (cumulative probability density function). The first of these, **PDF**, calculates binomial probabilities for Bernoulli experiments using Equation 3.6. It can be used to calculate the probability of either a specified number of successes or be used to generate a listing of the probabilities of each of all possible outcomes. Examples of its use are illustrated in Exhibit 3.1., where it is applied to the problem of round and wrinkled peas. The commands can be accessed through the menus by the route:

CALC > Probability Distributions > binomial

Having followed this route, a dialog box will present you with the opportunity to enter the number of trials, the probability of success, and whether you wish to generate a probability or cumulative distribution function.

---

* The answers are: P(0 wrinkled) = 0.116; P(1) = 0.312; P(2) = 0.336.

```
MTB > PDF;
SUBC> BINOMIAL 5 0.35.
```

Probability Density Function

```
Binomial with n = 5 and p = 0.350000

          x        P( X = x)
          0          0.1160
          1          0.3124
          2          0.3364
          3          0.1811
          4          0.0488
          5          0.0053


MTB > PRINT C1
```

Data Display

```
C1
    0    1    2    4

MTB > PDF C1;
SUBC> BINOMIAL 5 0.35.
```

Probability Density Function

```
Binomial with n = 5 and p = 0.350000

          x        P( X = x)
       0.00          0.1160
       1.00          0.3124
       2.00          0.3364
       4.00          0.0488

MTB >
```

Exhibit 3.1. Two examples of the use of the **PDF** command. In the first example, where no column is specified in the main **PDF** command, Minitab generates the probabilities for all possible outcomes of the experiment, ranging from no successes (X = 0) through to every trial being successful (X = 5). In the second example, a column is specified and Minitab calculates the probabilities for only those outcomes that are stored in the column. The subcommand in each case specifies that binomial probabilities are to be calculated, the number of trials in the experiment ($n = 5$), and the probability of a success ($p = 0.35$).

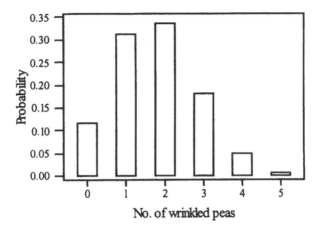

Figure 3.6. Plot of the probability density function for the wrinkled peas ($n = 5$, $p = 0.35$).

The probabilities of the various possible outcomes of such a binomial trial can be plotted in a bar chart (Figure 3.6.), which enables us to see very clearly the relative probabilities of each possible outcome. Since the bars each have similar widths, the probability of a specific outcome is proportional to both the height and the area of the bar relating to that outcome. The importance of this relationship between area of a graph and probability will be used extensively in the section on the normal distribution later in this chapter.

Looking at this histogram, we could say that in general we would expect 1 or 2 wrinkled pea plants from a Bernoulli trial of 5 plants, the peak probability is at 2 wrinkled pea plants. That is not to say that if we conducted a single trial we would necessarily expect about 1 or 2, but, that if many trials were performed, the average would be of that order. In fact, the mean of a binomial distribution is given by Equation 3.7.

$$mean\ (\mu) = np \tag{3.7}$$

Thus, in the case of the wrinkled peas, the mean would be:

$$mean = 5 \times 0.35 = 1.75\ \text{wrinkled pea plants,}$$

which is clearly reflected by the histogram in Figure 3.6. Although outcomes of binomial events can only be integers, averages can be real or decimal numbers. If we examined only 5 pea plants, we would not observe 1.75 pea plants that produce wrinkled seeds; but, if we examined many samples of 5 plants, we would expect that *on average* there would be 1.75 pea plants that produced wrinkled seeds. This is an example of the need to use statistics in biological research. For, while we can make mathematical predictions about the probable outcomes of experiments, we can only predict the *average* outcomes; there will always be some variation around the average.

It can also be shown that the standard deviation (σ) of a binomial distribution is given by Equation 3.8.

$$standard\ deviation\ (\sigma) = \sqrt{npq} \tag{3.8}$$

Therefore, the standard deviation of the wrinkled pea distribution is:

$$standard\ deviation = \sqrt{(5)(0.35)(0.65)} = 1.067\ wrinkled\ peas$$

In Equation 3.6, we calculated the probability of a number of successes, $P(X)$. On many occasions, we may be more concerned about the *proportion* of successes. For example, what proportion of a population that is exposed to an infection actually contracts the disease? What proportion of eggs laid hatch successfully? We calculate such proportions by dividing the number of successes (e.g., number contracting the disease, or hatching successfully) by the number of trials. Since we are dividing a binomial variable by a constant, such proportions can also be regarded as binomial probabilities. We will see later that proportions are very important in the biosciences. We can calculate the mean and variance of proportions very simply.

$$Proportion\ of\ successes = \frac{number\ of\ successes}{number\ of\ trials} = \frac{x}{n} = Prop_s$$

$$Proportion\ of\ failures = \frac{number\ of\ failures}{number\ of\ trials} = \frac{n-x}{n} = Prop_f$$

Mean proportion of successes is given by:

$$Mean\ of\ \left(\frac{x}{n}\right) = \frac{mean\ of\ x}{n} = \frac{np}{n} = p \tag{3.9}$$

and the variance of the proportion of successes is given by:

$$var\ of\left(\frac{x}{n}\right) = \frac{var\ of\ x}{n^2} = \frac{npq}{n^2} = \frac{pq}{n} \tag{3.10}$$

So far, the calculations have evaluated the probability of specific outcomes of binomial events. Sometimes, however, we may need to express the probability of obtaining, say, *less* than 3 pea plants that produce round seeds, i.e., the sum of the probabilities of getting 0, 1 or 2 successful outcomes. Here, we wish to calculate the probability of observing any one of a group of possible outcomes. The key word, *or*, is used so we need to apply the law of addition by calculating the sum of the probabilities of each of the individual outcomes (i.e., the sum of the probabilities of obtaining 0, 1 and 2 round pea plants).

$$P(0\ or\ 1\ or\ 2) = P(0) + P(1) + P(2)$$

If we look at the Minitab output in Exhibit 3.1., we can see that:

$$P(X < 3) = 0.1160 + 0.3124 + 0.3364 = 0.7648$$

Contrast this situation with the previous ones where we calculated the probability of individual, specific outcomes. The calculation above shows that this can be achieved using the **PDF** command and adding the relevant probabilities. Alternatively, we can use another Minitab command, **CDF**, which provides this function by calculating the *cumulative probabilities* for a specified number of successes in an experiment. Examples of this are shown in Exhibit 3.2., where we can see that the probability of obtaining less than 3 (i.e., $P(X \leq 2$ round pea plants)) successful outcomes is 0.2352. We may also use the law of complementarity in conjunction with **CDF** to calculate, for example, the probability of observing 4 *or more* round pea plants, which will be the same as the probability of a certainty (i.e., $P = 1$) minus the probability of obtaining less than 4 round pea plants:

$$P(X \geq 4) = 1 - P(X < 4) = 1 - 0.5716 = 0.4284$$

As with the probability density function, we can plot the cumulative distribution function in the form of a bar chart (Figure 3.7.).

```
MTB > CDF;
SUBC> BINOMIAL 5 0.65.
```

Cumulative Distribution Function

```
Binomial with n = 5 and p = 0.650000

          x       P( X <= x)
          0         0.0053
          1         0.0540
          2         0.2352
          3         0.5716
          4         0.8840
          5         1.0000

MTB > CDF 2;
SUBC> BINOMIAL 5 0.35.
```

Cumulative Distribution Function

```
Binomial with n = 5 and p = 0.350000

          x       P( X <= x)
       2.00         0.7648

MTB > CDF 3 K1;
SUBC> Binomial 5 0.65.
MTB > LET K2=1-K1 # K2=Prob.of 4 or more wrinkled pea plants
MTB > PRINT K1 K2
```

Data Display

```
K1        0.571585
K2        0.428415

MTB >
```

Exhibit 3.2. Three examples of the **CDF** command. The top example shows the complete c.d.f. for selecting round pea plants (probability of a success = 0.65) in 5 trials. The second shows the c.d.f. for 2 or less successes (in this example a success is a wrinkled pea plant, $p = 0.35$). Finally, the commands used to calculate the probability of obtaining 4 or more successes is shown; the law of complementarity is used, hence the need for the constant K2. The # symbol allows a note to be added at the end of a Minitab command line.

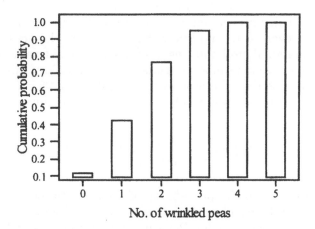

Figure 3.7. Cumulative probability distribution of the number of wrinkled peas (n = 5, p = 0.35).

The Poisson distribution and Poisson probabilities

In each of the preceding examples where trials were described by a binomial distribution, there was a maximum number of successes that could result from each trial. In Figures 3.6. and 3.7., for example, the maximum number of plants that could produce wrinkled peas was 5 because only 5 plants were produced; the upper limit of successes was fixed. In some areas of biology, as well as other fields of study, situations arise where the maximum number of successes is so large that we can consider that it is effectively limitless. Moreover, on many occasions, we may be unable to determine the maximum number of successes. For example, consider a study of the dispersion pattern of limpets (*Patella vulgaris*) on a rocky shore. We might throw quadrats (of area 0.25 m²) over a delineated area of the shore and count the number of limpets found in each quadrat. The probability of a particular limpet being found in a quadrat will be low because the total number of limpets is large compared to the area covered by the quadrats. However, because there are many limpets and the size of a quadrat is large compared to the size of a limpet (a limpet might occupy an area of only 0.002 m²), many quadrats may contain limpets. In statistical terms, the number of trials is large (i.e., the number of quadrats) and the probability of a success (a particular limpet being found in a quadrat) is small.

Since we cannot define the maximum number of successes, we will not be able to calculate the binomial probabilities using Equation 3.6. Furthermore, even if we did know the maximum, the actual calculation of the binomial probabilities would pose considerable problems when there are a large number of trials. As an example of this difficulty, the formula to calculate the probability of obtaining 25 successes from 100 trials with the probability of a success of 0.1 is given below.

$$P(25) = \frac{100!}{25!(100-25)!} 0.1^{25} 0.9^{75}$$

Fortunately, when the probability of a successful outcome is relatively small and the number of trials is large, the binomial distribution reduces to the much simpler Poisson distribution (named after the French mathematician, Siméon-Denis Poisson). As with the binomial distribution, the Poisson relates to data that are discrete counts or frequencies. Two other constraints also apply to the Poisson distribution, and they are:

(1) that each recorded object (in this example, limpets) is only a small fraction of the population from which the sample was taken, i.e., that many other limpets could have been found in the quadrats.

(2) that each observed object is independent of all others or, in other words, the presence of one limpet in a quadrat does not make the presence of another more or less likely; each *success* is a random event.

These two constraints have led to this distribution often being referred to as the distribution of rare or random events. It is used in many areas of biology, particularly where random events, or more often, deviations from randomness, are investigated. The Poisson distribution can be applied to situations where we are sampling objects in time or space. For example, we may wish to examine the frequency of release of packets of neurotransmitters from a synapse per unit time, the spacing behaviour of breeding birds, or the dispersion of plants over a defined area. The Poisson distribution can be utilised in the investigation of all of these problems.

In fact, in biology, we are generally interested in why dispersion patterns are not random. Such non-randomness is indicative of biological processes at work. If plants in a study area are clumped, this might indicate a heterogeneous distribution of soil variables or provide information regarding the mode by which a plant propagates or disperses. Epidemiologists might observe clusters of occurrences of a particular disease in a specific location, which could point to local environmental factors that contribute to the incidence of that disease. In contrast, regular spacing, where each individual is as far apart as possible from other individuals in the population, is indicative of competition, and this can be seen in the dispersion of nests of colonially nesting birds where space is limiting.

Having established the need for the Poisson distribution and where it might be applied, we can now move on to see how Poisson probabilities are calculated (Equation 3.11).

$$P(r) = \frac{e^{-\mu}\mu^r}{r!} \tag{3.11}$$

This equation will calculate the probability of a specific outcome of a process, which is described by a Poisson distribution (often known as a *Poisson process*), where r denotes the number of successes and μ denotes the mean number of successes. In terms of the quadrat example, $P(r)$ denotes the probability of obtaining a quadrat containing r limpets, and μ represents the mean number of limpets per quadrat (averaged over all quadrats in the survey). The mean is calculated very simply from empirical data (i.e., data gathered in the field). It is the mean number of successes per trial, or, in terms of the quadrat example:

$$\mu = \frac{total\ no.\ of\ limpets\ in\ all\ quadrats}{total\ no.\ of\ quadrats} \tag{3.12}$$

The symbol, e, a mathematical constant, which has a value of 2.3026 and is the base of natural or Napierian logarithms, is often referred to as the *exponential constant*. It is very important in many areas of science, and, in particular, it is a very important component of equations that describe growth processes in biology. We will see later that there is a Minitab function that can be used in calculations that incorporate e.

We can see that there is only one unknown parameter for the Poisson distribution, namely the mean. As a consequence, we can calculate Poisson probabilities with knowing only the mean, which, as

$$P(3) = \frac{e^{-\mu}\mu^r}{r!} = \frac{e^{-2}2^3}{3!} = \frac{(0.1353)(8)}{6} = 0.1804$$

Box 3.4. Calculation of the Poisson probability of 3 limpets in a quadrat, where the mean of the distribution is 2.

Equation 3.12 shows, is calculated from the sample. Therefore, the *mean* is the sole parameter of the Poisson distribution. A further characteristic of the Poisson distribution is that the variance of the distribution is equal to the mean. Thus, the calculation of Poisson probabilities is far simpler than those of the binomial distribution. An example calculation of Poisson probabilities based on this littoral survey of limpets is shown in Box 3.4. Try calculating the probability of obtaining 5 limpets per quadrat when the mean number of limpets per quadrat is 2 (check your answer against the Minitab example, Exhibit 3.3.).

As with the binomial distribution, the Minitab commands **PDF** and **CDF** can be used with appropriate subcommands to calculate Poisson probabilities as shown in Exhibit 3.3., and the corresponding probability histograms displayed in Figures 3.8. and 3.9. The commands can be accessed through the menus by a similar route to that used earlier for binomial probabilities:

CALC > Probability Distributions > Poisson

Once at the dialog box, you have to enter the mean, which is the parameter of the Poisson distribution.

As with the binomial distribution, we can calculate the frequencies of a particular number of successes that we would expect to observe by multiplying the probability of the number of successes by the number of trials. Therefore, in the limpet example, if 50 quadrats were used in the survey then we would expect the following number of quadrats to contain 3 limpets:

expected frequency = *(no. of quadrats)(probability of expected outcome)*
                  = *(50)(0.1804)*
                  = *9.02 quadrats*

How many quadrats would be expected to contain 4 limpets if 65 quadrats had been used in the survey?*

---

* Expected frequency = (65 quadrats)(0.0902) = 5.863 quadrats.

```
MTB > PDF;
SUBC> POISSON 2.
```

Probability Density Function

```
Poisson with mu = 2.00000

        x        P( X = x)
        0          0.1353
        1          0.2707
        2          0.2707
        3          0.1804
        4          0.0902
        5          0.0361
        6          0.0120
        7          0.0034
        8          0.0009
        9          0.0002
       10          0.0000
```

```
MTB > CDF;
SUBC> POISSON 2.
```

Cumulative Distribution Function

```
Poisson with mu = 2.00000

        x        P( X <= x)
        0          0.1353
        1          0.4060
        2          0.6767
        3          0.8571
        4          0.9473
        5          0.9834
        6          0.9955
        7          0.9989
        8          0.9998
        9          1.0000
```

```
MTB >
```

Exhibit 3.3. Examples of the use of the **PDF** and **CDF** commands in Minitab to calculate Poisson probabilities. The mean of the distribution, μ, is equal to 2.

The normal distribution and normal probabilities

The binomial and Poisson distributions describe discrete events. The sex of a piglet can only be male or female, and the number of limpets in a quadrat can only be measured in whole numbers; there are no possible intermediate values. Many variables in both biology and other subject areas, however, are continuous, i.e., the variables take values

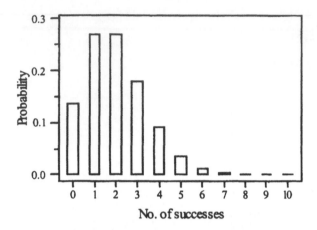

Figure 3.8. The probability density bar chart of a Poisson distribution with a mean (μ) of 2. The actual probability values are shown in the Minitab example, Exhibit 3.3.

that are real numbers. The *normal distribution,* which will be described in this section, is the most important distribution in statistics and it is a continuous distribution. In this section, I shall outline why the normal distribution is so important and how, under certain circumstances, the binomial and Poisson distributions approximate to

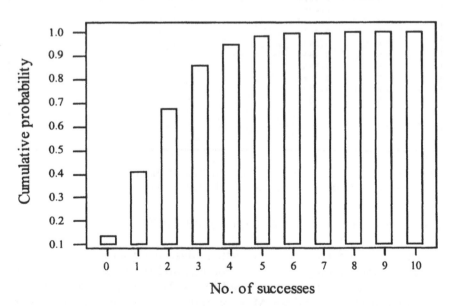

Figure 3.9. The cumulative frequency bar chart of a Poisson distribution (μ = 2).

the normal. Finally, we will see how to calculate and use normal probabilities.

In the discussion of measures of reliability and standard errors earlier in this chapter (Section 3.2), the central limit theorem was mentioned. Without going into mathematical detail, this theorem proves that if we take many samples from a population, the means of those samples will conform to a normal distribution, irrespective of whether the original population is normally distributed (as illustrated in Figure 3.2.). The underlying reason behind this relationship is that the central limit theorem states that any variable that is the sum of many random variables will be distributed normally. The sample means (which conform to a normal distribution) are each the result of a number of random events, namely the individual measurements that comprise each sample. In biology, this is a common occurrence, with many variables being the sum of many other random factors. For example, the height of an individual, the concentration of glucose in rats' blood, and the yields of crops are all the results of combinations of many genetic and environmental factors. Each case gives rise to biological variables that follow a normal distribution. It is for this reason that the normal distribution is not only extremely important in the biological sciences, but also central to many of the analytical techniques that are described in later chapters.

If we continue this line of argument where the sum of random events leads to a normal distribution, we can identify biological examples where the outcome of a binomial experiment will approximate to the normal distribution. An imaginary example based upon simple genetics illustrates this approximation very clearly.

Let us imagine that a particular phenotypic character is controlled by 5 unlinked genes. Assume also that each gene has 2 alleles, 1 of which is completely dominant over the other. If we know that, for each gene, the probabilities of occurrence of the dominant and recessive alleles are 0.2 and 0.8, respectively, then we can regard the phenotypic expression of this character in each individual organism as the result of a binomial experiment comprising 5 trials (the 5 genes). We can express the details of this experiment as follows:

n = no. of binomial trials = no. of genes = 5

p = probability of occurrence of a dominant allele = 0.2

q = probability of occurrence of a recessive allele = 0.8

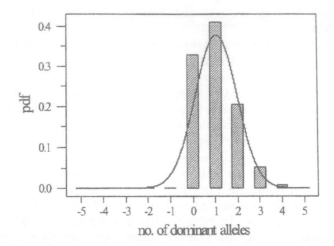

Figure 3.10. The binomial distribution (plotted as a histogram) that describes the number of dominant alleles that would be expected if there were 5 genes and the probability of a dominant allele was 0.2. Superimposed upon the binomial distribution is the expected normal distribution.

The mean and standard deviation of this binomial distribution are 1 and 0.89, respectively (as calculated by Equations 3.7 and 3.8). The binomial distribution that describes this experiment is displayed as a histogram in Figure 3.10. Superimposed upon this histogram is the theoretical normal distribution that has a similar mean and standard deviation. Although there is a close resemblance between the two distributions, there is a clear difference in terms of their symmetries. The normal distribution is perfectly symmetrical while the binomial is not.

The central limit theorem suggests that, with increasing numbers of random variables (or trials), the binomial distribution should more closely approximate to a normal distribution. Let us imagine, therefore, a similar genetic situation that differs in only one respect from the previous example, namely that 20 genes are involved rather than 5. Figure 3.11. illustrates the resultant binomial and normal distributions. The increased similarity between the two distributions is clear without having to delve into the mathematical theory underlying the central limit theorem.

Chatfield (1970) states that if the number of Bernoulli trials (in this example, genes) exceeds 20, then the two distributions match each other well for ranges of p (the probability of a success in the individual trials) between 0.3 and 0.7. The approximation is valid over a much larger

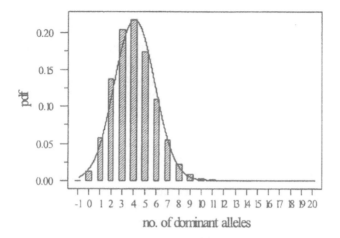

Figure 3.11. The binomial distribution (plotted as a histogram) that describes the number of dominant alleles that would be expected if there were 20 genes controlling the character. The probability of a dominant allele appearing is 0.2. Superimposed upon the binomial distribution is the expected normal distribution. There is a much closer match between the two distributions with 20 genes than with 5 genes as shown in Figure 3.10.

range if the number of trials is increased. Many authors (e.g., Chatfield, 1970; Clarke and Cooke, 1992) have offered rules of thumb for handling binomial probabilities, and these are summarised in Table 3.1.

Although these examples have been introduced to illustrate the relationship between the binomial and the normal distributions, they also demonstrate the fundamental biological process of continuous variation being derived from discrete events. If we assume that each occurrence of a dominant allele adds one unit of magnitude to the phenotypic character, then, with characters that are controlled by a

Conditions	Method and Distribution
(a) n large and p close to zero ($n > 50$ and $p \leq 0.1$)	Use Poisson approximation to the binomial with *mean* = *np*.
(b) n large and p not close to 1 ($np > 5$ and $nq > 5$)	Use normal approximation with *mean* = *np* and *variance* = *npq*.
(c) n is small	Use binomial distribution according to Equation 3.6.

Table 3.1. Rules of thumb governing the application of the binomial distribution and its approximation by the Poisson and normal distributions.

large number of genes, the contribution of a single gene will be relatively small, perhaps even smaller than the experimental error that
is intrinsic in measuring that character. Such genetic variation,
although discrete in its action, would thus appear as continuous phenotypic variation. Furthermore, the phenotype is generally the result
of interaction between genetic and environmental factors, and variation in the latter is continuous.

We saw earlier that when sample sizes are large the binomial
distribution can be closely approximated by the Poisson. It should come
as no surprise, therefore, to learn that the Poisson will approximate
to the normal distribution under certain conditions. In a similar way
to the binomial distribution, calculating Poisson distributions can
become very time consuming when large numbers are involved, particularly when the mean of the distribution is large. Conveniently, the
approximation to the normal is very good when the mean of the Poisson
distribution is large (greater than 10). This approximation is illustrated in Figure 3.12. for a Poisson distribution with a mean and
variance of 11.

On many occasions, we will take samples without knowing the form
of the distribution that describes them. However, the normal approx-

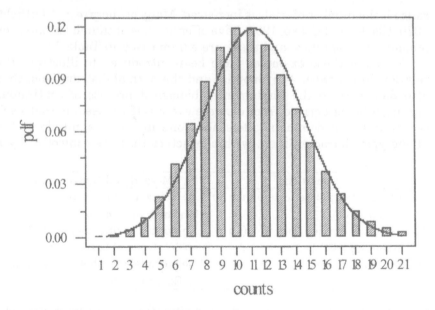

Figure 3.12. A histogram of a Poisson distribution (mean = 11) and the corresponding
normal distribution; there is a close similarity.

imations to the binomial and Poisson distributions provide an important general conclusion, namely that when we have large sample sizes, the distribution describing our measurements (irrespective of its form) will be approximated by a normal distribution. Therefore, by taking large samples, we can make a reasonable assumption that the distribution of our data set is consistent with a normal distribution. We shall see several examples of this conclusion in action in some of the analytical techniques described in later chapters.

As a consequence of the great importance of the normal distribution in quantitative science in general, many analytical techniques are based upon this distribution. Therefore, we need to gain a good working knowledge of normal probabilities and their interpretation in order to understand the key principles of conducting investigations in the biological sciences.

In the introduction to this chapter, I stated that the sample mean and variance were examples of statistics that we used in estimating population parameters. In the subsequent discussion of the binomial and Poisson distributions, the parameters by which these two distributions are characterised were introduced (the *number of trials* and *probability of a success* for the binomial; the *mean* for a Poisson distribution). All distributions can be characterised in a similar fashion. In the case of the normal distribution, the parameters are the mean, μ, and the standard deviation, σ. They can be seen in Equation 3.13, which describes the probability density function of the normal distribution for a specified value, X, of a measured variable (e.g., a specific concentration of cholesterol in the blood) and the mathematical constants, e and π. Fortunately, as biologists, we will never need to manipulate this equation directly. I include it here only for the sake of a complete record; we employ statistics packages to utilise this equation on our behalf!

$$pdf\ (X) = \frac{1}{\sigma\sqrt{2\pi}}\ e^{-(X-\mu)^2/2\sigma^2} \tag{3.13}$$

As with the binomial and Poisson distributions, we can calculate both the probability and cumulative density functions for a normal distribution. However, in contrast to the binomial and Poisson distributions, which relate to data that are in the form of discrete measurements or frequencies, the normal distribution describes continuous data. This fundamental difference requires a completely different approach to the way we view measurements of continuous variables

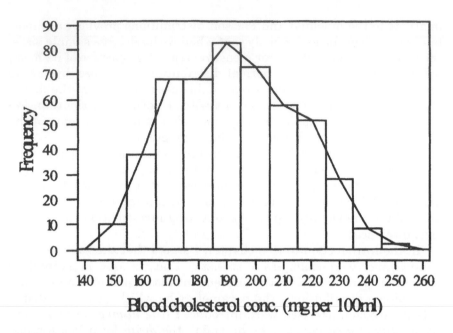

Figure 3.13. Frequency histogram and frequency polygon of 500 blood cholesterol concentrations (mg.100 ml⁻¹).

and consequently, in the way we handle probabilities relating to continuous variables.

For example, we can count very precisely how many limpets fall inside a quadrat* or the number of mites per leaf on an apple tree. However, we cannot say that the blood cholesterol level of an individual is *exactly* 180 mg.100 ml⁻¹ because the true level may be higher or lower by an infinitesimally small amount that is beyond the resolution of our measuring instruments. Furthermore, when we need to plot a frequency histogram of continuous data, we are presented with other difficulties. If we collected 500 blood samples and measured the cholesterol concentrations, we could plot a frequency histogram of the data (Figure 3.13.). However, the way in which we divide the x-axis into bars of a specified width is necessarily arbitrary because there are no natural divisions in continuous variables. As an alternative, we can plot the frequency data by joining the midpoints of the tops of the

---

* When using quadrat or grid square sampling methods, it is generally accepted that when an organism falls on the margin, we record it as inside the square if it is on one of two previously specified edges (e.g., the top or left) and outside the square if it is on either of the other two edges (e.g., bottom or right).

bars to produce a graph (or *frequency polygon*), also shown in Figure 3.13. The smoothness of the graph will improve with increasing volume of data and as the width of the bars decreases. Whatever we do, however, there will always be a certain element of subjectivity because we are applying techniques to continuous variables that have been designed for discrete variables.

Similarly, although we can calculate the probability of observing, say, 6 males in a litter of 8 piglets (the number of male piglets is a discrete variable), we cannot estimate the probability of randomly selecting an individual whose blood cholesterol level is exactly, say, 185 mg.ml^{-1}. However, this is not as catastrophic as it may appear since, if we cannot measure cholesterol levels exactly (or any other continuous variable for that matter), then there can be little practical value in being able to calculate the probability of finding an exact value of the variable under investigation. For continuous variables, therefore, we calculate the probability of selecting an individual who falls within a specified range of values. For example, we might calculate the probability of selecting an individual whose blood cholesterol concentration falls within the range 180–200 mg.100 ml^{-1}, or of selecting one whose concentration exceeded 190 mg.100 ml^{-1}.

In our consideration of binomial and Poisson probabilities, we used the cumulative distribution function to calculate the probability of ranges of outcomes (Minitab examples were illustrated in Exhibits 3.2. and 3.3.). The CDF is put to similar good use in calculating meaningful probabilities based upon the normal distribution.

We can see how cumulative distribution functions aid us in calculating useful normal probabilities in the following example based upon blood cholesterol levels. In this example, we have a normal distribution of cholesterol concentrations, the mean and standard deviation, being 193.13 and 22.3 mg.100 ml^{-1}, respectively. Let us imagine that we wish to calculate the probability of randomly selecting an individual whose cholesterol level falls within the range of 180–200 mg.100 ml^{-1}.

The cumulative density function of this population is shown in Figure 3.14. The shaded area in the graph covers the range of concentrations in which we are particularly interested. The y-axis of the CDF represents the probability of selecting an individual whose cholesterol level is less than or equal to a specified value. So, in this graph we can see that the probability of having a cholesterol level less than or equal to 180 mg.100 ml^{-1} is just under 0.3 (in fact, it is 0.278), and, for 200 mg.100 ml^{-1}, the probability is 0.621. If we wished to establish the probability of randomly selecting an individual whose blood cholesterol fell within this range, we could subtract the probability of

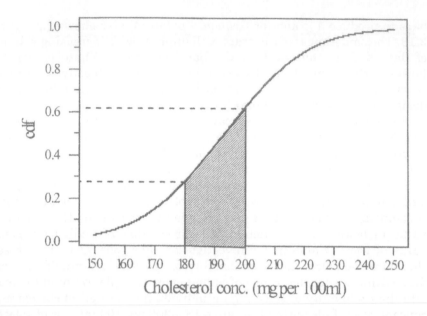

Figure 3.14. The cumulative distribution function of a normal distribution ($\mu = 193.13$ and $\sigma = 22.3$ mg.100 ml^{-1}). The shaded area represents the probability of selecting an individual from the range 180–200 mg.100 ml^{-1}.

selecting a person whose cholesterol was less than 180 mg.100 ml^{-1} from the probability of selecting someone whose cholesterol was less than 200 mg.100 ml^{-1}. The subtraction is shown in Exhibit 3.4. where the Minitab command **CDF** has been used to calculate the exact probabilities (the probability is 0.343). As with the binomial and Poisson distributions, probability and cumulative density functions can be obtained through the menu system very simply:

CALC > Probability Distribution > normal

The dialog box requires the mean and standard deviation to be entered as well as a column of data for which the PDF or CDF will be calculated (the x-axis of the distribution function).

There are an infinite number of normal distributions, the exact form of each depending upon the magnitude of its parameters (μ and σ). A standard normal curve has been devised to which every normal distribution can be reduced. By using this simple standardisation procedure, every normal curve can be made equivalent and thus be compared in a relatively straightforward way.

```
MTB > Note: The following CDF command calculates the cdf for
MTB > Note: cholesterol conc. of 180, and stores the ans in K1
MTB > CDF 180 K1;
SUBC> NORMAL 193.13 22.3.
MTB >
MTB > Note: The following CDF command calculates the cdf for
MTB > Note: cholesterol conc. of 200, and stores the ans in K2
MTB > CDF 200 K2;
SUBC> NORMAL 193.13 22.3.
MTB >
MTB > Note: The area between 180 and 200 is stored in K3
MTB > LET K3=K2-K1
MTB > PRINT K1-K3

Data Display

K1        0.278001
K2·       0.620986
K3        0.342985
MTB >
```

Exhibit 3.4. The Minitab procedure for calculating normal probabilities using the **CDF** command. The probability of randomly selecting an individual whose cholesterol is in the range 180–200 mg.100 ml^{-1} is 0.343.

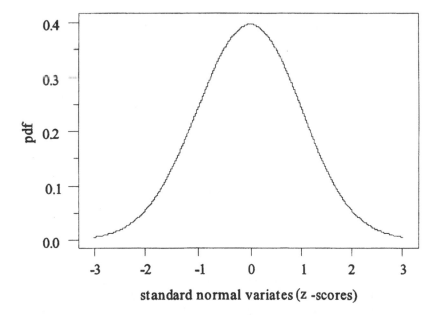

Figure 3.15. The probability density function of the standard normal distribution ($\mu = 0$, $\sigma = 1$).

There are several uses for this standard normal curve. For example, by being able to standardise every possible normal distribution, a single set of normal probability tables (Appendix A.1.) can be used for every normal distribution. Furthermore, we shall see in Section 3.4 that confidence intervals, which are an extremely important statistical tool, are calculated by employing the properties of the standard normal curve.

The standard normal curve is created by rescaling the horizontal axis of any normal distribution in terms of standard deviations from the mean. This process defines the mean of a standard normal curve as *zero*. The units of the *x*-axis of the standard curve are standard deviations and are generally referred to as *standard normal variates*, denoted by *z* (hence their other name of *z*-scores). The standard normal distribution is illustrated in Figure 3.15.

The formula to perform this standardisation is given by Equation 3.14.

$$z = \frac{X - \mu}{\sigma} \tag{3.14}$$

where, X denotes the raw measurement and μ and σ are the mean and standard deviation, respectively, of the population under consideration. Each raw measurement, once standardised, is expressed in terms of its distance from the mean in units of standard deviations or standard normal variates. Therefore, in terms of the distribution of cholesterol concentrations, where the mean and standard deviation were 193.13 and 22.3 mg.100 ml^{-1}, the standardised z-score of, for example, 180 mg.100 ml^{-1} is given by:

$$z = \frac{180 - 193.13}{22.3} = -0.59 \; \textit{standard normal variates}$$

In other words, a cholesterol concentration of 180 mg.100 ml^{-1} is 0.59 standard deviations below the mean.

Normal probability tables relate the area under a standard normal curve (which is equivalent to probability) to z-scores, and some of these are shown in Table 3.2. For example, 50% of the area under the curve is within 0.674 standard deviations on either side of the mean. In other words, there is a probability of 0.5 of randomly selecting an individual that is within 0.674 standard deviations of the mean.

Percentage of area under the curve	Number of standard deviations (z-scores) above and below the mean
50%	0.674
90%	1.645
95%	1.960
99%	2.576

Table 3.2. The limits (in terms of standard normal variates) which enclose specified percentages of area under a normal curve.

Although computer-based statistics packages are an easy route to obtaining probability values (as illustrated in Exhibit 3.4.), anyone who uses statistics has to refer to statistical tables at some time or another. There are many useful statistical distributions, and statistical tables can be presented in a variety of formats. Thus, it is important to gain a sound understanding of the basic principles of using statistical tables. A table of areas under the standard normal curve is presented in Appendix A.1. This normal distribution table expresses the area/probability *up to and equal to* the calculated values of z (i.e., in the shaded area of the distribution as shown below Appendix A.1., the table of normal probabilities).

In order to see how tables are used, let us examine two examples: $z = 0.0$ and $z = +1.96$. In the former case, the value in tables is given as 0.5, which indicates that half of the area under the standard normal curve lies below the value of 0. But remember that the mean of the standard normal distribution is 0. So the tables indicate that half of the area under the curve lies below the mean, which is quite logical. In other words, there is a probability of 0.5 that an individual selected at random will fall below the mean. By subtraction, we can also conclude that the probability of selecting an individual above the mean is 0.5 (using the law of complementarity).

In the second example, where $z = +1.96$, the probability value in tables is 0.975, i.e., the proportion of the area under the curve that is below $z = +1.96$. The remaining proportion of the total area must, therefore, be 0.025 (according to the law of complementarity: $1 - 0.975$), indicating that 2.5% of the population lies above a value equivalent to $z = +1.96$. According to Table 3.2., 95% of the area under the standard normal curve lies in the region between 1.96 standard deviations on either side of the mean, i.e., in the range -1.96 to $+1.96$ (as shown in Figure 3.16.). Or, in other words, 5% of the area lies outside that range: 2.5% below -1.96 and 2.5% above $+1.96$. These regions at the extremes of this (or any other) distribution are referred to as *tails*. This rela-

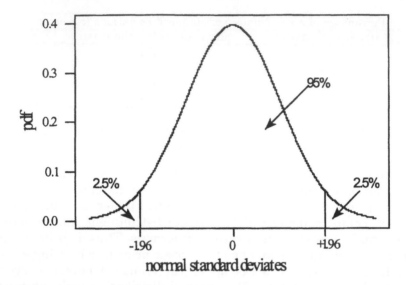

Figure 3.16. A standard normal distribution showing the proportions of the area under the curve within and outside the range –1.96 to +1.96 normal standard deviates.

tionship between standard errors, areas under the curve, and probability is possible because we are able to reduce all normal distributions to this standard normal distribution. The importance of this relationship will be illustrated in some detail in Section 3.4.

A further example of the use of z-scores and tables of normal probabilities is shown in Box 3.5. where the probability is calculated

---

Calculation of z-scores:

$z = (195 - 193.13) \div 22.3$ $= 0.084$
$z = (205 - 193.13) \div 22.3$ $= 0.532$

Using normal probability tables:

$P(z \le 0.084)$ $= 0.5319$
$P(z \le 0.532)$ $= 0.7019$
$\therefore P(195 \le X \le 205)$ $= 0.7019 - 0.5319 = 0.17$

Box 3.5. The probability of selecting an individual with a blood cholesterol concentration between 195–205 mg.100 ml^{-1} from a population with a mean and standard deviation of 193.13 and 22.3 mg.100 ml^{-1}, respectively, is equal to 0.17.

of finding an individual whose blood cholesterol concentration is in the range of 195–205 mg.100 ml⁻¹. The same computation using the Minitab **CDF** command is shown in Exhibit 3.5.

The slight difference in the answers between tables and Minitab occurs because the probabilities are calculated to more decimal places when using Minitab than when reading statistical tables.

Another example illustrates the additional aspect of calculating probabilities, namely where one value is below the mean and the other is above. In this second example, let us imagine that we wish to calculate the probability of randomly selecting an individual whose blood cholesterol concentration falls within the range 140–198 mg.100 ml⁻¹. Try to calculate the probability using Minitab by adapting the method illustrated in Exhibit 3.5. You will need to substitute 140 and 198 for 195 and 205, respectively.*

The calculation of this probability using z-scores and normal probability tables is shown in Box 3.6. In this calculation, we obtain a negative z-score for the lower cholesterol concentration (140 mg.100 ml⁻¹) since it is lower than the mean. We use the law of complementarity to calculate the probability of obtaining a z-score below −2.383; it is the equivalent of the probability of obtaining a z-score above +2.383.

Obviously, if we are considering two measurements, which are both below the mean of the population, then we would have to utilise the law of complementarity to obtain probability values for both measurements.

The preceding examples, based upon three important statistical distributions, have illustrated key relationships between raw measurement, mean, variance, standard error, and probability. The principles involved in these relationships are parallelled in other distributions. In later sections, we will encounter both calculations based upon similar relationships, and see how probability tables based upon these various distributions are used as part of hypothesis testing.

3.4 Confidence intervals

So far, we have seen the reasons *why* the standard error of the mean can be used as a measure of reliability, but not the methods of *how* it can be used. I have also mentioned that standard errors can be calculated for any population parameter. By bringing together the concept

---

* The answer is 0.5778.

```
MTB > CDF 195;
SUBC> NORMAL 193.13 22.3.
```

Cumulative Distribution Function

```
Normal with mean = 193.130 and standard deviation = 22.3000

        x        P( X <= x)
  195.0000          0.5334
```

```
MTB > CDF 205;
SUBC> NORMAL 193.13 22.3.
```

Cumulative Distribution Function

```
Normal with mean = 193.130 and standard deviation = 22.3000

        x        P( X <= x)
  205.0000          0.7027
```

```
MTB > Note: The overall probability is given by
MTB > Note: differences between two individual
MTB > Note: probabilities and is stored as K1.
MTB > NAME K1 'Prob'
MTB > LET K1= 0.7027-0.5334
MTB > PRINT K1
```

Data Display

```
Prob    0.169300
MTB >
```

Exhibit 3.5. Minitab calculation of normal probabilities using the **CDF** command. The probability of selecting an individual whose blood cholesterol concentration is between 195 and 205 mg.100 ml^{-1} is 0.1693.

of the standard error and the properties of the normal distribution that have just been introduced, we will be able to use this measure of reliability to estimate population parameters with a quantifiable degree of precision.

Confidence interval of a mean

Recall that when we took many samples and constructed a distribution of the sample means, the distribution was normal. Furthermore, the standard deviation of that distribution of sample means was the standard error of the mean. When we examined the properties of the normal distribution, we saw that 95% of the area under the standard

---

Calculation of z-scores:

$z = (198 - 193.13) \div 22.3 = +0.218$
$z = (140 - 193.13) \div 22.3 = -2.383$

Using normal probability tables:

$P(z \leq +0.218)$ $= 0.5871$
$P(z \leq -2.383) = 1 - P(z \geq +2.383) = 1 - 0.9913$ $= 0.0087$
$\therefore P(140 \leq X \leq 198) = 0.5871 - 0.0087$ $= 0.5784$

Box 3.6. The probability of selecting an individual with a blood cholesterol level in the range 140–198 mg.100 ml^{-1} from a population with a mean and SD of 193.13 and 22.3 mg.100 ml^{-1}, respectively, is equal to 0.5784.

normal distribution was found within 1.96 standard errors either side of the mean. It follows, therefore, that a range of 1.96 standard errors above and below the mean of this distribution of sample means will include 95% of all the sample means.

An alternative way of considering this is in terms of a sample mean, $\bar{x}$, of n individuals, and, more specifically, in terms of the ratio

$$(\bar{x} - \mu) / (\sigma / \sqrt{n})$$

which is the difference between the sample and population mean expressed as standard deviates. Since these standard deviates are distributed normally, 95% of them will be included within the range −1.96 and +1.96, and thus, we can say that there is a probability of 0.95 that this range, or *confidence interval,* includes the population mean. The upper and lower margins of the confidence interval are known, respectively, as the *upper* and *lower confidence limits.* Alternatively, we can say that, if we take many samples and calculate the 95% confidence interval for each one, we would expect 95% of them to include the true population mean. Up to now we have used the sample mean as a *point estimate* for the population mean. The confidence interval is an example of an *interval estimate.*

The formula to calculate the 95% confidence interval for a mean follows directly from this argument and is shown in Equation 3.15. Moreover, since we can calculate standard errors for any parameter, it follows that we can calculate confidence intervals for any population

---

95% Confidence Interval of the population mean = sample mean
± (1.96 × s.e. mean)

Upper Limit (95%) = 72.87 + 1.96 × 1.15 = 75.12
Lower Limit (95%) = 72.87 − 1.96 × 1.15 = 70.62

Box 3.7. Calculation of the 95% confidence interval for the mean resting pulse rate of adults.

parameter. In this section, I shall describe the methods for calculating confidence intervals for the mean and proportion of populations. In later chapters, we will encounter the calculation of other confidence intervals.

$$95\% \ conf. \ int. = sample \ mean \pm 1.96 \ std. \ err. \ of \ the \ mean \quad (3.15)$$

This particular formula can be used in two circumstances. If we know the standard deviation of the population, then this formula can be used, the standard error being calculated using the population standard deviation and the sample size. This particular circumstance would be highly unusual since, if we know the population standard deviation, then we would probably also know the value of the population mean, and would, therefore, not need to calculate a confidence interval for the mean.

The second circumstance in which we can use this formula is far more likely and occurs when we have large samples. Generally, we judge samples to be large if they include more than 30 subjects. In these cases, we use the standard error based upon the sample standard deviation in place of the population standard deviation. Box 3.7. shows how the 95% confidence interval of the resting adult pulse rate can be calculated. The data used in this example are stored in the Minitab data set **PULSE.MTW** that we encountered earlier in Chapter 2.

The confidence intervals calculated in this example enable us to say that we are 95% confident that the mean resting pulse rate of the population from which this sample was drawn lies between 70.62 and 75.12 beats.min^{-1}.

Conventionally, we frequently cite the 95% confidence interval, but we can use tables of areas under the standard normal curve to calculate confidence intervals that are wider or narrower (e.g., 99% or 90%). In such circumstances, we would use 2.576 (99%) and 1.645 (90%) instead of 1.96 in the general formula (Equation 3.15). Note that the width of the confidence interval increases as the level of confidence increases. Therefore, we have to trade off degree of confidence against usefulness since, at the ultimate extreme, we could have extremely high confidence but with intervals that are so wide that they are of no use at all.

Minitab incorporates a command to calculate confidence intervals based on this method and an example is shown in Exhibit 3.6. using the pulse data that we have encountered already. The command line is comprised of four components: the command itself (**ZINTERVAL**), the level of confidence required (e.g., 95% or 99%; the default value is 95%), the population standard deviation, and the column num-

```
MTB > Retrieve  'C:\MTBWIN\DATA\PULSE.MTW'
Retrieving worksheet from file: C:\MTBWIN\DATA\PULSE.MTW
Worksheet was saved on  5/31/1994
MTB > DESCRIBE C1

Descriptive Statistics

Variable      N      Mean    Median   TrMean    StDev    SEMean
PULSE1       92     72.87     71.00    72.61    11.01      1.15

Variable     Min       Max       Q1       Q3
PULSE1     48.00    100.00    64.00    80.00

MTB > ZINTERVAL 11.01 C1

Confidence Intervals

The assumed sigma = 11.0

Variable      N      Mean    StDev  SE Mean        95.0 % C.I.
PULSE1       92     72.87    11.01     1.15  (   70.62,    75.12)

MTB > ZINTERVAL 99.0 11.01 'PULSE1'

Confidence Intervals

The assumed sigma = 11.0

Variable      N      Mean    StDev  SE Mean        99.0 % C.I.
PULSE1       92     72.87    11.01     1.15  (   69.91,    75.83)

MTB >
```

Exhibit 3.6. Calculation of the 95% and 99% confidence intervals of the mean resting pulse rates (the default is 95%). Note that the level of confidence must precede the standard deviation.

ber(s). If the command is to be used on a large sample when the population standard deviation is unknown, then the *sample standard deviation* must be specified, and not the population standard deviation. As the example demonstrates, the standard deviation can be obtained from the output of the **DESCRIBE** command. In one of the examples, the required level of confidence is not specified, so the 95% confidence limits are calculated by default automatically by Minitab. The **ZINTERVAL** command can be executed via menus by the following route:

STAT > Basic Stats > 1-sample z

Basing calculations of confidence on the normal distribution when sample sizes are small reduces precision, particularly with very small samples. We can overcome such lack of precision by employing *Student's t-distribution,* which is a continuous distribution that is closely-related to the normal distribution. In fact, the *t*-distribution appears as a slightly flattened normal distribution, and when sample sizes are very large (i.e., infinite) the two distributions are identical. The calculation of confidence intervals using the *t*-distribution follows a similar pattern to the procedure we used with the normal distribution. The only difference in the process is that for a given level of confidence (e.g., 95% or 99%) there is not a single value by which the standard error is multiplied as there was with the normal distribution (e.g., $1.96 \times SE$ for 95% confidence intervals). Instead, we multiply the standard error by a value of *t,* which is dependent upon the size of the sample from which the confidence interval is calculated. Values of *t* are large for small samples, decreasing in magnitude as sample size increases until sample size is infinite, when *t* will be the same as for the normal distribution. Consequently, the limits calculated using the *t*-distribution are wider than those based upon the normal distribution. The empirical logic of this is clear: larger samples should give a more reliable estimate of the population mean than smaller samples.

In order to calculate a confidence interval based upon the *t*-distribution, we need to use either a statistics package, or to derive the appropriate value of *t* from *t*-tables (Appendix A.2.). In these tables, the header row refers to the probability (0.05, 0.025, and 0.01) and the row labels down the left hand side are related to sample size. They are called the *degrees of freedom* and are equal to the *sample size* − 1. We have already encountered the degrees of freedom in Chapter 2 as

Mean   = 253.93 mg.100 ml^{-1}
Standard deviation   = 47.71 mg.100 ml^{-1}
Sample size (n)   = 28
Degrees of freedom   = n − 1 = 27
Standard error   = 47.71 ÷ √28 = 9.016 mg.100 ml^{-1}

99% confidence interval*   = mean ± t_{27} (SE)
                           = 253.95 ± 2.7707(9.016) = 253.95 ± 24.98
99% confidence interval   = 228.92 to 278.88 mg.100 ml^{-1}

Box 3.8. Calculation of 99% confidence intervals for the mean cholesterol concentration based upon student's t-distribution for a small sample (n = 28). *The subscript attached to t indicates the degrees of freedom.

the denominator of the equation that calculated the sample variance and the sample standard deviation.*

As an example, we can examine the Minitab data set, CHOLEST.MTW, in which only 28 measurements of blood cholesterol were taken two days after the subjects had suffered a heart attack. Box 3.8. illustrates the method by which 99% confidence intervals are calculated.

Minitab provides a command, **TINTERVAL,** to calculate confidence intervals based upon the t-distribution. This command uses the sample standard deviation to calculate the standard error and, by default, calculates the 95% confidence intervals. In its simplest form, therefore, we only have to provide the command and the column in which the data are stored. An example is shown in Exhibit 3.7. where the 95% confidence intervals are calculated for the cholesterol data. In this exhibit, I have also included the calculation of confidence intervals using the **ZINTERVAL** command in order to illustrate the difference between using the normal and t-distributions. Notice that the interval is larger when the t-distribution is employed. Since the sample size is below 30, it is the t-distribution that should be used in these circumstances, and not the normal distribution. The **TINTER-VAL** command can be executed via menus by the following route:

* You may ask why we examine tables with headers marked 0.05 when we want to calculate 95% confidence intervals. The answer to that question is as follows. Statistical tables such as those for the t-distribution contain *critical values,* which indicate points at the left and right ends of the horizontal axis beyond which lies 5% of the area under the curve. Alternatively, we can say that the critical values indicate that 100–5% of the area lies within these limits. We shall see in later chapters why it is more convenient for them to be formatted in this way.

```
MTB > Retrieve  'C:\MTBWIN\DATA\CHOLEST.MTW'
Retrieving worksheet from file: C:\MTBWIN\DATA\CHOLEST.MTW
Worksheet was saved on  5/31/1994
MTB > INFO
```

Information on the Worksheet

```
Column   Name           Count  Missing
C1       2-DAY             28        0
C2       4-DAY             28        0
C3       14-DAY            28        9
C4       CONTROL           30        0

MTB > TINTERVAL 95.0 '2-DAY'
```

Confidence Intervals

```
Variable     N      Mean     StDev  SE Mean        95.0 % C.I.
2-DAY       28    253.93     47.71     9.02  (  235.42,   272.43)

MTB > ZINTERVAL 95.0 47.71 '2-DAY'
```

Confidence Intervals

```
The assumed sigma = 47.7

Variable     N      Mean     StDev  SE Mean        95.0 % C.I.
2-DAY       28    253.93     47.71     9.02  (  236.25,   271.61)

MTB >
```

Exhibit 3.7. The calculation of a 95% confidence interval based upon the t-distribution using Minitab. Contrast the width of the confidence intervals calculated using the t-distribution with those calculated using the (inappropriate) normal probability distribution.

STAT > Basic Stats > 1-sample t

Confidence interval of a proportion

Frequently, we need to make statements concerning the proportion of a population that exhibits particular characteristics. For example, breeding success can be expressed as the proportion of offspring born that survives until fledging or weaning; alternatively, we might wish to describe the proportion of the population that is male. In such circumstances, there is obviously benefit to be gained from being able to calculate a confidence interval associated with our estimate of the population proportion.

During the discussion of binomial probabilities earlier in this chapter, we saw that proportions can often be regarded as binomial variables with a mean and variance given by Equations 3.9 and 3.10, respectively. We can use this fact to develop a formula that calculates confidence intervals for proportions. Since the binomial distribution approximates to the normal distribution under certain conditions (Table 3.1.), we can adapt the formula (Equation 3.15) that we used to calculate confidence intervals based upon the normal distribution to provide an equivalent formula to calculate confidence intervals of a proportion (Equation 3.16).

95% conf. int. for a prop. = obs. prop. ± z (sd of prop.)   (3.16)

The method by which the confidence interval is calculated is illustrated in the following example. Let us imagine that we wished to estimate the proportion of mice that will develop liver cancer after exposure to a suspected carcinogen. An experiment was conducted in which 50 mice were exposed to the suspect substance. After a period of time, the livers of the mice were examined and 20 mice were found to have developed liver tumours. The calculation of the confidence interval for this proportion is shown in Box 3.9. There is no specific

Number of mice	= 50
No. of mice with tumours	= 20
Proportion of livers with a tumour	= 20 ÷ 50 = 0.4

variance of
    proportion
	= (prop. with tumours) (prop. without
	    tumours) ÷ no. of mice
	= (0.4)(0.6) ÷ 50 = 0.0048

SD of proportion	= √0.0048 = 0.069
95% confidence interval of proportion	= 0.4 ± 1.96 (0.069) = 0.4 ± 0.14

Box 3.9. Calculation of 95% confidence interval of the proportion of mice with liver tumours.

Minitab command to calculate confidence intervals for proportions, but, as the calculation above illustrates, the arithmetic is very straight-forward and so a manual calculation is not very taxing.

An estimation of a confidence interval for a proportion is dependent upon the normal approximation to the binomial distribution. Table 3.1. detailed the conditions under which this approximation holds true. Where those conditions are not satisfied, we are not justified in adopt-ing such a method. Therefore, when designing experiments upon which such confidence intervals will be based, these constraints should be taken seriously, and we must ensure that sample sizes are large enough. Obviously, with such constraints, we cannot utilise the t-dis-tribution when sample sizes are small.

3.5   Graphical presentation of reliability and confidence

One doesn't have to read many papers in scientific journals before coming across a graph, histogram, or bar chart that includes standard deviations, standard errors, or confidence intervals in the form of *error bars* (an example is shown in Figure 3.17.). In both of the preceding

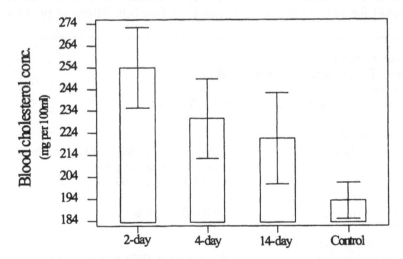

Figure 3.17. A bar chart of the mean concentration of blood cholesterol (mg.100 ml^{-1}) for a group of healthy adult males (*Control*); and, for a group of adult males who have suffered a heart attack (measured 2, 4, and 14 days after their heart attack). The bars represent the mean concentration on three different days after their attack. The vertical error bars represent the 95% confidence interval of the means.

chapters, I have emphasised the importance of data description; a clear picture communicates information to the reader in a far more approachable and memorable way than a series of equations or paragraphs of text. By adding vertical bars that represent standard deviations, standard errors, or confidence intervals to any graph, the reader can gain an immediate impression of the variability, reliability, or precision of the data being presented. Moreover, in Part II (Questions of Comparison), we will see how confidence intervals can be used in hypothesis testing.

The addition of the confidence intervals demonstrates very clearly both the magnitude of difference between the 2-day cholesterol concentrations and those of the control group, and it shows that the width of the confidence intervals of the 3 heart attack groups are very similar. We shall see in later chapters that the comparability of variability is an extremely important prerequisite to many statistical tests. Clearly, therefore, confidence intervals have a multitude of important roles in hypothesis testing and statistical methods.

3.6   Exercises

The solutions of some of these problems can be found using Minitab. However, greatest benefit will be gained from finding the solutions both by diagrammatic methods (such as tree and Venn diagrams), manual calculations, and also by using Minitab.

1. A botanical survey was carried out using quadrats. The results showed that 20% of the quadrats contained Species A, 33% contained Species B, and 8% contained both Species A and B. Draw a Venn diagram to describe these results. Calculate the probability of a quadrat containing:
   (a) neither Species A nor Species B.
   (b) Species A, or Species B, or both species.
   (c) Species B, but not Species A.
   (d) Species A, or Species B, but not both.

2. In a population, the probability of being dark haired is 0.52 and light haired is 0.48. Furthermore, the probabilities of dark and light eye colours are 0.15 and 0.85, respectively. Using a tree diagram, calculate the probability of an individual being both light haired and dark eyed. If the population comprised 4800 individuals, how many light-eyed/dark-haired people would you expect to find?

3.   It is estimated that 1.5% of a population is infected with a particular virus and the effectiveness of a national screening program for this virus is being examined. It has already been established that, if an individual is infected, then the probability of testing positive is 0.975, and the probability of an uninfected individual being diagnosed correctly is 0.925. If 50,000 individuals are tested,
   (a) how many would be expected to test positive?
   (b) how many uninfected individuals would be incorrectly diagnosed?
   (c) what is the probability of a person who had tested negative being uninfected?

4.   If the probability of a male or female piglet being born is equal, what is the probability of a sow giving birth to a litter of 6 males and 2 females?

5.   *Nasonia vitripennis* is a parasitoid wasp that lays its eggs in the larvae of blowflies. If the probability of a larva being parasitised is 0.37, what is the probability that more than 4 larvae will be parasitised out of a sample of 20?

6.   A sample of blood was examined and the proportion of cells that possessed a binding site for a particular antigen was 0.45. If we take a sample of 12 cells at random, what is the probability of:
   (a) finding 3 cells with the binding site?
   (b) finding fewer than 5 cells with the binding site?
   (c) finding between 6 and 8 cells with the site?

7.   The number of packets of the neurotransmitter, acetyl choline, released at a synaptic endplate by a nerve impulse is described by a Poisson distribution with a mean of 2.33. What is the probability of:
   (a) 3 packets being released?
   (b) 7 or more packets being released?
   (c) between 3 and 5 packets being released?
   (d) less than 4 packets being released.

8.   The number of red blood cells found in grid squares on a haematocrit are dispersed according to a Poisson distribution with a mean of 1.913 cells per grid square. What is the probability of finding:
   (a) 2–4 cells in a grid square?
   (b) more than 4 cells in a square?

9.   The data listed are the toe lengths of two closely-related species of birds. Calculate the 95% confidence intervals of

the means for both species, and plot the means with their confidence intervals on a bar chart. If a normal distribution is assumed, what is the probability of finding:
(a) an individual of *Species A* with a toe longer than 3.0 cm and
(b) a specimen of *Species B* with a toe shorter than 3.4 cm?

Species A
2.47, 2.56, 2.64, 2.70, 2.71, 2.81, 2.81, 2.81, 2.85, 2.85, 2.88, 2.90, 2.90, 2.91, 2.91, 2.94, 2.95, 3.00, 3.02, 3.02, 3.03, 3.08, 3.09, 3.09, 3.12, 3.13, 3.14, 3.14, 3.15, 3.16, 3.17

Species B
3.28, 3.28, 3.32, 3.33, 3.37, 3.47, 3.54, 3.58, 3.61, 3.62, 3.70, 3.77, 3.80, 3.83, 3.83, 3.87, 3.92, 3.93, 3.97, 3.98, 4.02, 4.04, 4.04, 4.07, 4.08, 4.11, 4.12, 4.17, 4.26, 4.27, 4.28

10. The length and breadth of a sample of 243 cuckoo eggs (in mm) are tabulated below (Latter, 1902). Assuming that these data have been drawn from a normally-distributed population, what is the probability of randomly selecting an egg whose length is greater than 23.55 mm? Calculate the 95% confidence intervals for the means of both the length and breadth of the eggs.

Length (mm)	Frequency	Breadth (mm)	Frequency
19	1	14	1
19.5	1	14.5	1
20	7	15	5
20.5	3	15.5	9
21	29	16	73
21.5	13	16.5	51
22	54	17	80
22.5	38	17.5	15
23	47	18	7
23.5	22	18.5	0
24	21	19	1
24.5	5		
25	2		

Courtesy of Latter, O.H. (1902) The egg of *Cuculus canorus*, *Biometrika*, 1, 164–176. With permission.

11. The following data are the weights (g) of the hippocampis of 30 rat brains.

```
0.136560 0.111750 0.117649 0.093217 0.114261 0.144439 0.107676
0.123443 0.131794 0.150556 0.138949 0.133171 0.114115 0.124783
0.117288 0.133854 0.126733 0.151998 0.114656 0.102751 0.162781
0.131275 0.140805 0.181277 0.156171 0.160017 0.140801 0.119567
0.149358 0.174512
```

Assuming that the sample has been drawn from a normal distribution, calculate the 99% confidence interval for the mean. What is the probability of finding a hippocampus that weighs between 0.15 and 0.16 g?

12. The heights of maize plants (cm) were measured for two groups (fertile and sterile pollen). Produce a bar chart showing the means and 95% confidence intervals for these data.

Height — Fertile (cm)	Height — Sterile (cm)
233.68	220.98
271.78	223.52
248.92	248.92
246.38	238.76
241.3	236.22
238.76	236.22
233.68	248.92
243.84	218.44
248.92	200.66
264.16	228.6
246.38	238.76
226.06	231.14

13. An experimenter conducting investigations of animal preferences lets the animals choose to turn left or right in a t-maze. Prior to the experiments starting, preliminary trials were conducted to identify and eliminate individuals that were biased in favour of turning in a specific direction (left or right-handedness). For one particular animal, the proportion of times it turned left was 0.65 in 40 trials. Construct 95% confidence intervals for this proportion and comment on whether the animal demonstrated any evidence of bias.

3.7   Summary of Minitab Commands

PDF (with subcommands for each statistical distribution)
CDF (subcommands similar to **PDF**)
ZINTERVAL
TINTERVAL

3.8   Summary

1. The precision or reliability with which a population parameter has been estimated by a sample statistic is measured by a standard error.
2. The binomial, Poisson, and normal distributions have been described and the relationship between them has been illustrated. The binomial and Poisson, under certain conditions, can be approximated by the normal distribution.
3. Interval estimates in the form of confidence intervals have been introduced for means and proportions.

CHAPTER 4

Sampling

4.1 Introduction

The previous chapters have all emphasised the need to take samples from a population. In an ideal world, a sample provides a true reflection of the population from which it is derived. Therefore, practical techniques that enable us to take representative samples are of critical importance to good investigations. Given the importance of sampling in quantitative investigations, many textbooks have been devoted solely to the subject of sampling techniques (e.g., Cochran, 1977; Barnett, 1991). In this chapter, I introduce the basic principles of sampling and some of the main practical techniques that are employed in the biological sciences. In the final section, I have included a brief outline of some of the techniques that enable us to determine the size of the sample that we need to gather in order to satisfy the objectives of an investigation.

Biological populations not only have inherent variability (since no two individuals are identical), but they also comprise many diverse sub-groups. For example, in many animal species, marked differences may be manifested between the sexes, ages, and life-history stages in terms of size, behaviour, and preferred habitat or food. Therefore, if we do not take care in the way we sample, we may select samples that appear to estimate the population reliably (because they are composed of very similar individuals), but are, in fact, biassed in favour of a particular section of the population. Samples have to be selected in such a way that they accommodate this variability in a population. We must ensure that all relevant sections of the population are included in our sampling regime, otherwise we run the risk of gathering biassed samples. In the first section of this chapter, I shall describe the main sampling techniques utilised by biologists that minimise the potential for sampling bias.

Logic would suggest that the likelihood of selecting biassed samples should decrease with sample size simply because very large samples are more like the population than very small samples. Therefore, we might suppose that in order to avoid inaccurate sampling, we merely have to take very large samples. However, if we take samples that are unnecessarily large, we will waste valuable resources and we may also damage the population that we are sampling (there are also some other potential problems that are detailed below). At the same time, taking samples that are too small may lead to real effects remaining undetected or to our obtaining a biassed view of the population because we have failed to sample the population comprehensively. When deciding upon the size of sample that we need to take from a population, we have to strike a balance between experimental expediency on the one hand, and resource and ethical constraints on the other. In the final section of this chapter, I shall discuss some of the ways in which we can determine the appropriate size of a sample.

The combination of the key elements of sampling methodology that will be described in this chapter and the principles of probability that were introduced in Chapter 3, provide the basic tools to perform biological investigations. The remaining chapters are devoted to describing how these tools can be applied and adapted to different types of questions that arise in the biosciences.

Before discussing specific sampling techniques, let us consider in more detail the reasons why we take samples as opposed to censussing entire populations, and the general properties that all samples must possess. In doing so, the importance of, and reasons underlying, the key principles of sampling methodology will become clearer. There are four main reasons for using samples as opposed to entire populations:

(i)   *unknown population size:* In the majority of biological investigations, we are not in a position to specify the entire population and so we have to take samples. For example, if we wished to examine the age-structure of a population of limpets on a seashore, there is no possibility of the population being specified; we can view it as essentially infinite.

(ii) *resource limitations:* Every measurement that we make in an investigation takes time and costs money in terms of manpower and equipment requirements as, indeed, does the analysis that follows it. Resource constraints require us to take samples rather than perform an entire census of the population, even if the latter is possible. Inevitably, resource limitations are the major determinants controlling the bounds of any investigation. Good planning prior to starting the practical stages of a project will ensure that we use the available resources as efficiently as possible.

(iii) *disruption to populations or habitats and other ethical considerations:* Many measurements require the destruction of the subjects from which they are derived. Examples include removal of whales and other cetaceans from the oceans; sacrificing laboratory rodents in order to examine tissue development; extraction of soil samples; and the destruction of plant material in order to estimate dry matter content.

Clearly, if we take too large of a sample, we may endanger a natural population. This is certainly a vigorously debated topic in terms of the global population of sea mammals; it can also be a major consideration on a smaller scale when dealing with either sensitive habitats or locally rare species. Similarly, with soil samples we could potentially be contributing to the destruction of habitats that are important to species that may not even be the focus of the main investigation. We must always, therefore, take a very broad view when designing experiments that are to be performed in natural habitats.

Although there is no danger of the extinction of the species in those experiments where the animals that are sacrificed have been bred specifically for laboratory use, we do have a moral obligation to min imise the number of animals used. Again, detailed planning will help us to achieve this goal. Moreover, ethical committees in universities and research institutes will undoubtedly require that a full justification of the proposed sample sizes be provided for scrutiny prior to such experimentation.

The final example of dry matter estimations illustrates that at the outset of any experiment we must be aware of how many measurements are needed if every piece of sampled material has to be analysed destructively. As was mentioned briefly in Chapter 1, in such circumstances we must ensure that our sampling regime leaves sufficient material for physical or chemical analysis in the later stages of an investigation, and, at the same time, does not alter in any significant way the remaining environment (e.g., does not allow more light to fall onto the remaining plants, or modify root competition by increasing space). Furthermore, you must ascertain *before* the experiment starts how much material is required for chemical analysis. Sadly, I have had occasion to inform researchers that they have not gathered enough material for analysis. Although I have used a botanical example here, the principle is equally applicable in any other area of biology (e.g., medical trials using laboratory rodents).

    (iv) *sampling may be more accurate than a total population census:* This reason for needing to take samples is, per-haps, the most surprising. Why should a sample of mea-surements provide a better judge of a population than a census of the entire population itself?

Whenever we take a series of measurements, there is always the potential for errors of two sorts: *sampling errors,* which are those due to the way we choose the subjects (or sampling units), and *non-sampling errors,* which are those due to other factors, and which will increase in line with the volume of data. For example, with a large experiment several researchers may have to share the observational tasks. Although each may take the measurements in a consistent and repeatable manner, they may also differ slightly from each other in their technique, thereby introducing systematic errors. Similarly, when transcribing observations from recording sheets or notebooks, errors are made. If one transcription error is made for, say, every 200 observations, then the number of errors in a complete census of a population comprising several thousand individuals will be very large in comparison to a sample. On many occasions errors such as these may be far greater than sampling errors. Therefore, a well-organised sample that does not over-extend resources may, paradoxically, provide a better representation of a population than a complete census.

Having concluded that we must take samples, how do we ensure that a sample is representative of the population? In Chapter 3 we saw that consistency was important and that we could place greater confidence on our estimates of the population parameters by taking a larger sample (remember the standard error decreases in magnitude as the sample size increases). Therefore, one component in our efforts to ensure that our sample faithfully represents the population is *replication.*

The second problem that we have to avoid is the introduction of bias. If we wish to select 20 sheep for measurement, we might simply "point at random" to select them. However, inadvertently or subcon-sciously we may be selecting in a non-random way. We may, for exam-ple be selecting large individuals, or ones that are grazing on longer grass than the rest of the flock, or ones with more black on their noses. We must ensure that we do not introduce any bias of our own into the sampling procedure, which we can achieve by using a randomisation process that is independently constructed. Therefore, in order to obtain an unbiassed representative sample that gives repeatable results, we must take *replicated* sets of *randomly selected* measurements.

The first stage of this process of selection requires that we define very carefully the *sampling units* of our investigation, or the specific elements of the population on which we wish to record observations. The choice of sampling units largely depends upon the nature of the question that we are attempting to answer and the data type (nominal, ordinal, interval, and ratio as defined in Chapter 1) required for that investigation. Nominal data types dictate fundamentally different sampling units to the other three data types. When dealing with nominal data, we *count* units, whereas with the other three, we *measure* characteristics of units.

When we are dealing with counts or frequencies, we need to know how many individuals are contained in a defined unit area (e.g., plants per 0.25 m^2 quadrat, birds per nest), volume (e.g., fish larvae per m^3 of seawater, red cells per ml of blood), time period (eggs laid per minute or neurotransmitter vesicles released at a synapse per second), or if the unit is the whole or part of an individual organism (e.g., numbers of parasites per host or aphids per leaf). Thus with counts, the sampling units are not the biological subjects themselves, but the space, volume, or time interval in which they are observed, and the population is the total number of units that could have been sampled (such as the limpet example in Chapter 3 when the Poisson distribution was discussed). In contrast, when we measure characteristics of biological material (whole organisms or parts), the sampling units will be defined in terms of those groups, individuals, or parts of individuals. For example, we can measure the mean pulse rate for a human, the daily energy requirement of a sheep, or the speed at which an action potential travels along a nerve fibre. The human, the sheep, and the bundle of nerve fibres would be the sampling units in these examples and the population from which they are drawn would be all the relevant individuals that could have been sampled. In the pulse example, that is not necessarily the entire world population, for we may be interested only in, say, a specific group of individuals who are suffering from a respiratory disease. The clear definition of both the sampling unit and population from which it is drawn is of paramount importance at the outset of an investigation; without such definitions, errors and confusion will arise.

In a classical paper, Morris (1955) reviewed the essential characteristics of sampling units. Although he discussed these principles in the context of a particular insect pest (the spruce budworm), his paper provides so much valuable information about sampling techniques in fieldwork that, along with the texts by Southwood (1978) and Krebs (1989), it should be required reading for all field ecologists. Likewise, Altmann's (1974) paper and the book by Martin and Bateson (1993) should be held in similar regard by behavioural scientists. Although Morris was specifically discussing investigations involving counts, several of the points are equally relevant to sampling units relating to measurement.

1. All sampling units in the population have an equal chance of selection. Bias will be introduced if we do not select randomly.

2. The unit must be stable or, if changing, it must be easily monitored. If the unit changes shape, size, or condition (e.g., growth and foliage changes in trees), then the counts would be affected. If a sampling unit becomes more or less favourable to the individuals being counted during the sampling period (e.g., the immune status of a host animal altering during early development), then comparisons based upon those units will be invalid.

3. The proportion of the population found in a unit must remain constant. If rhythms, such as circadian or seasonal variations, cause changes in the population in specific areas, then the sampling regime must accommodate those rhythms. For example, the sleep:wake cycle of mobile animals and the vertical movement of plankton in the ocean on a daily basis will have a substantial impact on the outcome of sampling.

4. If we are dealing with units for counting, then we must be able to define the unit of area, volume, or time interval.

5. The sampling unit must be clearly identifiable in practice. If we are counting parasites on hosts, then we must ensure that we avoid counting hosts twice. If we are observing the behaviour of individual animals then they must also be identifiable (perhaps through individual marking techniques). Permanent transects or grids must be identifiable in the field. In time-based observations, we must have a method of time keeping so that we can distinguish between one time period and the next. Overlapping sampling units would lead to multiple measurements.

6. The sampling unit should be of a reasonable size. If it is small, the variance of counts will be large, but, since the cost increases with size, a balance has to be struck between variance and cost.

7. The size of the unit must be relatively large compared to the size of the organism being counted, thereby reducing errors due to edge effects. Bias can be introduced if many organisms are found on the edges of a sampling unit.

8. The size of sampling units used to investigate mobile subjects should be approximately the same size as the range of their movement. If the unit is too small, individuals may move from one unit to another. Clearly, this is a particular danger when we are using contiguous grid squares. Counting twice can be a very serious problem.

Implicit in several of the points listed by Morris is the requirement that sampling units are completely independent of each other; that is to say that a measurement or count derived from one sample is unaffected in any way by the measurement or count derived from any other sample. We shall see in later chapters that independence is a critical assumption of many analytical techniques and is, therefore, one of the key principles in quantitative investigations. This assumption is widespread, but, sadly, frequently violated. All scientists must have a thorough understanding of the principles of independence in order to avoid the potential pitfalls. There are two primary areas where independence may be threatened: *multiple measurement* and undue *influence through association*.

The dangers of multiple measurement are relatively easy to envisage. If an individual is counted more than once or contributes more than one measurement to an analysis, then it will clearly exert undue influence on the final results. Unfortunately, erroneous multiple measurement can sometimes be mistaken for valuable replication. We can define replication as: *sampling from more than one sampling unit,* and we have already seen in Chapter 3 that it is essential for precision. So, as a simple rule, we should use only one measurement of a particular variable from a sampling unit in any one analysis. This problem of multiple measurements should not be confused with measuring several variables on the same subject; for example, recording the heights and weights of individuals (data of this sort provide the material for discussion in Chapter 8 concerning sequential relationships). If we are to avoid multiple measurements, then we have to ensure that sampling units such as grid squares or time intervals do not overlap. Therefore, the fifth point identified by Morris (listed above) is critical: we must be able to identify sampling units, otherwise the potential for errors due to repeated measurements is great. There are some obvious amendments to this rule, which we shall encounter in later chapters, one being developmental studies, where we actually set out to investigate the dependence of events in one time period on events that occurred previously.

Multiple measurements of a single variable from individual sampling units are not, themselves, a problem. Rather, we should only use a single datum from each sampling unit in an analysis. This distinction can be illustrated by an example from endocrinology. The concentrations of various hormones in the blood follow a diurnal pattern (e.g., plasma cortisol concentration in humans may vary as much as three-fold during its daily cycle). How do we take meaningful samples in such situations? One way forward in these circumstances is to take several blood samples from each individual at different times of day and then pool the measurements from each person into a mean, thus providing a single datum for each person (i.e., each sampling unit) for statistical analysis.

Sometimes we may be in danger of selecting the subjects that are closely associated to each other and this can lead to another violation of the independence principle. For example, if we were interested in the birthweights of starlings (*Sturnus vulgaris*) and we examined 5 nests, weighing a total of 21 nestlings, could we justify regarding the individual nestlings as sampling units? All those birds found in a nest would be genetically related and their individual weights would not, therefore, be totally independent. In fact, the weights would not only depend upon genetic factors, but also upon the food gathering abilities of their parents. Consequently, we must adopt the *nests* rather than the *nestlings* as sampling units, and use the mean weight per nest as the sample statistic for our analysis. There are many similar examples of this phenomenon throughout biology where we have to use the subject group as the sampling unit because the response of the individual organism is dependent upon the group as a whole. Another frequently encountered example would be the growth of potted plants. The growth rates of each plant will be interdependent, just like caged laboratory mice, because dominance relationships have been shown to cause feeding, and hence body weight effects. In order to avoid violating independence, we must define the sampling units clearly and unambiguously. Their definition should follow directly from the question that we are attempting to answer and the way in which we construct the null hypothesis. You must, therefore, be prepared to spend time defining the sampling units before commencing the practical work.

In many studies you will need to pose several questions based upon the same biological subjects, but each question may require different sampling units. The following example illustrates the variety of sampling units that may be required in related questions.

Compare three questions concerning the behaviour and milk production of dairy cows. We might wish to examine the social behaviour of cows, particularly their spacing while grazing. To do so we could divide a field into grid squares and, on each observation, count the number of cows within each grid square. The sampling unit here would be the grid square, and the measurements would be the number of cows per square. If we were to investigate the dominance relationships between the cows in the herd, we would need to observe the interactions between pairs of cows to record the winner and loser of each interaction in order to compile an interaction matrix. The population of sampling units here is every possible pair of cows in the herd, where we would record the number of interactions per pair. Finally, for milk production, we could use each individual as the natural sampling unit, measuring the daily milk yield of each cow.

Sampling units are drawn from the population that we are investigating. However, the biological population in which we have an interest may be so large that we cannot attempt to sample widely and give every individual in that population an opportunity of being selected. We could not, for example, contemplate taking samples without any constraint from the entire world human population; we would have to confine ourselves to sampling from a sub-section of the total population. Similarly, if we used quadrats to investigate the distribution of particular woodland plant species, we might be constrained from sampling across the whole study area. When we take measurements on a herd of cows, the herd is a sub-set of the entire biological population of cows. In each case, we define a *statistical population* from which we will take samples that are smaller than the biological population that we are investigating. This statistical population is known as the *sampling frame*.

Having identified the need to clearly define both the sampling units and the sampling frame from which the samples will be drawn, we must now determine the particular procedures by which we actually select the sampling units from the sampling frame. There are many ways in which we can select a sample, each of which has its own particular attributes that suit different types of problems. There is insufficient space here to discuss all the available methods in detail and since there are several excellent texts that cover this topic in extensive detail (e.g., Cochran, 1977; Barnett, 1991), I shall outline just three of the most frequently encountered procedures. These three methods illustrate the major principles of sampling and, should you need to use other methods, they should enable you to become familiar enough with the basic terminology to adapt methods from monographs on this subject to suit your own purpose. Also remember that during the planning stages of any investigation, you should consult the available literature on the topic area; this will help you to identify the appropriate sampling methods suitable for your particular needs.

4.2   Sampling techniques

Simple random sampling

At its simplest, random sampling gives every member of the sampling frame the same probability of being selected. For example, imagine that we conducted a simple experiment to compare two groups of animals in terms of their growth rates, where each group is fed a different diet (Treatment A or Treatment B). If we have 20 animals of broadly similar age and weight, we should need to allocate 10 animals to each treatment. In this case, we can use *simple random sampling* or *simple random allocation;* each individual has an equal chance of being allocated to a particular treatment group. In experiments where simple random sampling is used, all individuals would be subjected to identical conditions except the experimental treatment itself. Therefore, the animals would be in a single enclosure and differ only with respect to the feed that they are given.

There are many ways of allocating individuals randomly to a particular treatment, but all incorporate a structured approach to random selection. This is not a contradiction in terms, but rather a precaution against subconscious selection. The big question is how do we select?

We can adopt low technology methods such as numbering all the animals, writing the numbers on pieces of paper, and then drawing these from a jar. Alternatively, we can use random number tables or generate random numbers using a computer (e.g., using a computer-based statistics package such as Minitab).

Random number tables and random numbers generated by a computer package are not, strictly speaking, random. They are all produced by computer programs and, as such, are generated according to some definable, and thus non-random, rules. However, the sequence of numbers will repeat itself only after many hundreds of thousands of numbers have been generated so they can effectively be regarded as random. Specifically, they are called *pseudo-random numbers* to distinguish them from *true* random numbers.

A set of random number tables is provided in Appendix A.3. Notice that the random numbers are set out in blocks of five digits with row numbers to aid us in using the tables. There is nothing magical about the blocks of digits; the arrangement is just for convenience. If we need to select numbers between 1 and 10, we need to use only the first 2 numbers of each block. Whenever we use random number tables, we must select numbers in an orderly pattern; if we do not, we compromise the pseudo-randomness of the tables. So, starting at a fixed point, e.g., the top left corner, we move in a fixed direction (vertically, horizontally, or diagonally across the page). We can start at any point on the page (even selecting the location by pointing blindly with a pin), but we have to proceed from that point in a fixed direction. Once the required number of individuals have been selected, mark the position in the tables with a pencil mark. On the next occasion that you use the table, you can start from the position where you finished last time. This last point is very important because if we continue to start our selection of random numbers at the same position in the tables on each occasion, we will introduce systematic errors into the selection procedure by always selecting from similar locations in a study site or even using the same individuals.

To see how random selection actually works, let us imagine that our animal experiment requires us to select 20 animals from a total stock of 200. If the animals have been numbered, then we use the random number tables to select the 20 that will be used. For the sake of this illustration, let us start at the top left and move vertically down the first column of random numbers. Since we are selecting from a total of 200 animals, we need to look at the first 3 digits of each block of the table. Whenever we encounter a number outside our required range, we simply move on to the next one. If we find an acceptable number on more than one occasion, we ignore the second selection and move on to another until we have the necessary sample size composed of unique numbers within the required range. When we reach the bottom of the first column, we can start selecting from the top again, which, in this case, would be 527 (52 from the first block and 7 from the second). Our selection procedure would be as follows:

We would reject 494, 553, 862, ..., 370 because they are outside the required range; 063 would be accepted and allocated to Treatment

```
MTB > RANDOM 20 C1;
SUBC>    INTEGER 1 200.
MTB > PRINT C1


C1
    124    125     58    156     48    101    111     91    154     35    169
      3     16    143    130     95     35    158    104     31

MTB >
```

Exhibit 4.1. An example of Minitab generating 20 random numbers from a range of 1 to 200 and storing them in Column 1. Notice that 35 is repeated. Therefore, always generate more numbers than required so sufficient, unique numbers are produced.

A, 197 to Treatment B, 150 to A, 092 to B, and 032 to A in the first vertical pass through the table. Start again at 527 and continue with this sampling procedure to complete the allocation of animals to the 2 experimental groups.*

We could use random numbers just as easily to generate coordinates within grid squares for taking soil cores by selecting a pair of random numbers to represent the x and y coordinates of the position within the grid for the soil core.

The process of using a computer program to generate random numbers is very similar to that of using tables, but with the added advantage of being able to specify only numbers that are within the required range. An example is shown in Exhibit 4.1. where, like the manual example using tables, a number has been selected twice. Therefore, when using a computer to generate random numbers, always generate more than required.

The menu-based route to generating random numbers is as follows:

CALC > Random Data > Integer

Simple random sampling is a valuable technique for populations that are uniform in their dispersion and that do not tend to be composed of sub-groups that differ in any respect and that may be influential to the investigation. For example, different age classes of the bivalve mollusk, *Macoma baltica,* are found at different levels of the shore, and a simple random sample may by chance lead to individuals being selected more from one area than from others, thus giving a

---

* Treatment A will have animals: 63, 150, 32, 136, 140, 97, 112, 55, 88, 198
  Treatment B will have animals: 197, 92, 52, 9, 68, 45, 166, 73, 193, 95

sample biassed in favour of one age class. If a study population is heterogeneous in its dispersion or composition, then simple random sampling cannot be guaranteed to produce a truly representative sample, a clear violation of its underlying principle. Furthermore, an additional problem caused by heterogeneously dispersed populations is that of a large variance. As a result of such difficulties, this sampling technique is often used to allocate similar individuals to treatment groups in experiments, but is not often used in field experimentation where we frequently need to adopt more elaborate sampling methods.

Stratified random sampling

Prior to the commencement of some investigations we may already possess some insight into the study population or its location as a result of earlier research, such as with the *Macoma* example. Similarly, we may wish to estimate the population density of a species of orchid on an area of sloping ground that we know increases in soil moisture from top to bottom. Such variation could have significant implications for the distribution of orchids. We can profit from such knowledge by taking this variability into account when designing the sampling regime. By dividing the area into several sub-sections (known as *strata*) according to soil moisture content (e.g., low, medium, and high) and taking simple random samples from each stratum, we can be more confident of obtaining a reasonably representative sample. This method is called *stratified random sampling* and is illustrated by the soil moisture example in Figure 4.1.

If the strata have been chosen appropriately, then the variability between samples within each stratum will be small compared to the variation between strata. This will result in the estimate of the population density having a smaller standard error than if simple random sampling had been employed, i.e., stratification can give a more precise estimate.

There are other potential advantages to stratification in addition to increased precision. By dividing the population, we are in a position to draw more detailed conclusions about the population or the area over which it is distributed. Thus we would be able to comment with greater authority about the relative density of orchids at the three different levels of soil moisture.

A further advantage of stratification concerns the size of the strata. In the orchid example, the strata were constructed by the investigator to be of similar size, but this need not always be the case and physical

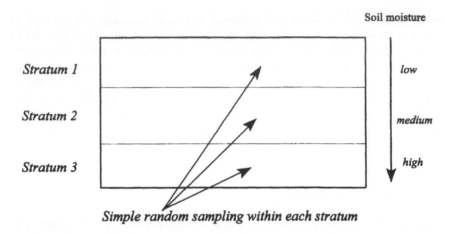

Figure 4.1. A diagrammatic representation of how stratified random sampling could be used to estimate population size on a study area where there was a soil moisture gradient.

constraints can sometimes preclude this possibility altogether. We can, however, account for strata of dissimilar sizes in subsequent analysis. This is illustrated in the following worked example. Let us imagine that we wish to estimate the number of trees in an area of woodland and that there are four plots (A, B, C, and D) within the wood from which we take samples. The plots (strata), however, are not of equal size, being 6, 8, 8, and 10 ha, respectively. Within each stratum we count the trees within a number of grid squares, which each measure 0.25 ha in area; these grid squares are the sampling units. The locations of each grid within the strata are determined by simple random sampling (a diagram of the wood and the sampling regime is illustrated in Figure 4.2.). The grid squares from each stratum have to provide as true a reflection of the population within the stratum as possible. It follows, therefore, that we may need to sample more grid squares in larger strata than in smaller ones. A straightforward approach to determining how we should divide our sampling effort is by *proportional allocation*. In the following worked example, 20 grids can be sampled, which is about 15% of the total number of sampling units, thus allowing 4, 5, 5, and 6 squares to be taken from A, B, C, and D, respectively, in strict accordance with their relative sizes.

The raw data are tabulated in Table 4.1. along with the mean and standard deviation of the density of trees per grid square within each plot using the formulae that were introduced earlier (Equations 2.2, 2.4, and 2.7).

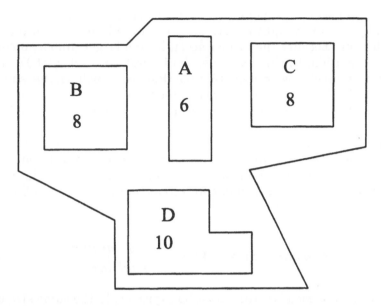

Figure 4.2. Four strata (A, B, C, and D) are defined in the woodland. The number of grid squares sampled from each stratum (4, 5, 5, and 6, respectively) is proportional to the relative sizes. The location of each grid square within a stratum is determined using simple random sampling (random numbers used to select x and y-coordinates of each grid).

Stratum area (ha)	A	B	C	D
	6	8	8	10
Grid square				
1	8	14	7	6
2	15	12	4	3
3	9	15	8	7
4	12	9	6	5
5		13	9	6
6				4
Mean no. of trees per grid	11.00	12.60	6.80	5.17
Variance	10.00	5.30	3.70	2.17

Table 4.1. Counts of trees in the four strata (A, B, C, and D). The number of grids sampled per stratum was determined by proportional allocation, 15% of the total area of each stratum was sampled. The mean density and variance for each 0.25 ha grid square are detailed.

In estimating the population size from a stratified sample, we should give more weight to estimates from the larger strata than the smaller ones because the former account for a larger proportion of the total population. In this example, the largest stratum, Plot D, seems to have a much lower density than Plot A, which is the smallest stratum (5.17 versus 11.00). If we treated the results from all the strata equally, then the samples drawn from the smallest plot would have the same influence on our estimate of the population mean as those taken from the largest stratum. The calculation of the overall population density takes account of the relative sizes of the different strata by using a weighting factor, which is proportional to the relative size of each stratum, and the formula for its calculation is given in Equation 4.1.

$$Weighting\ factor = W_i = \frac{area\ of\ stratum\ i}{total\ area} \qquad (4.1)$$

The area of each stratum must be expressed in the same size units as the sample units. In this example, the grid squares (the sampling units) are 0.25 ha in area, and thus, the area of each stratum must be expressed in units of 0.25 ha (not SI or any other set of units). The areas of each stratum are as follows:

$$Plot\ A = 6\ ha \div 0.25\ ha = 24\ grids$$
$$Plot\ B = 8\ ha \div 0.25\ ha = 32\ grids$$
$$Plot\ C = 8\ ha \div 0.25\ ha = 32\ grids$$
$$Plot\ D = 10\ ha \div 0.25\ ha = 40\ grids$$

We can calculate the overall mean density of trees per grid square using Equation 4.2. This is our estimate of the population density. The full calculation is shown in Box 4.1.

$$Mean\ pop.\ density = \frac{\Sigma(stratum\ area)(stratum\ mean\ density)}{total\ area} \qquad (4.2)$$

or, in other words,

$$Mean\ pop.\ density = \Sigma(W_i\ stratum\ mean\ density) \qquad (4.3)$$

The variance of the population mean is calculated by combining the weighting factor and variance for each stratum as shown in Equation 4.4, where n_i is the sample size (the number of grids sampled in each stratum).

$$Variance \ of \ population \ mean = \Sigma \left(\frac{W_i^2 s_i^2}{n_i} \right) \qquad (4.4)$$

Equation 4.4 shows that the calculation of the overall variance of a stratified sample depends only upon the variances of the individual strata, not the variability between the strata. This gives us some control over the overall precision of our estimates. For, if we can stratify in such a way that the variance within each stratum is low, then we can reduce the overall variance. We shall see in Chapter 7 that this principle forms the basis of a very important experimental design, which is known as the *randomised block design*.

	Stratum Area (no. of grid squares)	Mean density	Variance	Weighting factor, W_i	Sample size
A	24	11.0000	10.00	0.1875	4
B	32	12.6000	5.30	0.2500	5
C	32	6.8000	3.70	0.2500	5
D	40	5.1700	2.17	0.3125	6

$$\bar{x}_{ST} = \frac{(24)(11) + (32)(16) + (32)(6.8) + (40)(5.17)}{128}$$

$$= \frac{264 + 403.2 + 217.6 + 206.8}{128} = 8.528$$

$$Variance = \frac{(0.1875)^2 10}{4} + \frac{(0.25)^2 5.30}{5} + \frac{(0.25)^2 3.70}{5} + \frac{(0.3125)^2 2.17}{6}$$

$$= 0.088 + 0.066 + 0.046 + 0.035$$

$$= 0.235$$

Box 4.1. Calculation of the sample mean and variance of the density of trees, based upon the stratified sample detailed in Table 4.1.

The sampling units in this example are grid squares, but stratified random sampling is not restricted to a single type of sampling unit. We could just as easily stratify a sampling frame of human subjects according to gender, age class, social, or medical background. Such stratification is used extensively in both social sciences and epidemiological studies.

Systematic sampling

Both simple random and stratified random sampling techniques, as their names suggest, rely on random selection procedures. In these two methods, each member of the population or stratum has the same probability of being selected as all the others in the population or stratum. Systematic sampling, on the other hand, is not truly random although it does include an element of random selection.

We can use systematic sampling wherever each member of the population can be numbered or where there is a definable area or period over which the population is to be observed (such as experimental plots, grid squares on a microscope slide, or specified periods of time). In principle, systematic sampling requires us to sample each ith individual within the population. For example, we might wish to sample 5% of the plants growing within an experimental plot. Using this method, we would select every 20th plant. We would randomly select a number between 1 and 20 to determine the first plant to be sampled, say 12, then proceed to select the following plants: 32, 52, 72, 92, 112, and so on throughout the plot. Another key example of the use of systematic sampling is transect sampling applied both to small mammal trapping and to botanical surveys (Southwood, 1978; Krebs, 1989).

The advantage of systematic sampling over simple random sampling is that it is quicker; we need to consult random number tables only once. Furthermore, with this method, we take samples evenly throughout the population, thus avoiding any vagaries of simple random selection that could conceivably lead us to take a sample from a restricted portion of the population, which may or may not be representative. We can even use systematic sampling within strata so that the advantages of stratified sampling are not lost. In fact, Snedecor and Cochran (1956) suggested that systematic sampling methods often provide more accurate results than simple random sampling.

The potential disadvantage of systematic sampling occurs if there is regular variation within the population that has the same frequency as the sampling frequency. In such a circumstance, we could obtain a biassed view of the population. This disadvantage is not present when we are using a systematic sampling method to sub-sample from a sampling unit that has been selected randomly, such as when we use point quadrats to sub-sample within grid squares in vegetation studies. Krebs (1989) cites several ecological references that suggest that systematic sampling is a valuable and fairly robust technique to use in field biology. The important rule when considering systematic sampling is that we should be reasonably familiar with the subject matter of the investigation. For example, if we intended to use this method for sampling wheat plants across a field, we would need to be familiar with any systematic variability of the field (e.g., soil characteristics), and scrutinise the data for any periodic variability that might be indicative of regular variations in the field conditions.

As an illustration of using systematic sampling techniques, the following example is based upon the pulse rate data that we have already encountered. For the sake of argument, let us consider those 92 individuals as a population from which we wanted to take samples in order to estimate the mean resting pulse rates.* First of all, we need to determine the sequence in which we will sample. In this case, we can systematically work our way through the sample row by row. If we want a sample of, say, 10% of the entire group, we need to start at a suitable row and then use every tenth row after that point. The way we select the starting point is as follows. Since there are 92 individuals and we wish to take a 10% sample, we need to select 92/10 individuals, which, rounded down, is 9. We then select a random number between 1 and 9 as our starting point, and then select every tenth row after that point. In the following exhibit (4.2.), the random starting point was 5 and the **COPY** command is being used to select the required rows of data. Perform this operation and, using appropriate commands, obtain a value for the means and standard deviations of the whole group and the systematic sample.**

4.3   Sample size determination

The first and second sections of this chapter have, respectively, described the main reasons *why* we sample and also the basic princi-

---

* I am treating this group here as a population purely for the purposes of this example; in Chapter 3, the 92 individuals were, quite correctly, regarded as a sample of a much larger population.

** The means are 72.87 and 72.44, respectively, for the whole group and the sample; the standard deviations are: 11.01 and 12.20, respectively.

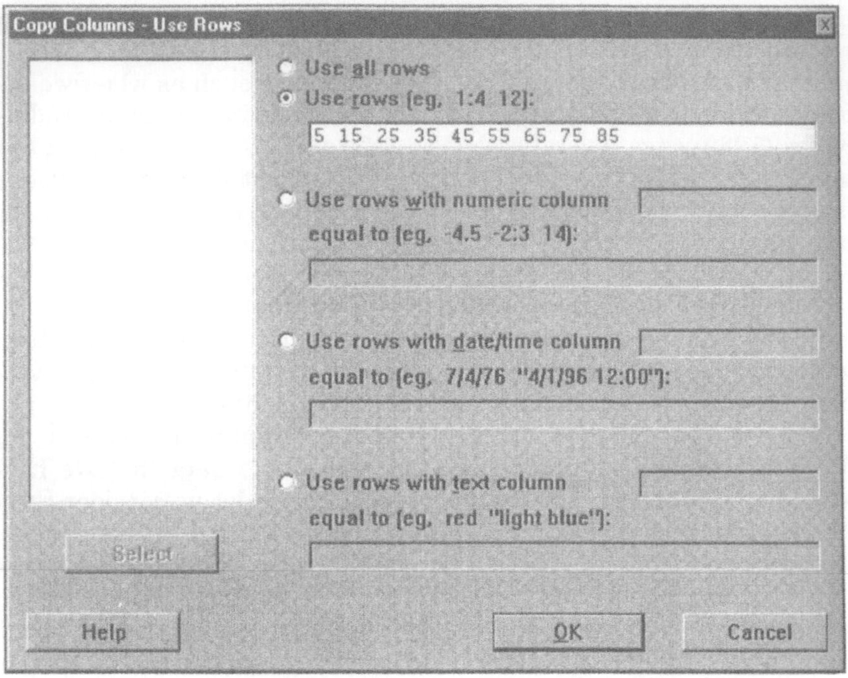

Exhibit 4.2. The dialog box of the **COPY** command that is used to specify the rows that will be sampled systematically.

ples underlying three of the most important sampling strategies, thus illustrating *how* we sample. We have yet to decide, however, upon the size of the samples that we need to take. The methods by which minimum sample sizes are calculated vary considerably according to the purpose for which we are sampling, but, although there are a large number of methods available, they each share some common principles. Hence, the objective of this final section is to introduce the principles of sample size determination by describing two specific methods. Those that I will describe are the methods used when we wish to use samples to estimate either a population mean or a population proportion. An understanding of the principles underlying these three methods will prepare you to adapt other methods for more elaborate situations as future needs arise.

How large a sample is necessary when estimating a population mean?

We can develop a simple method to answer this question by adapting the formulae we used in Chapter 3 to calculate a confidence interval (a worked example was shown in Box 3.8.). Recall that a confidence interval enables us to make statements like, *we are 95% confident that the true population mean lies between the lower and upper confidence limits.* Here the confidence interval is evaluated using Equation 4.5.

$$CI = \bar{x} \pm t_{0.05, n-1} \frac{\hat{\sigma}}{\sqrt{n}} \qquad (4.5)$$

In this formula, $\bar{x}$ is the sample mean; n represents the sample size; t is the value in t-tables for 95% confidence (i.e., $P = 1 - 0.95$)* and $n - 1$ degrees of freedom; $\hat{\sigma}$ and is an estimate of the population standard deviation. We could restate this equation by saying that we estimate the population mean with a specified margin of error, say w units, on either side of the sample mean, where w is given by Equation 4.6:

$$w = t_{0.05,\ n-1} \frac{\hat{\sigma}}{\sqrt{n}} \qquad (4.6)$$

Now, by rearranging this formula, we arrive at an expression (Equation 4.7) that will enable us to calculate the minimum sample size (n) required to estimate a population mean within a specified margin of error (w).

$$n = \frac{t_{0.05, n-1}^2 \hat{\sigma}^2}{w^2} \qquad (4.7)$$

Before we examine how this formula can be used, however, we must ask ourselves a number of questions, some of which are more biological than statistical.

---

* Remember that, in Chapter 3, we used the probability $P = 1 - 0.95 = 0.05$ column of t-tables to indicate that there was only a 5% chance of the true population mean being outside the confidence interval.

First, how precise an estimate of the population mean is really needed? If we examine Equation 4.7, we can see the importance of this question since the required sample size (n) is dependent upon w^2. So, as the acceptable margin of error (w) increases, the minimum sample size decreases in line with the square of w; a small increase in the acceptable margin of error would lead to a large reduction in the sample size requirements. Therefore, during the planning stages of an investigation, we have to decide how precise an answer we need. For example, in estimating pulse rates, would we be satisfied with a margin of error of ±10, ±5, or ±2 beats.min^{-1}?

Second, the sample size is also dependent upon the value of t, which is determined by the level of confidence we require: a high confidence level (e.g., 99%) leading to a larger minimum sample size than lower levels of confidence (e.g., 95%). So, what level of confidence do we require in our conclusions? The answer to this question will also depend upon the subject matter under examination and the use to which the conclusions will be put.

Third, the sample size will depend upon the population variance, with very large variances requiring us to take large samples. The population variance will be unknown before we start our investigation (if we already knew the variance, then it is likely that we would also know the population mean), and so we need to estimate it somehow. Therefore, a prerequisite of using any method to estimate sample size requirements is some prior knowledge or estimate of the population variance. This might be acquired from previous work reported in the literature or by means of a small-scale pilot study, but some estimate is absolutely essential. The questions that we have to ask are, what is already known about our subject material? What information is available in the literature? What preliminary work is necessary before starting the main investigation? Since we will be using a sample rather than the population variance, the expression is rewritten more correctly as Equation 4.8 with the sample variance (s^2) replacing the estimate of the population variance ($\hat{\sigma}^2$).

$$n = \frac{t^2_{0.05,\,n-1}s^2}{w^2} \qquad (4.8)$$

The value of t in this expression depends not only upon the level of confidence that we require, but also upon the sample size itself. For instance, compare the following 2 values of t for sample sizes of 21 and 11: $t_{0.05,\,20} = 2.0860$ and $t_{0.05,\,10} = 2.2281$. This means that we cannot simply substitute into Equation 4.8 values for t^2, s^2, and w^2 to obtain an absolute value for n. Instead we have to use a trial-and-error or *heuristic* approach. In simple terms, we make an initial guess as to what the required sample size might be and use that to obtain a value of t from tables (the t-distribution tables are in Appendix A.2; although, as a rule of thumb, one could obtain a reasonable first value for n by letting $t = 2$ because the value of t will not be far from this value). Then we use Equation 4.8 to calculate a value for n and use the resultant value of n to obtain a new value for t and then repeat the calculation. By repeating the procedure, you will very quickly arrive at the required sample size. An example illustrates this process in Box 4.2., in which you can see that although a wildly inaccurate first guess was used, a final estimate for the sample size was obtained after only 3 rounds of calculation.

The main principles used in this calculation are repeated in all other methods of estimating minimum sample sizes. We have to know something about the variability of the subject material, and we have to specify both the degree of confidence and the margin of error that we are prepared to accept.

How large a sample is necessary when estimating a population proportion?

We answer this question using a method that closely parallels that used to determine the minimum sample size needed to estimate the population mean. Just as when we calculated the sample size necessary to estimate the population mean, the appropriate expression for the proportion is based upon the calculation of the confidence interval, which was given in Chapter 3 (Equation 3.16). We need to specify both an acceptable margin of error and the required level of confidence, and we also require an estimate of the population variance of the proportion.

The margin of error is a straightforward quantity to define. For example, we might need to estimate a proportion, such as breeding success or incidence of infection, to within ±0.1, ±0.02, or some other proportion, the actual quantity depending upon the subject matter and the use to which the result will be put.

With regard to the confidence interval for a proportion, you will recall from Chapter 3 that we assumed that our data met conditions that allowed the binomial distribution to be approximated by the normal distribution. If those conditions were not met, then we could not legitimately calculate a confidence interval for a proportion. A

How large a sample is necessary to estimate the mean resting pulse rate with a margin of error of ±3 beats.min^{-1} with a 95% confidence?

A sample of ten pulse rates gave a sample standard deviation (s)
$= 10.03$ beats.min^{-1}

∴ sample variance (s^2) $= 100.06$

Acceptable margin of error (w) $= ±3$ beats.min^{-1}

Let initial guess of required sample size (n) $= 15$
$t_{0.05,\ 14}$ $= 2.1448$

Substituting in Equation 4.8,

$$\therefore n = \frac{(2.1448)^2 (10.03)^2}{3^2} = 51.42$$

Using $n = 51$, we obtain a value of $t_{0.05,\ 50} = 2.0086$ which we substitute in Equation 4.8 and recalculate n:

$$\therefore n = \frac{(2.0086)^2 (10.03)^2}{3^2} = 45.10$$

Again, using the new value of $n = 45$, we obtain a new value for $t_{0.05,\ 40} = 2.0211$ and recalculate. The t-tables available to us do not have an entry for 44 degrees of freedom, so we use the nearest conservative value (i.e., lower) degrees of freedom which, in this case, is 40.

$$\therefore n = \frac{(2.0211)^2 (10.03)^2}{3^2} = 45.66$$

We have now had two successive values of n that are similar (i.e., their integer parts are identical: 45), we then round up to the nearest integer and use a *sample size of 46 pulse rates.*

Box 4.2. Calculation of the minimum sample size needed to estimate the resting pulse rate of humans with a confidence of 95% and a margin of error of ±3 beats.min^{-1}.

similar assumption has to be made in order to determine the minimum sample sizes for estimating proportions. Table 3.1. summarised the necessary conditions: the sample size must be large; the probability of a success (p) must not be close to 1; and both np and nq (where q is the probability of a failure) should both be greater than 5. Since the assumptions require that the sample size must always be large, we use the normal standard deviate (z-score) rather than the t-distribution to define the appropriate level of confidence.

The variance of a proportion was described by Equation 3.10, and is given below:

$$Variance \ of \ proportion \ of \ successes = \frac{pq}{n}$$

where p is the proportion of successes and q (or $1 - p$) is the proportion of failures. By substituting z^2 for t^2 and $\hat{p}(1 - \hat{p})$ into Equation 4.8, we obtain an expression to calculate the minimum sample size (Equation 4.9) in which w again represents the acceptable margin of error.

$$Min. \ sample \ size = \hat{p}(1 - \hat{p}) \frac{(z - score)^2}{w^2} \tag{4.9}$$

where, since we do not know the proportion prior to sampling, we can only estimate the value of p, hence the use of $\hat{p}$ in the expression. If previous work provides an estimate, we can use that; or we could perform a pilot study if resources were available. Alternatively, we can obtain a conservative estimate (i.e., an overestimate) of the necessary sample size because the value of $\hat{p} = (1 - \hat{p})$ is at its maximum when they are equal (i.e., $\hat{p} = [1 - \hat{p}] = 0.5$), the product equalling 0.25. Therefore, a conservative estimate of the minimum sample size would be given by:

$$n = 0.25 \frac{(z - score)^2}{w^2}$$

Two examples of the calculation are illustrated in Box 4.3. In these examples, the researcher wished to calculate the minimum sample size necessary to estimate the proportion of colourblind individuals in the male population. The first calculation makes a conservative estimate of $\hat{p}(1 - \hat{p})$, while the second calculation makes use of prior

How large a sample is necessary to estimate the proportion of colourblind males in the population with a margin of error of ±0.1 with 95% confidence?

(a) No knowledge of the proportion in the population exists, therefore the conservative estimate of $\hat{p}$ and $1 - \hat{p}$ will be used.

$$n = 0.25\frac{(1.96)^2}{(0.1)^2} = 96.04 \approx 97$$

(b) A previous sample found a value of $\hat{p}(1 - \hat{p}) = 0.2$

$$n = 0.20\frac{(1.96)^2}{(0.1)^2} = 76.83 \approx 77$$

Box 4.3. The calculation of minimum sample sizes necessary to estimate a population proportion. Note the difference in the sample sizes using the two methods emphasising the value of previous knowledge, the larger of the two sample size recommendations exceeding the smaller by 26%.

knowledge. Although the values used in the two calculations are not very different (0.25 and 0.2, respectively), the minimum sample sizes do differ considerably: 20 more subjects are required according to the conservative estimate. Clearly, gathering too many data is a waste of valuable resources. So, the moral of this tale is that knowledge of the subject matter is of paramount importance.

4. 4   Exercises

1.   The data in Table 4.2. comprise the body weights (kg) of a population of 20 males and 80 females.
     (a) Using both random number tables and Minitab, select simple random samples of 10 males and 10 females and calculate the means and 95% confidence intervals for the means for both sexes.
     (b) Using both random number tables and Minitab, select simple random samples of 15 males and 15 females

Male						
80.16	76.72	77.19	71.96	76.07	73.53	72.12
73.93	72.09	76.28	71.82	83.31	78.53	81.38
78.29	74.81	65.89	82.83	75.26	74.58	

Female						
64.47	62.98	59.07	70.53	65.72	63.18	55.37
63.40	68.95	58.65	55.59	63.39	57.28	61.42
64.84	64.86	65.34	67.13	67.93	69.09	63.74
61.47	69.35	62.09	57.89	59.51	55.18	62.91
68.30	63.62	59.67	59.20	63.91	61.07	53.88
66.68	67.90	56.70	59.40	64.16	56.54	66.74
63.93	64.37	58.20	62.01	63.65	65.46	66.39
56.66	65.08	57.40	69.95	57.07	66.26	59.29
67.13	51.30	54.95	62.92	63.23	62.71	64.96
62.15	67.78	51.95	62.43	72.69	60.84	66.63
56.82	65.45	63.96	61.32	59.56	66.04	58.61
68.32	54.77	66.28				

Table 4.2. The weights (kg) of a population of 20 males and 80 females.

and calculate the means and 95% confidence intervals for the means for both sexes.

(c) Comment upon the accuracy of the two different sample sizes in estimating the mean weights of all individuals of the two sexes (the mean weights are: males = 75.84 kg; females = 62.47 kg) and comment on the precision of the confidence intervals.

2. Calculate estimates of the mean weight of the population listed in Table 4.2. based upon:
   (a) the samples of 10 females and 10 males of Exercise 1;
   (b) the samples of 15 females and 15 males of Exercise 1;
   (c) using a stratified random sample of 4 males and 16 females.
   (d) Which sampling strategy provides the most accurate estimate of the mean weight of the population? (The overall population mean is actually 65.144 kg)

3. Illustrating your answers with appropriate examples, describe the circumstances in which you would use simple random sampling, systematic sampling, and stratified

random sampling, justifying in each case your decision to use a particular sampling strategy.

4. A researcher wishes to estimate the mean rate of water turnover (ml.g^{-1} of bodyweight.day^{-1}) for ostriches. Previously, a sample standard deviation of 0.01 ml.g^{-1}.day^{-1} has been recorded. A meaningful margin of error would be 0.005 ml.g^{-1}.day^{-1} and the researcher requires an answer with 95% confidence. What sample size would you recommend?

5. Using a sample standard deviation of 10.03 beats.min^{-1}, calculate the minimum sample size needed to estimate the mean pulse rate with a margin of error of 5 beats.min^{-1} and with 99% confidence.

6. An estimate of the mortality rate for nestlings of a bird species is required (as measured by percentage mortality of nestlings not reaching fledging age). An earlier study suggested that the proportion dying was about 0.13. The researcher will accept a margin of error of ±0.1 and wants 99% confidence. How many broods would you suggest that he observes?

7. The density of goose droppings (number of droppings per m^2) provides an index of population density. How many fields in which geese roost should be sampled in order to estimate the mean density of droppings? Earlier studies suggested a standard deviation of 6.5 m^{-2}, and the ecologists require a margin of error of ±2 with a confidence level of 95%.

8. The percentage of crude protein in the common reed (*Phagmites australis*) is to be estimated. The researchers have some concerns about the quality of an earlier estimate of the percentage which was 17.5%. How great a difference is there between the minimum sample sizes required to estimate the proportion of protein using both of the available methods of calculation? (Assume that ±5% is an acceptable margin of error and 95% confidence is required.)

4.5   Summary of Minitab commands

COPY

RANDOM (with subcommand **INTEGER**)

4.6   Summary

1. The reasons why we need to take samples were reviewed along with the definitions of sampling units and sampling frame.
2. Three sampling strategies (simple random, stratified, and systematic) were introduced along with a description of the situations in which each can be applied.
3. Methods for calculating the minimum sample size necessary to estimate a population mean and a population proportion were described.

PART II

Questions of Comparison

General Introduction

The previous chapters have outlined the basic principles of statistics that are fundamental to all biological investigations. These principles enable us to pose and answer biological questions despite the problems caused by biological variability and the intrinsic heterogeneity of the environment within which we conduct our studies. The function of the remaining chapters is to describe in practical detail how we employ these principles in answering different types of biological questions. In Chapter 1, I categorised problems into four basic groups. Chapters 5–7 are devoted to the first of these categories: *Questions of Comparison.*

The need to compare biological data is one of the most commonly encountered types of questions in biology. For example, ecologists compare the relative abundance of populations in different habitats; geneticists compare gene frequencies in different populations; agronomists compare crop yields under various cultivation regimes; ethologists and psychologists compare the behaviour of individuals under different social contexts; cell biologists compare growth rates of tissues; and epidemiologists compare the spread of diseases. Moreover, in addition to the wide-ranging occurrence of questions of comparison, the variety of possible types of comparisons is also extremely wide.

If we consider the examples cited above, we can see that they could all be expressed in the form of a general question like:

Is group A greater/faster/more complex than group B?

In other words, we can compare two groups against each other. But there are many other forms of comparison that occur frequently in the biological sciences. Consider the following examples:

(i)   We may wish to compare a group of individuals against a standard value, such as testing whether the pulse rates of a group of individuals conform to a normal rate of 72 beats.min^{-1}.

(ii) Two groups may be compared to determine, for example, incubation periods of insect eggs at 2 temperatures. This type of comparison involves independent groups; individuals in one treatment are totally separate from the other group within the experiment.

(iii) In contrast to comparisons of independent groups, there are occasions where advantages can be gained if subjects appear in more than one experimental group. Imagine, for example, that we wish to assess the effectiveness of a particular drug. Individuals may vary considerably in their responses and so we have to take account of such individual variability in conducting the investigation. This can be achieved by taking two measurements from each subject, one before and another after the administration of the drug. Clearly, in this sort of comparison we have two related samples, each subject appearing in both the *before* and *after* groups.

(iv) One may, perhaps, wish to compare more than two groups, e.g., what is the respiration rate in 50% sea water of three species of shrimp of a particular genus, one from fresh water, one from an estuarine environment, and the third from the sea? In this situation, we would be comparing *three* independent groups.

(v) Alternatively, in agricultural research, we may wish to assess the effects of several different concentrations of nitrate and phosphate in a fertiliser on crop yields. In such an investigation, we would compare the variable, crop yield in response to two separate *factors* (nitrate and phosphate), each of which having been administered at several concentrations (*levels*).

Added to this variety of comparisons, the biological questions that we pose may give rise to different data types: frequencies, proportions, or measurements on ordinal, interval, or ratio scales. Every one of these different types of comparisons requires its own particular approach, and the methods by which we tackle each one will be described in the following chapters. In Chapter 5, I will describe the simplest of comparisons, one set of measurements contrasted with a standard measurement. In Chapter 6, I will illustrate those comparisons involving two samples. Comparisons involving more than two samples are outlined in Chapter 7, and more complex problems in which we compare the effects of several factors simultaneously are also considered in Chapter 7.

Before proceeding on to the more detailed descriptions of various types of comparisons, you should recognise four key questions that have to be addressed in all comparisons:

(a) What characteristics of the samples should be compared?
(b) What other characteristics of the samples will have to be taken into consideration?
(c) How do we actually decide if one group differs from a standard value or from another group?
(d) If we do find a difference, how do we assess the biological significance of the result?

The answers to the first two questions determine what we need to measure in order to perform a comparison. The third question concerns how we interpret the statistical results. The fourth relates the results back to the original objective of the investigation, the biological context. The answers may vary according to the nature of both the data under examination and the comparison itself. However, by considering each of these questions in the context of the following chapters, you should gain a better understanding of the common principles of comparison.

Prior to examining different comparisons in detail, we can make some general comments about these key questions. First of all, we have to determine what property of the sample we should use in a comparison. In Chapter 2, we saw that the mean and the median are the natural choices of measure to use in comparisons, the mean generally being used when the data are distributed symmetrically, and the median being particularly appropriate when the data seem to be drawn from a skewed distribution.

In answering the first question, we have also provided a partial answer to the second; the decision to use either the mean or the median will be based upon the distribution of the data. Are the data distributed symmetrically? Or, in more statistical language, are the data consistent with a normal distribution? In addition to the variability of the data, we also have to pay attention to the reliability or precision of the sample, which, after all, is representing the population in which we have an interest. Therefore, on every occasion that we use a sample, we need to gauge its reliability as a model of the population, and this is achieved by using the appropriate standard error.

The third question relates to the results of statistical analysis and how we use them to decide whether to accept or reject the null hypothesis. Statistical inference allows us to calculate the probability of the observed results occurring if the null hypothesis is true. Statistical analysis does NOT provide the probability of the null hypothesis being true. Certain conventions have been established to help us with this decision. If the probability of obtaining our observed set of measurements falls below 0.05 (i.e., less than a 1 in 20 chance), we reject the null hypothesis and accept the alternative hypothesis. This criterion is the point at which we usually start to consider rejecting the null hypothesis, but, in critical situations, we may require stronger evidence so that the risk of being wrong is much smaller. This convention applies across all types of investigations, not just questions of comparison. Another way of regarding this measure of probability is by understanding that it is the probability of wrongly rejecting a true null hypothesis. This type of error is referred to as a *Type I error.* This is why we must always report the probability to a reasonable degree of precision; we are reporting the probability of a Type I error (critical values of statistical tables generally enable us to quote probabilities to 3 decimal places). The converse error is when we wrongly accept a false null hypothesis. This is called a *Type II error.*

The answer to the fourth question, concerning the biological significance of statistical results, will always be specific to each biological investigation. I will, therefore, confine my discussion of this question to the context of specific examples. The most important general remark that can be made about this question is that the answer to it forms the basis for the conclusions of every biological investigation, not just questions of comparison.

CHAPTER 5

Single Sample Comparisons

5.1 Introduction

The simplest form of a comparison that we might encounter is when we need to compare some data of our own against a standard value. Such standards may originate from a variety of sources, such as a speculative hypothesis or a specific practical need. For example, when calibrating analytical equipment, we might measure the concentration of a standard solution on several occasions and then contrast that sample of measurements with the known concentration. If the measurements made by the equipment agree with the actual concentration, then we know that the calibration has been successful.

The choice of method used to perform this or any other type of comparison depends on many biological and statistical factors. As we saw in Chapter 3, the size of sample was critical to the choice of method to calculate confidence intervals. This is also an essential consideration when we compare a sample against a standard value; the sample size plays a major part in determining the methods available for use. I have followed this distinction in grouping the methods available for this type of comparison, those methods suitable for large samples, and those that are appropriate for small samples. As with confidence intervals, we regard large samples as having more than 30 measurements. A third section details the methods that we can employ to analyse data that do not conform to a normal distribution.

5.2 Comparisons of a large sample (or where the population standard deviation is known)

There are various approaches that we can adopt to perform this comparison, our choice of method depending largely on whether we use the mean or the median of our sample to compare against the standard

0.3466	0.3418	0.3446	0.3449	0.3424	0.3427	0.3515
0.3499	0.3547	0.3453	0.3498	0.3503	0.3526	0.3517
0.3468	0.3514	0.3475	0.3496	0.3539	0.3416	0.3435
0.3599	0.3384	0.3446	0.3513	0.3482	0.3556	0.3537
0.3526	0.3483	0.3457	0.3435	0.3406	0.3496	0.3531
0.3502	0.3576	0.3512	0.3468	0.3485	0.3478	0.3441
0.3505	0.3455	0.3562	0.3511	0.3450	0.3543	0.3458
0.3497	0.3540	0.3495	0.3393	0.3461	0.3464	0.3436
0.3484	0.3422	0.3548	0.3532	0.3512	0.3491	0.3518
0.3551						

Table 5.1. Eggshell thickness measurements (mm).

value. Below, I describe each of the most frequently encountered methods in the context of examples, and outline how we decide which is the most appropriate method to use in particular contexts.

In the first example, the mean shell thickness of eggs produced by hens in a large poultry unit was known to be 0.35 mm. After a change in feed supplier, the manager wished to determine if the change in feed had led to any change in the shell thickness. A random sample of 64 eggs was taken and the thickness of the shell of each was recorded (the data are listed in Table 5.1.). Care was taken to ensure that each egg was collected from a different hen, thus ensuring that the data were independent measurements of the shell thickness. The importance of independence was discussed in the context of sampling procedures in Chapter 4. In terms of these eggshell data, the comparison would be meaningless if the sample comprised measurements of several eggs from a small number of individuals rather than single eggs taken from individual hens. Data independence is a fundamental principle of hypothesis testing, and it will appear frequently in the discussion of many different types of investigations throughout this book.

The objective of this investigation is to discover if there has been any change in the shell thickness as a consequence of the change of food. In other words, does the shell thickness differ from 0.35 mm? Our first task in the process of answering this question is to formulate a testable null hypothesis and a complementary alternative hypothesis:

H_N: The thickness of eggshells after the food change is equal to 0.35 mm

H_A: The thickness of eggshells after the food change is not equal to 0.35 mm.

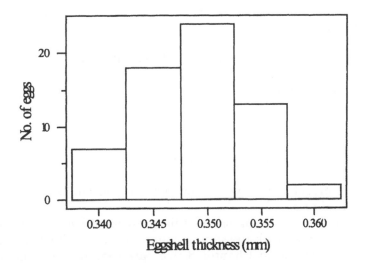

Figure 5.1. The frequency histogram of eggshell thickness measurements.

Before considering the question quantitatively, we should first examine the data visually (Figure 5.1.). Clearly, there is a certain amount of variability in the shell thickness, but the measurements seem to be distributed fairly symmetrically around a peak of about 0.35 mm, and there does not appear to be any obvious evidence of extreme values. From this visual evidence, we are probably justified in concluding that the sample was consistent with a normal distribution, and, consequently, that the *mean* shell thickness, rather than the *median*, is the appropriate sample statistic to use. In this particular example, the mean and median are actually very similar, 0.34871 and 0.3493 mm, respectively. We can now refine the hypotheses to include the fact that we are using the *mean* in the comparison:

H_N: The *mean* thickness of eggshells after the food change is equal to 0.35 mm.

H_A: The *mean* thickness of eggshells after the food change is not equal to 0.35 mm.

In statistical terms, the null hypothesis of this particular example could be viewed in the form of the question:

Is this sample of shell thickness measurements consistent with a population mean of 0.35 mm?

Thus the null and alternative hypotheses can also be expressed in algebraic notation:

$$H_N: \mu = 0.35 \text{ mm} \qquad H_A: \mu \neq 0.35 \text{ mm}$$

Since the alternative hypothesis would be accepted if we found that our mean was either greater than or less than 0.35, it can be expanded as:

$$H_A: (\mu < 0.35 \text{ mm OR } \mu > 0.35 \text{ mm})$$

where μ represents the mean thickness of the shells of the entire population of eggs.

One simple way of approaching this question utilises the confidence interval associated with the sample mean. Recalling the discussion of confidence intervals in Chapter 3, we saw that we can calculate a confidence interval that defines a range around the sample mean within which we are 95% confident of finding the true population mean. Therefore, one way of answering this question is to calculate the 95% confidence interval around our sample mean to see if it includes the standard value of 0.35 mm. If the 95% confidence interval does include 0.35 mm, then our sample would seem to be consistent with the null hypothesis (according to the convention of hypothesis rejection and acceptance that was outlined in the General Introduction in Part II). Conversely, if the confidence interval does not include 0.35 mm, we would reject the null hypothesis and accept the alternative.

The calculation of the confidence interval for this sample of eggshell data is detailed in Box 5.1. Here 0.35 mm does not fall within the confidence interval. Therefore, we can conclude that the mean thickness of eggshells is statistically different from the standard value of 0.35 mm.

The statistical conclusion that we can draw from this calculation is that there is only a small probability that this sample of eggs has been drawn from a population whose mean equals 0.35 mm. In other words, if the null hypothesis is true, then the probability of finding these results is less than 0.05. Therefore, since the probability is so low, we reject the null hypothesis and conclude that the eggshell thickness with the new food is statistically different; in fact the egg shells are thinner than before the change in food. The Minitab method for calculating confidence intervals was described in Chapter 3 and the calculation of the 95% confidence interval for the eggshell data is shown in Exhibit 5.1.

Mean shell thickness = 0.34871 mm
Standard deviation = 0.00468 mm
Sample size = 64
∴ Standard error = 0.00468 ÷ √64 = 0.00059

$$95\% \ confidence \ interval = \bar{x} \pm 1.96 \ sem$$

Upper confidence limit = 0.34871 + 1.96(0.00059) = 0.3499
Lower confidence limit = 0.34871 − 1.96(0.00059) = 0.3476

Box 5.1. Calculation of the 95% confidence interval of eggshell thickness (based upon the normal distribution).

While it is perfectly acceptable to reject the null hypothesis and to express the probability of obtaining these results as less than 0.05, we should wherever possible express this probability as precisely as we can. For example, if the probability is less than 0.001 and we report it as such, rather than merely as less than 0.05, we provide the reader with a much more informed view of the confidence of the result. In

```
MTB > Describe  'eggshell'.

Descriptive Statistics

Variable        N      Mean    Median   Tr Mean    StDev    SEMean
eggshell       64   0.34871   0.34930   0.34871   0.00468   0.00059

Variable       Min       Max        Q1        Q3
eggshell   0.33840   0.35990   0.34508   0.35177


MTB > ZInterval 95.0 0.00468 'eggshell'.

Confidence Intervals

The assumed sigma = 0.00468

Variable       N      Mean     StDev   SE Mean        95.0 % C.I.
eggshell      64   0.34871   0.00468   0.00058   ( 0.34756, 0.34985)
```

Exhibit 5.1. Calculation of the 95% confidence interval of eggshell thickness using Minitab. We use the **DESCRIBE** command to obtain a value of the sample standard deviation to use as an estimate of the population sd (sigma). This calculation is based upon the normal distribution for large samples (n > 30).

order to obtain these more precise probability values, we could calculate a series of confidence intervals of varying sizes (such as 95%, 99%, 99.9%, and so on), but that would be a rather tedious and time-consuming procedure. However, one should remember that, just because the level of confidence is increasing it does not mean that we are increasingly confident that the mean equals 0.35 mm. Rather, it means we are increasingly confident that the population mean lies within the defined limits (100% confidence intervals include all possible values). Therefore, although we have managed to test this null hypothesis successfully by means of a confidence interval, the primary function of confidence intervals is not hypothesis testing, but rather using samples to estimate the size of population parameters such as the mean. Thereby we answer such questions as, what is the mean of a population based upon a sample drawn from that population? Hypothesis testing is really just a convenient by-product of the confidence interval. In general, though, if we wish to test a null hypothesis, we should use statistical tests that specifically calculate the probability of obtaining the observed results if the null hypothesis is true. The term *statistical test* describes a procedure that tests a null hypothesis.

When dealing with large samples like the eggshell data, we can utilise the formula for calculating z-scores (Equation 3.14 applied to the sample mean) to calculate the required probability in a direct manner. Referring specifically to the eggshell example, we can express the difference between the sample mean (0.34871 mm) and the standard value (0.35 mm) in terms of standard errors. We use this difference to evaluate the probability of our sample having been drawn from a normally distributed population with a mean of 0.35 mm. In other words, the statistical test simply calculates z according to Equation 3.14 (the calculation for this example is shown in Box 5.2.). Then, by using a table of normal probabilities, it directly determines the probability of obtaining a value of z that is equally as large. As with the calculation of confidence intervals, this method is only appropriate under two conditions: when we already know the population variance, or when the sample size is large (conventionally when $n > 30$). This testing procedure is known as a *one-sample z-test*.

The calculation shown in Box 5.2. evaluates the difference, in terms of standard errors (z-scores), between our sample mean and the standard value of 0.35 mm. Examples in Chapter 3 illustrated how z-scores and normal probability tables can be used to calculate the probability of selecting particular individuals from a normally distributed population. We can use those same procedures in this type of comparison to calculate the probability of selecting a random sample that has a

$$z = \frac{\bar{x} - \mu}{\sigma_{\bar{x}}} = \frac{0.34871 - 0.35}{\dfrac{0.00468}{\sqrt{64}}}$$

$$z = \frac{0.34871 - 0.35}{0.000585}$$

$$z = -2.21$$

Box 5.2. Calculation of z, the number of standard errors between the sample mean ($\bar{x}$) of 0.3487 mm and the hypothesised mean (μ) of 0.35 mm. The denominator ($\sigma_{\bar{z}}$) is the

standard error of the mean and is estimated by $\dfrac{s}{\sqrt{n}}$.

mean that differs from 0.35 mm by at least as much as that of our own sample (either greater or smaller). In other words, we can calculate the probability of selecting a random sample that has a mean either *less than or equal to 0.34871 mm* or *equal to or greater than 0.35129 mm.* We have to calculate the probability of the mean being greater *or* smaller because that is the way that we constructed the null and alternative hypotheses. We posed the question in such a way to test the data for a difference in *either* direction (larger or smaller). We shall return to the subject of framing hypotheses later in this section.

By converting, or *standardising,* the measurements of the shell thickness to standard normal deviates, we will ask the question, what is the probability of obtaining a value of z, where

$$z < -2.21 \quad \text{or} \quad z > +2.21$$

which can be rewritten as[*]

$$|z| > 2.21$$

Normal probability tables (Appendix A.1.) tell us that the probability of selecting a sample with a mean less than or equal to +2.21 z-scores is 0.9864 (as shown in Figure 5.2.). By the law of complemen-

[*] The notation | | denotes the absolute value or *modulus* of a number, i.e., ignores a minus sign. Therefore, $|-2| = 2$.

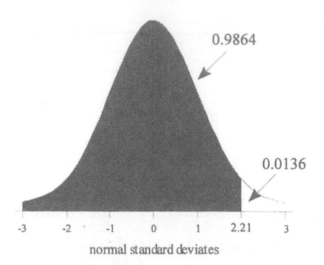

Figure 5.2. The area under the standard normal distribution *below* z = +2.21 represents 0.9864 of the total area, leaving 1 − 0.9864 above z = +2.21.

tarity (described in Chapter 3), the probability of selecting a sample with a mean greater than +2.21 z-scores is given by:

$$P(z > +2.21) = 1 - P(z \leq +2.21) = 0.0136$$

But, we require

$$P(z > +2.21 \quad \text{or} \quad z < -2.21)$$

which, according to the law of addition (Chapter 3) can be expressed as:

$$P(z > +2.21) + P(z < -2.21) = 0.0136 + 0.0136 = 0.0272$$

This probability is shown graphically in Figure 5.3.

Therefore, the *statistical* conclusion that we can draw about this sample of eggshells is that there is only a very small probability (P = 0.0272) that such a sample was drawn from a population with a mean of 0.35 mm, which is a much more precise conclusion than if it were drawn directly by calculating 95% confidence intervals.

A one-sample z-test is performed using the Minitab command **ZTEST,** which is to be found in the same dialog box as the calculation of the confidence interval by the route:

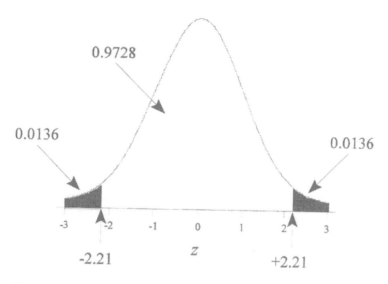

Figure 5.3. The normal curve illustrating the proportions of the area above $z = +2.21$ and below $z = -2.21$. These proportions represent the probability of randomly selecting a sample of eggs that differs by at least 2.21 standard normal deviates from the mean.

STAT > Basic Stats > 1-sample z

However, in order to test the hypothesis rather than calculate a confidence interval, we must select the *Test mean* option rather than *Confidence interval*, which is illustrated in Exhibit 5.2.

The Minitab output of the **ZTEST** command is shown in Exhibit 5.3. The structure of the output comprises three different elements: the hypothesis under test, descriptive statistics, and the results of the analysis. When reporting the results of this investigation, all three components should be included in the report.

The statistical conclusion that can be drawn from this analysis is that a reduction in eggshell thickness has occurred following the change in diet ($z = -2.21$; $P = 0.027$). In reporting these results, we would include these statistical conclusions (the test statistic, z, and the probability of observing these results if the null hypothesis is true) along with descriptive statistics such as the mean, sample size, and either the standard deviation or standard error of the mean. These various components must all be included in the *results* section of the report. We have to convey the question addressed by the investigation in very clear terms to the reader (hence the need to refer to the null hypothesis). We also need to describe or summarize the data in order

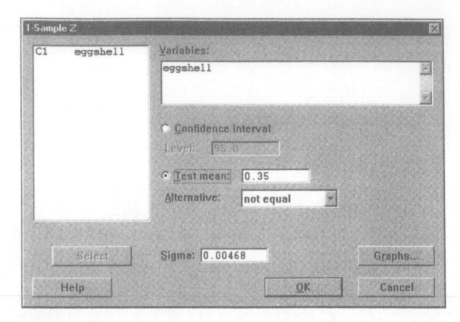

Exhibit 5.2. The Minitab dialog box to conduct the z-test on the eggshell data. Note that the *Test mean* rather than the *Confidence interval* option must be selected, specifying the hypothesised mean against which the sample mean is being compared (in this case, 0.35). Sigma is the sample standard deviation (found using descriptive statistics) that is used as an estimate of the population standard deviation.

```
MTB > ZTest 0.35 .00468 'eggshell';
SUBC>    Alternative 0.

Z-Test


Test of mu = 0.350000 vs mu not = 0.350000
The assumed sigma = 0.00468

Variable      N      Mean     StDev    SE Mean        Z    P-Value
eggshell     64  0.348706  0.004683   0.000585    -2.21      0.027
```

Exhibit 5.3. The output of the Minitab command to perform a z-test. The command line can either be generated by means of the dialog box or typed directly. The hypothesised mean is entered before the estimate of the population standard deviation. The subcommand **Alternative** relates to the alternative hypothesis and will be discussed in detail later in this chapter. The output includes a statement of the null (mu = 0.35) and alternative hypotheses (mu ≠ 0.35); descriptive statistics (sample size, mean, standard deviation, and standard error); and the results of the analysis (z and P).

to report the magnitude and direction of any observed difference. Plus, we have to report the results of the analysis, i.e., whether we should accept or reject the null hypothesis.

In the *discussion* section of the report, we would discuss the results in the context of the objectives of the investigation. In other words, this is where we evaluate the *biological* significance of the results. As biologists, we may be interested in knowing what component of the food has caused the change, and the physiological mechanism through which the change was mediated. Such an interest would lead to new testable hypotheses, and, perhaps, further experimentation. The *discussion* section of a report, therefore, is where we answer Question (d) that was identified in the General Introduction to Part II: the biological significance of the results in the context of both the objectives of the investigation and the relevant literature.

Before leaving the eggshell example, let us consider it from another perspective, that of an egg producer. For example, if eggshells are known to start cracking during normal egg collection procedures when the shells become thinner than 0.33 mm, then the producer would view the observed decrease in shell thickness to 0.348 mm as totally *insignificant* in practical terms. The eggshells, although reduced in thickness, would still not pose a threat to economic production. Therefore producers, in contrast to nutritionists, would draw totally different conclusions from these results. The contrast between the perspectives of the producer and the nutritionist shows us that a single data set can be utilised in two very different ways. In fact, the biologist and the producer are really posing two wholly different questions, with the producer asking:

Is the thickness of eggshells greater than or equal to 0.33 mm (i.e., thick enough) or less than 0.33 mm (too thin)?

If we re-examine our original null and alternative hypotheses, we can see that they are not specifically focused on the needs of the egg producer, whose sole concern is whether the shell thickness has dropped below 0.33 mm. In fact, the needs of the producer would not even be satisfied by hypotheses modelled on those with 0.33 mm substituted for 0.35 mm, but would instead require a directional element to be included. An appropriate pair of hypotheses for the egg producer would be:

H_N: The mean thickness of eggshells after the food change is *greater than or equal to* 0.33 mm.

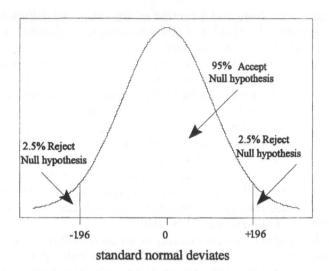

standard normal deviates

Figure 5.4. The acceptance and rejection areas of a two-tailed hypothesis, based upon a normal distribution.

H_A: The mean thickness of eggshells after the food change is *less than* 0.33 mm.

Like all pairs of hypotheses, these account for every possible outcome of the analysis and they are specifically tailored to match the objectives of the investigation. The key question that we have to address is how will the testing procedure for this uni-directional hypothesis differ from that of the earlier bi-directional hypothesis?

Figure 5.4. shows a standard normal distribution with the acceptance and rejection regions for the bi-directional hypothesis. If the observed mean is greater than 1.96 standard normal deviates away from the hypothesised or standard mean (greater or less than), then we reject the null hypothesis. In other words, we reject the null hypothesis if the observed mean falls within either tail of the distribution. Consequently, bi-directional hypotheses are known as *two-tailed hypotheses*. Since we have adopted the convention of rejecting the null hypothesis if the probability of it being true is less than 5%, then the rejection region comprises 5% of the area under the curve divided equally between each tail of the distribution.

If we now consider the question posed by the egg producer in which we are only interested in whether the shell thickness is less than 0.33 mm, we can see that the rejection region should be confined to the left-hand tail of the distribution. The rejection and acceptance

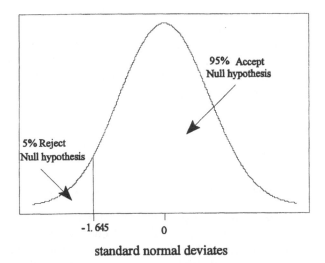

standard normal deviates

Figure 5.5. The acceptance and rejection areas of a one-tailed hypothesis, based upon a normal distribution.

regions for this uni-directional or *one-tailed hypothesis* are shown in Figure 5.5. Since we continue to follow the convention of 5% as the criterion for rejection or acceptance of the null hypothesis, the rejection region occupies 5% of the area under the curve, but in one tail only. The area below 1.645 standard normal deviates (examine normal probability tables in Appendix A.1. to verify the origin of this value).

In order to see how we actually test one-tailed hypotheses in practice, imagine that the feed for our flock of hens has been changed once again. A sample of 65 eggs has been selected at random and the mean and standard deviation of the sample were calculated as 0.32815 and 0.004841 mm, respectively (the measurements are shown in Table 5.2.).

Using Equation 3.15 to calculate z-scores, we find that the sample mean is 3.083 standard normal variates below the threshold shell thickness of 0.33 mm:

$$z = \frac{0.32815 - 0.33}{\dfrac{0.004841}{\sqrt{65}}} = -3.083$$

Referring back to Figure 5.5., we can see that this value of z falls well within the rejection region for this one-tailed null hypothesis.

0.3254	0.3334	0.3327	0.3284	0.3176	0.3301	0.3287
0.3216	0.3276	0.3245	0.3280	0.3376	0.3320	0.3290
0.3260	0.3209	0.3334	0.3215	0.3269	0.3321	0.3296
0.3317	0.3297	0.3337	0.3211	0.3241	0.3246	0.3384
0.3291	0.3307	0.3291	0.3314	0.3280	0.3262	0.3237
0.3268	0.3418	0.3306	0.3312	0.3215	0.3267	0.3277
0.3306	0.3176	0.3216	0.3235	0.3310	0.3263	0.3272
0.3255	0.3300	0.3256	0.3242	0.3297	0.3317	0.3198
0.3268	0.3299	0.3267	0.3289	0.3373	0.3370	0.3258
0.3273	0.3312					

Table 5.2. Eggshell thickness (mm) from a sample of 65 eggs.

Furthermore, by using normal probability tables, we obtain the more precise value of 0.001 as the probability* of observing such a result if the null hypothesis is true. Therefore, we can conclude that after the second food change, the shell thickness decreased ($z = -3.083$; $P < 0.001$) below the threshold thickness of 0.33 mm at which shells will crack under normal husbandry practice. Urgent remedial action would be necessary on the part of the egg producer.

The major difference between testing one and two-tailed hypotheses is in defining the rejection and acceptance regions. These, in turn, determine the calculation of probability, which is a frequent point of confusion for students. One technique that I have found helpful in this respect is to sketch diagrams like Figures 5.4. and 5.5., marking the acceptance/rejection regions as well as the location of the observed value of z.

A question that is often asked relates to the choice of one or two-tailed tests. If, for example, the result of a two-tailed test is such that we accept the null hypothesis, but the decision is a very fine one because the probability is, say, 0.06, could we not reassess the data in terms of a one-tailed test in order to be able to reject the null hypothesis? The answer to this question is an unequivocal NO! There are two major reasons why we cannot re-analyse a two-tailed null hypothesis as if it were one-tailed.

The first is that the null and alternative hypotheses are dependent upon the original question of the investigation. We have seen in the two eggshell examples that the nutritionist and the egg producer were asking two totally distinct questions that required different hypotheses. Therefore, if we change the hypotheses, we change the objective of the investigation.

---

* $P(z < -3.083) = P(z > +3.083) = 1 - P(z < +3.083) = 1 - 0.9990 = 0.001$.

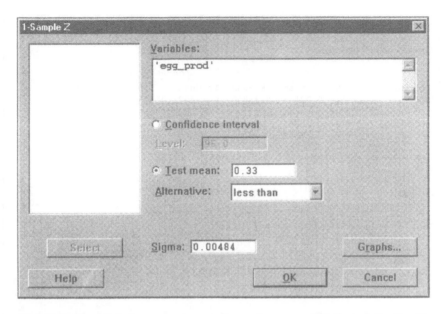

Exhibit 5.4. The Minitab menu screen used to perform a one-tailed z-test: the direction of the alternative hypothesis has to be selected. The data listed in Table 5.2. are stored in column **C2.** The value of *Sigma* (the population standard deviation) was found by using the **DESCRIBE** command (see Exhibit 5.5. for full output of the analysis).

The second reason is that if we have conducted a two-tailed test, our rejection region for the null hypothesis comprises 2.5% of the area in each tail of the distribution, which is a total of 5% (Figure 5.4.). By switching to a one-tailed test, we effectively add a further 2.5% to the rejection region in one tail. By doing so, we have created a rejection region of 7.5% — the criterion for rejection would be P = 0.075, not 0.05.

I would actually go further and argue that making post hoc changes to the hypotheses indicates that perhaps we are more interested in rejecting null hypotheses than rigorously searching for answers to biological questions. The simple rule for hypothesis testing is to formulate the null and alternative hypotheses *before* gathering any data, and to make sure that they match the objectives of the investigation. Do not modify hypotheses after data collection or analysis.

Minitab provides a method for performing one and two-tailed z-tests (based upon the same menu illustrated in Exhibit 5.2.) where the direction of the alternative hypothesis is specified as *not equal* (two-tailed); *less than* or *greater than* (one-tailed). Contrast Exhibit 5.4. where a one-tailed test is performed with Exhibit 5.2., where a two-tailed test is performed.

```
MTB > Describe  C2.
```

Descriptive Statistics

```
Variable          N     Mean   Median  Tr Mean    StDev  SE Mean
C2               65  0.32815  0.32800  0.32809  0.00484  0.00060

Variable        Min      Max       Q1       Q3
C2          0.31760  0.34180  0.32545  0.33110

MTB > ZTest .33 .00484  C2;
SUBC>   Alternative -1.
```

Z-Test

```
Test of mu = 0.330000 vs mu < 0.330000
The assumed sigma = 0.00484

Variable      N      Mean    StDev  SE Mean          Z          P
C2           65  0.328154 0.004841 0.000600      -3.08     0.0011

MTB >
```

Exhibit 5.5. Minitab output of the result of a one-tailed z-test. Note that the null and alternative hypotheses are included in the output. Descriptive statistics are included to obtain the value of the standard deviation that has to be included in the z-test command. The **ALTERNATIVE** subcommand indicates that the alternative hypothesis shows that the mean is *less than* 0.33.

We have seen how the z-test enables us to compare the mean of a single large sample against a standard value. The distribution of measurements that comprise our sample approximate to a normal distribution if the sample is large. In the next two sections, respectively, we will see how similar comparisons can be performed on small samples and on data that we cannot assume to have been drawn from a normal distribution.

5.3   Comparisons of a small sample

When we considered confidence intervals in Chapter 3, we had to use the t-distribution if the sample size was small (i.e., less than 30). Likewise, when comparing small samples against a standard we have to use the t-distribution, rather than the normal distribution. But, other than that difference, the principles of testing a hypothesis using a small sample are very similar to those used for a large sample. We can see how the one-sample t-test can be used to perform one and two-tailed tests by re-examining the maize data that were featured in Exercise 12 in Chapter 3. These data comprised height measurements of two sets of maize plants, one with fertile pollen and the other with sterile pollen. Let us pose two simple questions:

(a) Does the height of maize plants with fertile pollen differ from 240 cm? and

(b) Are the maize plants with sterile pollen shorter than 240 cm?

The first question leads to two-tailed hypotheses of the form:

H_N: The mean height of maize plants with fertile pollen does not differ from 240 cm.

H_A: The mean height of maize plants with fertile pollen does differ from 240 cm.

In order to test this hypothesis, we have to calculate how much the mean height differs from the standard height of 240 cm. As with the z-test, we measure this difference in terms of standard errors, but now we use the t-distribution, rather than the normal distribution, to calculate the probability of obtaining the observed results. The formula that we use to perform this conversion is a direct parallel to Equation 3.14 that we used for the z-test:

$$t = \frac{\bar{x} - \mu}{s.e.} \tag{5.1}$$

where $\bar{x}$ and $s.e.$ represent the sample mean and standard error of the mean, respectively, and μ represents the standard value with which the sample is being compared.

Before performing this calculation, let us define the acceptance and rejection criteria of the null hypothesis in terms of the t-distribution. You will recall from the discussion of confidence intervals based upon the t-distribution (Chapter 3), that the similarity between the t and normal distributions diminishes with decreasing sample size. We used degrees of freedom to take account of the sample size and we follow

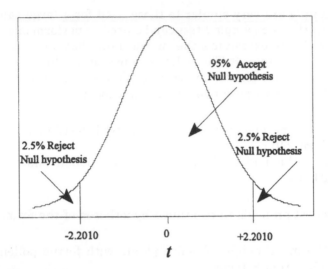

Figure 5.6. The acceptance and rejection areas of a two-tailed hypothesis, based upon a t-distribution with 11 degrees of freedom.

the same procedure with this test. The number of degrees of freedom is calculated in exactly the same way as it is for confidence intervals:

$$degrees\ of\ freedom = sample\ size - 1 \qquad (5.2)$$

Thus, in the case of the maize plants, there are 11 degrees of freedom. We can see in the t-table (Appendix A.2.) that with 11 degrees of freedom 5% of the area under the curve lies equally divided in the two tails, below $t = -2.2010$ and above $t = +2.2010$. Therefore, the rejection/acceptance regions are defined according to Figure 5.6.

Having established the acceptance and rejection criteria, we can now proceed to compare the heights of the maize plants (mean = 245.32 cm; s.d. = 12.74 cm; n = 12) with height of the specified mean of 240 cm, and this calculation is shown in Box 5.3.

The calculated value of t falls within the acceptance region for the null hypothesis. If we examine the t-tables, we can see that if the null hypothesis is true, the probability of observing such a result is greater than 0.1. In reporting such a result, we would cite the value of t, the degrees of freedom, and the probability (in this case, $t = 1.45$; df = 11; $P > 0.1$). By accepting a null hypothesis, we are saying that it provides an acceptable description of the situation under examination; we are not saying that it is true. Furthermore, remember that the sample is

$$t = \frac{\bar{x} - \mu}{s_{\bar{x}}} = \frac{245.32 - 240}{\dfrac{12.74}{\sqrt{12}}}$$

$$t = \frac{245.32 - 240}{3.68}$$

$$t = 1.45$$

Box 5.3. Calculation of t, the number of standard errors between the sample mean ($\bar{x}$) of 245.32 and the hypothesised mean (μ) of 240 cm. The denominator ($s_{\bar{x}}$) is the standard error of the mean.

being used as a representation of the population, so the hypotheses should be framed in terms of the population and not the sample.

Let us now turn to the second question that relates to the sterile maize plants. Are they shorter than 240 cm? The required null and alternative hypotheses are clearly one-tailed because the question includes the key words *shorter than*. Whenever a question includes a directional element, a one-tailed hypothesis will be required. For this example, the hypotheses will be:

H_N: Sterile maize plants are taller than or equal to 240 cm.

H_A: Sterile maize plants are shorter than or equal to 240 cm.

As with the two-tailed tests, we calculate the difference between the observed mean and the standard value (240 cm) and express this in terms of standard errors:

$$t = \frac{230.93 - 240}{3.95} = -2.3$$

Again, there are 11 degrees of freedom, but, since this is a one-tailed test, interpreting the t-tables will differ slightly from the method used for the two-tailed test, in which 5% of the area under the curve was divided equally between each tail (2.5% below -2.2010 and 2.5% above $+2.2010$). In contrast, for the one-tailed test, we require 5% in one tail. The $P = 0.1$ column of t-tables indicates 10% of the area under

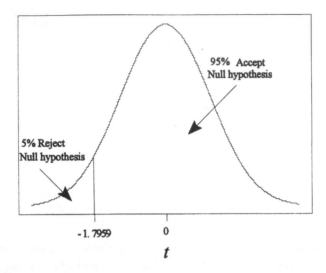

Figure 5.7. The acceptance and rejection areas of a one-tailed hypothesis, based upon a *t*-distribution with 11 degrees of freedom.

the curve spread between the two tails, i.e., 5% in each. Therefore, the appropriate critical value for this one-tailed test with 11 degrees of freedom will be −1.7959 (it's negative because we are interested in the left-hand tail) as shown in Figure 5.7.

Extending this reasoning across the entire table, the critical probability levels at the head of each column should be divided by a factor of two for one-tailed tests. There is no set convention as to how statistical tables are presented. The *t*-tables in Appendix A.2. have been designed for two-tailed tests, hence the necessary manipulation for one-tailed tests. Before using any statistical tables, establish whether they are set up as one or two-tailed tables. This is often stated explicitly, or, as in the case of the *t*-tables in this book, by the diagram at the foot of the tables indicating both tails.

Returning to the results of this comparison, the calculated value of t (−2.3) falls within the rejection region, and looking at the *t*-table, we can see that it lies between 2.2010 and 2.7181. Therefore, we can conclude that the sterile maize plants are significantly shorter than 240 cm ($t = -2.3$; df = 11; $P < 0.025$).

The Minitab command to perform the one-sample *t*-test is **TTEST** and can be found via the same menu route as the one-sample *z*-test:

STAT > Basic Stats > 1-sample t

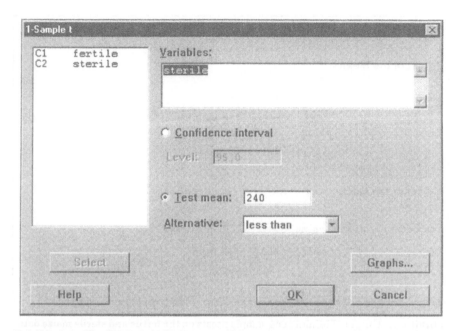

Exhibit 5.6. The Minitab dialog box to perform a one-sample *t*-test. Note that *less than* has been selected in the *Alternative* box, thereby selecting a one-tailed test.

5.4   Comparisons of samples that have not been drawn from normal distributions

The *t*-test is based upon two fundamental assumptions. The measurements or observations comprising a sample must be independent of each other (as discussed in Chapter 4 and in the context of the eggshell example) and the populations from which the samples are drawn should follow a normal distribution.

The assumption of normality is very important, and if it is not satisfied, conclusions may be invalid. There are two broad approaches that we can adopt to overcome this problem. One approach is to use statistical tests that do not rely upon this assumption, and it is this approach that will be the subject of this section. The other approach is to perform an arithmetic transformation that converts the measurements into a form that does conform to a normal distribution. We shall see later that normality is only one of several assumptions that are implicit in many statistical tests. From time to time such assumptions are not satisfied and many data transformations have been

```
MTB > TTest 240   'fertile';
SUBC>   Alternative 0.
```

T-Test of the Mean

```
Test of mu = 240.00 vs mu not = 240.00

Variable      N       Mean     StDev    SE Mean        T          P
fertile      12     245.32     12.74       3.68     1.45       0.18

MTB > TTest 240   'sterile';
SUBC>   Alternative -1.
```

T-Test of the Mean

```
Test of mu = 240.00 vs mu < 240.00

Variable      N       Mean     StDev    SE Mean        T          P
sterile      12     230.93     13.68       3.95    -2.30      0.021
```

Exhibit 5.7. Output of Minitab one-sample *t*-tests on the fertile and sterile maize data. The first is a two-tailed test and the second is a one-tailed test. Note that the descriptive statistics (mean, standard deviation, and standard error) are all included in the output as well as clear statements of the null and alternative hypotheses. It is also worth noting that Minitab provides an exact probability value (0.18 and 0.021 here), whereas if we use *t*-tables we will be less precise with estimates of >0.1 and <0.05.

established to cope with such circumstances. Transformations are also used when the sample of data being analysed contains values that may differ by several orders of magnitude (e.g., in counts of bacterial populations). Since data transformations are so widespread throughout biology, and because great care is required in their use and the reporting of transformed results, they are the subject of a detailed discussion in Section 7.4.

There is a group of statistical techniques that do not require the distribution of the data to conform to a particular statistical distribution. They are called *non-parametric* or *distribution-free* tests. Several excellent books are devoted to this group of tests (e.g., Siegel and Castellan, 1988; Sprent, 1993), and other general texts (e.g., Zar, 1996) include this group alongside the parametric tests that do make assumptions regarding statistical distributions.

Generally, non-parametric tests are not as powerful as the equivalent parametric tests when the assumption of normality is satisfied (i.e., they are more likely to lead to Type II errors than the equivalent parametric test). Therefore, if it is valid to use parametric tests, they should provide a more powerful analysis. Sometimes, however, we do need to use non-parametric tests, and so some of the more frequently used techniques will be described in the context of investigations where they are relevant.

Before proceeding to outline the non-parametric methods that we can use to perform comparisons, let us consider what sort of biological data might not conform to a normal distribution.

In Chapter 3, we saw that biometrical variables such as height were the result of a summation of many random variables (both genetic and environmental), and, according to the central limit theorem, they could be expected to conform to a normal distribution. In fact, many biological variables measured on a ratio scale would be expected to fall into this category, including eggshell thickness. However, there are many other biological variables that could be categorised as data types other than ratio. For example, there are nominal data such as sex, species, or phenotypic characters (e.g., round or wrinkled peas), or ordinal data such as position in a dominance hierarchy. Often we would not be able to assume that such data were drawn from populations that were distributed normally. A further example of non-normal data applies to variables that can be measured on a ratio scale, but, because of the way that they are recorded, normality cannot be assumed. This latter situation is where we have to truncate the scale of measurement in some way. For example, imagine that we were counting the number of aphids per leaf. We might record the aphid numbers according to the following scale:

$$0, 1, 2, 3, 4, 5+$$

On this scale, any value above 4 aphids is recorded as 5+. This type of scale is referred to as a *censored scale*, and the data are known as *censored data*. Situations of this type arise when the investigator runs out of time or other resources and has to curtail the measurements in some way. This frequently occurs in survival studies. In pharmacological investigations, for example, the time taken for an individual to respond to a treatment or die is frequently recorded, but, after a specified period, observations cease and all individuals who haven't responded to treatment or who are still surviving are grouped together in a single category and given a value of *more than x time units* (paralleling 5+ aphids per leaf). By ascribing this arbitrary cut-off value to several subjects, we have recorded data to a lower level precision for these individuals compared with the other subjects in the sample. As a consequence, we have a limited and variable knowledge of the population from which the data have been drawn.

I shall describe three non-parametric techniques (*binomial*, *sign*, and *Wilcoxon one-sample tests*) that we can use where we have to compare a single sample against a standard value. The first and second can be applied to all data types including nominal variables, i.e., those data that fall into categories, such as male/female, yes/no, and left/right. The sign test is, in fact, a special case of the binomial test, and we will also encounter it in later sections. The Wilcoxon test can be used where we have been able to measure the magnitude of the variable of interest to some degree, even if it is only on an ordinal scale. Whereas the one-sample *z* and *t*-tests enabled us to compare a sample mean with a standard value, the non-parametric tests that I describe here compare sample medians with a specified standard. The reason that we use the median rather than the mean is that it is not affected by outliers as much as the mean. For example, the probability of a value being less than the median is always 0.5, whereas the probability of a value being less than a mean depends upon the distribution involved.

Comparing nominal data against a standard value

In psychological investigations, animals are often presented with choices. We may, for example, wish to examine the preference of a rat or chick for a novel food or nest design. Such investigations are frequently performed by placing the subject in a T-shaped maze. The animal walks up the central corridor and is forced to turn left or right at the T-junction with the choices being presented in the arms of the T-maze. An experiment comprises several trials for each subject with the food, for example, being switched at random between the left and right arms between trials. Over the years, however, many animal species have been observed to exhibit lateral preferences, a tendency to favor one side (left or right) over the other. The following example describes a pre-experiment test, which was used to screen out rats who exhibit such a bias.

In this test, rats were introduced into the T-maze and the direction in which they turned was recorded (there was no physical difference between the two arms of the maze). Since we are interested in detecting bias in either direction, we conduct a two-tailed test, and the null and alternative hypotheses would be:

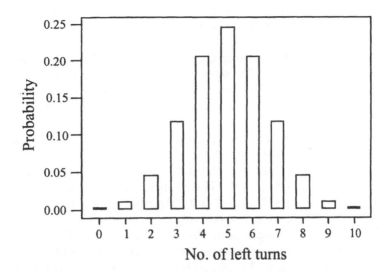

Figure 5.8. The probability of a binomial distribution describing the unbiassed turning behaviour of rats. The probability of the rats turning left is the same as turning right ($P = 0.5$), and the number of trials is 10.

H_N: A rat will turn left and right to an equal extent.

H_A: A rat will turn left and right to an unequal extent.

Since there are only two possible outcomes of each trial (turning left or right), this is, in effect, a binomial experiment comprising several trials. The probability of all possible outcomes will be described by the binomial distribution, with the probability of turning in either direction being equal, i.e., 0.5. Figure 5.8. shows the probability of turning to the left in an experiment comprising 10 trials.

In an experiment with a single rat, the investigators observed it taking 1 turn to the left and 9 to the right. The question is whether this result is consistent with the null hypothesis, or whether the rat should be removed because it is exhibiting a bias toward one direction?

Following the same reasoning that we used for one-sample z and t-tests, we need to calculate the probability of obtaining results with an equal or greater degree of bias as our observations. Since we are conducting a two-tailed test, this requires us to calculate the probability of the rat taking 0, 1, 9 or 10 left turns, for which we can use Equation 3.6:

$$P(0, 1, 9, \text{ or } 10 \text{ left turns}) = 0.000977 + 0.009766$$
$$+ 0.009766 + 0.000977 = 0.021486$$

The probability of obtaining these results and the null hypothesis being true is less than 0.05, so we would reject the null hypothesis that this rat was unbiased, and remove it from the experiment. This is a very simple test that can be applied in any situation where there are two alternative outcomes for which the respective probabilities are known. Since it is based upon the binomial distribution, it is known as the *binomial test*. We can calculate the probability of a particular outcome using the Minitab probability distribution commands that were introduced in Chapter 3. The results for this rat example are illustrated in Exhibits 5.8. and 5.9. (the dialog box to perform the command and the results, respectively).

Although it is a simple test, the calculations can become rather tedious (as we saw in the calculation of binomial probabilities in Chapter 3) and thus it is advisable to use the computer wherever possible.

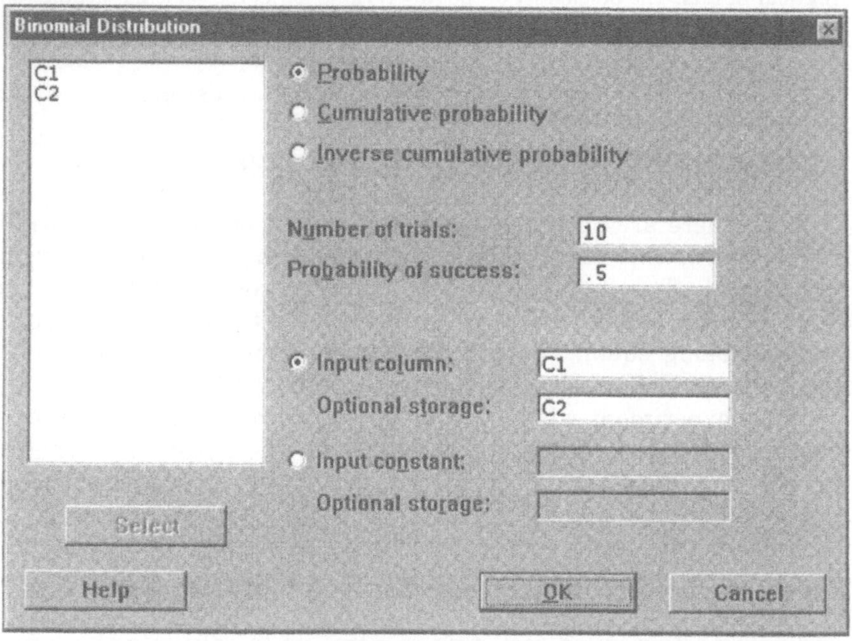

Exhibit 5.8. The dialog box to calculate binomial probabilities. The data in column **C1** are the outcomes for which probabilities are required (in the rat example, the data are 0, 1, 9, and 10). The calculated probabilities are stored in column **C2**.

```
MTB > PDF C1 C2;
SUBC>    Binomial 10 .5.
MTB > PRINT C1 C2
```

Data Display

```
 Row    C1          C2

   1      0   0.0009766
   2      1   0.0097656
   3      9   0.0097656
   4     10   0.0009766

MTB > SUM C2
```

Column Sum

```
     Sum of C2 = 0.021484
MTB >
```

Exhibit 5.9. Minitab output of the calculation of binomial probabilities. In this particular example, in which the probability of turning left is 0.5 and the number of trials is 10, we need to calculate the probability of a rat turning left on 0, 1, 9, or 10 occasions. The probabilities for each of the four outcomes is stored in column C2, and the sum is calculated as 0.021484.

In the specific situation where the probabilities of the two outcomes are equal (i.e., both are 0.5), we can use another simple test that is based upon the binomial distribution, the *sign test*.[*] This test is so called because it is used where a trial can lead to a positive or negative outcome. In order to perform this test, we merely count the number of the most frequent outcomes and compare that number with the appropriate section of sign test probability tables, of which a set is included as Appendix A.4. This table has been constructed by calculating binomial probabilities for all possible outcomes of binomial experiments where the number of trials is less than or equal to 35. This table shows two-tailed probabilities; if a one-tailed test is being performed, the probabilities at the head of each column should be divided by 2. In this particular example, with the rat turning to the left on only 1 occasion and turning to the right in the 9 other trials, we can see from the tables the probability of turning in one particular direction on 9 or more occasions is less than 0.05, but more than 0.02 (as shown below in the extract of the tables):[**]

[*] We shall also see this test in Chapter 6 when we consider comparisons of related pairs of data.

[**] If we had constructed a one-tailed hypothesis, the probability of obtaining these results would have been less than 0.025, but greater than 0.01.

n	0.2	0.1	0.05	0.02
10	8	9	9	10

Therefore, in this pre-experiment trial, we can report that if the null hypothesis is correct (i.e., the rat is unbiased), then the results are not very likely ($P < 0.05$) and we can reject the null hypothesis and remove this rat from the experiment. The sign test has given the same result as the binomial test, but not to the same level of precision with regard to the magnitude of the probability.

Minitab provides a command to perform the one-sample sign test, and the main dialog box is found within the *nonparametric* section of the **Stat** menu (and is illustrated in Exhibit 5.10.):

Stat > Nonparametrics > 1-Sample Sign

In this example of rat choices in a T-maze, unlike previous examples, the experimental observations are not in the form of a measured variable, but are frequencies of each of two nominal categories (left

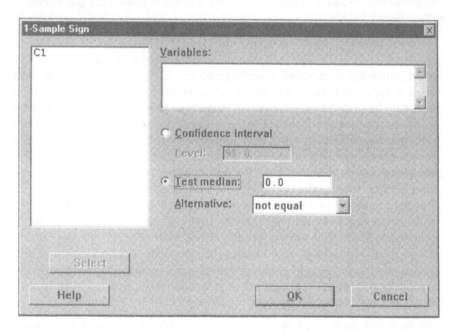

Exhibit 5.10. The dialog box for a one-sample *sign test*. It is very similar to that of the one-sample *t*-test.

```
MTB > PRINT C1

Data Display


C1
   -1    1    1    1    1    1    1    1    1    1

MTB > STest 0.0 C1;
SUBC>    Alternative 0.

Sign Test for Median


Sign test of median = 0.00000 versus  not =  0.00000

                    N  Below  Equal  Above       P   Median
C1                 10     1      0      9   0.0215    1.000
MTB >
```

Exhibit 5.11. The data and results of a sign test, analysing the turning behaviour of a rat in a T-maze. In 10 trials, the rat turned left on only 1 occasion. The probability of observing such behaviour if the null hypothesis of no bias is true is 0.0215 (in agreement with the outcome of the binomial test).

and right). Therefore, we have to code the observations in some way before we can perform the analysis. A straightforward method is to enter the left turns as −1 and the right turns as +1 (or left and right). We can then test the null hypothesis that the median value equals 0 since we would expect a similar number of left and right turns (i.e., the same number of −1s as +1s). The data and output of the Minitab sign test for this example are shown in Exhibit 5.11.

In this particular example, the rat could only turn left or right and thus, no observation could actually be the same as the hypothesised median (in this case 0). However, we could have applied this test to ordinal data where the observations could be the same as the median. In such circumstances, we remove those records from the analysis. For example, imagine that we wanted to test whether the following sample of measurements differed from a median of 25.

$$18, 20, 25, 32, 26, 29, 21, 23$$

We would remove the measurement that is the same as the hypothesised median, and conduct the sign test only on those data that are either above or below the median value. Minitab also follows this procedure and we can see in Exhibit 5.11. that Minitab reports the number of measurements that fall below, are equal to, or are above the hypothesised value.

In Chapter 3 we saw that the binomial distribution can be approximated by the normal distribution when sample sizes are large. We can make use of this approximation when we wish to compare a large sample with a known standard. Since both the binomial and sign tests are based upon the binomial distribution, this approximation applies to both tests. The mean and variance of a binomial distribution are, respectively, given by:

$$\mu = np$$

and

$$\sigma^2 = npq$$

We can then apply these to Equation 3.14. to calculate z-scores (as illustrated earlier in Box 5.2.). However, errors can occur as a consequence of using a continuous distribution (the normal) to approximate a discrete distribution (the binomial). Thus, a *continuity correction* is applied to the formula used to calculate the z-scores. The correction factor reduces the difference between the observed number of successes and the mean value (μ) as shown in Equation 5.3.

$$z = \frac{(x \pm 0.5) - \mu}{\sigma} \tag{5.3}$$

In order to reduce the difference, we would add 0.5 to x (the number of successes) if it is lower than μ and subtract 0.5 if the number of successes is larger than μ. As an example of the normal approximation, imagine that we performed another experiment with a rat in a T-maze, but the rat had to perform 35 trials. We observed it turning to the right on 8 occasions (the observed number of successes), and the calculation is detailed below in Box 5.4.

Table 3.1. showed the criteria determining when we should use the normal approximation to the binomial distribution. In the context of the sign test, some authors suggest that we should use this approximation when the sample size exceeds 35 (Siegel and Castellan, 1988), while Minitab only invokes the normal approximation when the sam-

The probability of a rat turning left $= \ p \ = \ 0.5$
The probability of a rat turning right $= \ q \ = \ 0.5$
The number of trials $= \ n \ = \ 35$
The number of turns to the right $= \ \ \ = \ 8$

$\therefore$ the mean $= np \ = (35)(0.5) \qquad = 17.50$
and the variance $= npq = (35)(0.5)(0.5) = \ 8.75$

and so,

$$z = \frac{(no. \ of \ right \ turns \pm 0.5) - \mu}{\sqrt{npq}} = \frac{(8 + 0.5) - 17.50}{\sqrt{8.75}}$$

$$= -3.042$$

Referring to normal probability tables, we can see that the two-tailed probability of obtaining this result, or one that is even more extreme, is 0.0024.

Box 5.4. Calculation of the probability of a rat turning to the right on 8 occasions in 35 trials; the calculation is performed using the normal approximation to the binomial distribution. The continuity correction is applied to the formula, and, since the number of turns is lower than the hypothesised mean, 0.5 is added to the observed number of turns, thus reducing the difference. Since the probability of obtaining these results if the null hypothesis is true is 0.0024, we reject the null hypothesis and conclude that the rat is biased towards the left.

ple size is greater than 50. I have constructed the binomial probability tables in Appendix A.4. for sample sizes up to and including 35.

Comparing measured data against a standard value

The sign test is very useful when we have observations that are not quantitative. However, because the test does not take into account the *magnitude* of any difference between an observation and the hypothesised median, it is not a particularly powerful test when compared with those that do incorporate the magnitude of any difference. For this reason, wherever we have quantitative data, we should avoid using the sign test if possible. If the measurements do not conform to a normal distribution, or are from censored data, then we should use an alternative test, the one-sample Wilcoxon test, which was named after the statistician who devised it.

The Wilcoxon test, often known as the *Wilcoxon signed-rank test*, takes into account both the magnitude of a measurement as well as its sign. So, in the example below, in which retention times for a drug have been measured, we can compare the times against a standard value, not only in terms of whether they are longer or shorter, but also the degree to which they differ from a specified retention time. In this respect, the Wilcoxon test is a direct parallel to the t and z-tests. It is a far more powerful tool than the sign test, but it is less powerful than the t and z-tests. Given this relative power, it should be used in preference to the sign test, wherever possible.

The procedure to conduct a one-sample Wilcoxon test is best explained by means of an example. The experiment described here examined the length of time that a drug remained at an effective concentration in the blood of patients. For this drug to be effective, it must be retained in the blood at a minimum concentration for at least 12 hours. The drug was administered orally as a capsule and the duration for which it remained at the necessary concentration in the blood was recorded (in hours). Recording continued for a maximum of 24 hours after administration. Seventeen individuals were tested, and, by the end of the monitoring period, the concentration had fallen below the critical level in 14 patients. The retention times for the 3 remaining patients were recorded as 24+ hours; the data set is clearly censored. The question that is posed in this investigation is, do these data indicate that this mode of administration satisfies the clinical need of a minimum retention time of 12 hours? The null hypothesis to be tested will be one-tailed:

H_N: The median retention time will be less than or equal to 12 hours.

H_A: The median retention time will be greater than 12 hours.

Since the data are censored at one end of their distribution, we cannot assume that they are symmetrically distributed, so we have to compare the median rather than the mean. The data for this experiment are shown in Table 5.3. and the calculation is detailed in Box 5.5.

The procedure for the Wilcoxon one-sample test comprises four stages:

(i)   Find the difference between each measurement and the hypothesised median.

6.00	7.00	8.80	9.20	11.20
12.20	12.50	14.00	17.00	19.20
20.50	22.00	22.60	23.20	24+
24+	24+			

Table 5.3. Retention time (hours) of the drug in the blood.

(ii) Rank the differences in order of absolute size, i.e., ignoring the sign of the difference.

(iii) For each group (the positive and the negative differences), calculate the sum of the ranks.

(iv) The test statistic, T, is the larger of the two rank sums and should be compared against the critical values in the Wilcoxon tables (Appendix A.5.). This table provides critical values for two-tailed probabilities, and thus, for this particular example, the probabilities must be halved in a similar way to the sign test example.

If a measurement coincides precisely with the median against which the data are to be compared, that particular datum is omitted from the analysis and the number of data is reduced accordingly. Notice also in Box 5.5. that some differences are identical and thus are given the same rank (known as *tied ranks* or *ties*). The procedure for allocating ranks to identical measurements is quite straightforward. The drug retention data include two sets of tied measurements: 7 and 17 hours. Each differs from the hypothesised median by 5 hours, and the three longest retention times have all been recorded as 24+ hours, which occupy the 15th, 16th, and 17th positions. The first of these sets of ties occupies the seventh and eighth positions in the sequence of measurements. Their rank is calculated as the mean of the two positions that they occupy: 7.5. Similarly, for the other group of ties, they are each given a rank of 16, their mean position in the sequence of measurements. This method of allocating ranks to ties scores is used throughout statistics, and we will encounter it on several further occasions in this book.

In this particular test we need to find the larger of the rank sums, whereas in other tests we have to calculate the smaller sum. Sometimes, the number of subjects in each group is widely different, thus making the calculation of the smaller rank sum considerably simpler than that of the larger group. We can reduce the effort required for these calculations by utilising the following expression, which calculates the total rank sums for the entire set of measurements:

$$total\ rank\ sum = \frac{N(N+1)}{2}$$

where N is the total number of measurements in the sample. Using this expression along with the rank sum for the smaller group, we can calculate the larger rank sum by subtraction:

$$rank\ sum\ of\ larger\ group = \frac{N(N+1)}{2} - rank\ sum\ of\ smaller\ group$$

The distribution of the Wilcoxon test statistic can be approximated by the normal distribution in a similar way as the binomial and sign tests when sample sizes are large. Sprent (1993) states that for sample sizes of 20 or more, the normal approximation can be used where the mean and standard deviation are given by Equations 5.4 and 5.5, in which N represents the number of values that differ from the hypothesised median.

$$\mu = \frac{N(N+1)}{4} \tag{5.4}$$

and

$$\sigma = \sqrt{\frac{N(N+1)(2N+1)}{24}} \tag{5.5}$$

We must apply a continuity correction to the equation used to calculate the z-score:

$$z = \frac{(|T-\mu|-0.5)}{\sigma} \tag{5.6}$$

A further complication arises when using the normal approximation if we have tied ranks that require us to include a further term in Equation 5.5 to account for the number of sets of tied ranks in our set of measurements, and this is shown in Equation 5.7.

The required retention time for this drug to be effective is 12 hours. This calculation, therefore, must test the null hypothesis that median retention time is less than or equal to 12 hours.

Retention time (hours)	Difference from hypothesised median	Rank
6.0	−6.0	9.0
7.0	−5.0	7.5
8.8	−3.2	6.0
9.2	−2.8	5.0
11.2	−0.8	3.0
12.2	0.2	1.0
12.5	0.5	2.0
14.0	2.0	4.0
17.0	5.0	7.5
19.2	7.2	10.0
20.5	8.5	11.0
22.0	10.0	12.0
22.6	10.6	13.0
23.2	11.2	14.0
24+	12.0	16.0
24+	12.0	16.0
24+	12.0	16.0

Sample size   = 17
The sum of the ranks for the positive differences = 122.5
The sum of the ranks for the negative differences = 30.5

Comparing the larger of these two values with the Wilcoxon tables in Appendix A.5., we see that the one-tailed probability of observing such a rank sum is less than 0.025 (the critical values for one-tailed probabilities of 0.025 and 0.01 in the tables are 119 and 126, respectively). Therefore, we reject the null hypothesis and accept the alternative hypothesis: the drug retention time is greater than 12 hours and the method of administration is acceptable.

Box 5.5. Calculation of the one-sample Wilcoxon test for the drug retention data shown in Table 5.3.

$$\sigma = \sqrt{\dfrac{N(N+1)(2N+1) - \dfrac{\Sigma t}{2}}{24}} \qquad (5.7)$$

In this equation, Σt relates to the groups of tied ranks within the set of measurements. In this drug example, there are two sets of tied ranks, one with the rank of 7.5 and the other with the rank of 16. In the first of these groups, there are 2 measurements (7 and 17 hours), and in the second, there are 3 (each recorded as 24+ hours). In the following expression, which describes the calculation of Σt, the number of ties in each group (2 and 3) are denoted by t_i. For each group of ties, the term $(t^3 - t)$ is calculated and then summed together to produce a value for Σt. Thus, for the drug retention data, Σt is given by:

$$\Sigma t = \Sigma(t_i^3 - t_i)$$

$$= (2^3 - 2) + (3^3 - 3)$$

$$= 30$$

Minitab provides a procedure to perform the Wilcoxon test, and, like the sign test, it can be accessed through the non-parametric section in the **Stat** menu. Any tied values are taken into consideration by Minitab. Exhibit 5.12. illustrates the results of the analysis of the drug retention data.

The Wilcoxon test is far more powerful than the sign test because it takes into consideration the magnitude of the differences between the raw measurements and the hypothesised median; in fact, it is almost as powerful as the one-sample t-test. We will see in Chapter 6 that this test (like the one-sample t-test) can be utilised in another context, that of comparisons between two related samples. In that chapter we will also see how to compare two independent samples of measurements.

```
MTB >.WTest 12   'Duration';
SUBC>    Alternative 1.
```

Wilcoxon Signed Rank Test

```
Test of median = 12.00 versus median > 12.00

              N for   Wilcoxon             Estimated
         N    Test    Statistic      P      Median
Duration 17    17       122.5      0.016    17.10
MTB >
```

Exhibit 5.12. Minitab output from a Wilcoxon one-sample test performed on the drug retention data. In this output, N represents the total number of measurements in the sample; N *for test* would be reduced if any measurements were identical to the median value against which the measurements were being compared. As with the z, t, and *sign* tests, the **Alternative** subcommand enables a one or two-tailed test to be performed. The *Estimated Median* represents the estimated median of the population from which the sample has been drawn.

5.5 Exercises

1. In each of the following examples, select an appropriate statistical test for a two-tailed hypothesis that compares the sample against a specified value:

   (a) The data are symmetrically distributed, the sample size is 25, and the population standard deviation is unknown.

   (b) The data are the frequencies of males and females of a particular species in a sample taken from a small area of woodland. If there is an equal probability of being male or female, what test would you perform to establish that the sample was consistent with a 50:50 male:female ratio?

   (c) The sample size is 65 and the data are normally distributed.

   (d) The ratio of yellow to green flowers in the F_1 generation of a plant breeding experiment should be 3:1 according to the predictions of Mendel's laws. If you were provided with the actual results (i.e., the numbers of each colour), how would you calculate the probability of observing that particular outcome of the experiment?

(e) Body condition in sheep is graded on an ordinal scale 1–5 (1 being poor, and 5 being fat). If you have a sample of 20 measurements, what test would you use to determine whether the population from which the sample of sheep was drawn differed from an ideal grade of 3?

2. In Exercise 8 of Chapter 2, the total leaf areas of samples of four varieties of soya were tabulated. Using a one-sample t-test, compare the samples against an expected mean of 250 cm^2 for each of these varieties. Which varieties, if any, differ from the expected mean? Why is a t-test appropriate?

3. Using the data relating to toe length that were listed in Exercise 9 of Chapter 3, test the following null hypotheses:
    (a) that the mean toe length of *Species A* is greater than or equal to 3.1 cm
    (b) that the mean toe length of *Species B* is equal to 3.9 cm
    Explain your choice of statistical test.

4. One can expect a growth rate of 0.2 kg.day^{-1} during the first 4 weeks of a pig's life. The following data were recorded for 9 piglets. Is there any evidence to suggest that they are not growing at the expected rate? A one-sample t-test is the appropriate test to use.

    0.180, 0.185, 0.215, 0.178, 0.180, 0.200, 0.181, 0.180, 0.201

    Reanalyse these data using a sign test and a Wilcoxon test. Do these further tests identify whether the median growth rate differs from 0.2? What do these three sets of results demonstrate about the relative power of the different tests?

5. The Blackcap (*Sylvia atricapilla*) is a small bird that generally becomes fully fledged (i.e., has full flight feathers) after a mean period of 11 days in the nest. The following data have been collected and represent the time (in days) between hatching and fledging. Are they consistent with the known fledging period? Explain your reasons for the choice of test:

    11, 14+, 11, 10, 12, 13, 14+, 13, 14+, 14+, 12, 13, 14+, 9, 10

6. In the experiment described above in Exercise 1(d), 16 yellow flowers and 2 green flowers were recorded in the F_1 generation. Does this result indicate that more yellow flowers were produced than would have been predicted by the ratio of 3:1?

5.6 Summary of Minitab commands

ZTEST (with subcommand **ALTERNATE**)

TTEST (with subcommand **ALTERNATE**)

PDF (with subcommand **BINOMIAL**)

STEST (with subcommand **ALTERNATE**)

WTEST (with subcommand **ALTERNATE**)

5.7 Summary

1. The principal elements of hypothesis testing have been introduced in the context of single sample comparisons. In other words, comparing a single sample of measurements against a specified or hypothesised value.
2. Hypotheses can be one-tailed or two-tailed. Two-tailed hypotheses are in the form of:

H_N: The mean (or median) is equal to the standard value.

H_A: The mean (or median) is not equal to the standard value.

One-tailed hypotheses incorporate a directional element and are of two basic forms:

H_N: The mean (or median) is greater than or equal to the standard value.

H_A: The mean (or median) is less than the standard value.

or

H_N: *The mean (or median) is less than or equal to the standard value.*

H_A: *The mean (or median) is greater than the standard value.*

The key word that signals whether the hypotheses are one- or two-tailed are *less than, greater than* (for one-tailed), and *equal to* (for two-tailed).

3.   The acceptance/rejection regions for one and two-tailed hypotheses were illustrated with respect to the normal and t-distributions.

4.   Once the question has been formulated, the choice of statistical test is determined by sample size and whether the data conform to a normal distribution.

Normally distributed data	large samples ($n > 30$) or small samples and population standard deviation known	one-sample z-test
	small samples ($n \leq 30$), population standard deviation unknown	one-sample t-test
Non-normally distributed data	nominal data falling into two categories (i.e., binomial data and P known)	binomial test
	binomial data and P unknown (assumed $P = 0.5$)	sign test
	magnitude of difference from hypothesised value measured (on an ordinal scale at least)	Wilcoxon test

5.   When samples sizes are large, alternative forms of all the non-parametric tests can be used that are based upon the normal distribution.

CHAPTER 6

Comparing Two Samples

6.1 Introduction

The basic principles of hypothesis testing using statistical techniques were introduced in the last chapter in the context of very simple comparisons, comparing a single sample of measurements against a standard or hypothesised value. Those principles apply equally well across all types of biological questions, and, in this chapter, I shall describe the methods by which they can be applied to problems where we need to compare two samples of measurements. However, in the General Introduction to Part II, I described two types of two-sample comparisons, and, although it is very important to know how to *perform* a particular test procedure, it is far more important to understand how to *recognise* which particular type of two-sample comparisons is appropriate to answer our specific question. (In fact, this ability to recognise the true nature of a question is one of the most important skills that a biologist must possess, and this skill is not confined only to comparisons of two samples.) Therefore, before describing the various ways that we can compare two samples, I shall detail the different types of two-sample comparisons that we might encounter.

Two-sample comparisons can be divided into those relating to *independent* or *unpaired* samples and those that are concerned with *dependent* or *paired* samples. All biologists (in fact, all scientists) must be able to distinguish between questions that fall into these two groups. Comparisons of this type are so common in biology that this ability to discriminate is absolutely essential. To illustrate the difference, I shall use the pulse rate investigation that was featured in a number of earlier examples. Two groups of individuals were involved in this particular investigation, and, at the start, the resting pulse rates of all the individuals in the two groups were measured. One group then exercised while the other group continued to rest. After exercise, the pulse rates of all individuals (exercisers and non-exercisers) were recorded. This experiment provides many pulse rate measurements that enable us to perform several different comparisons by contrasting different groups of measurements.

1. We can compare the pulse rates of the members of the exercise and non-exercise groups before any exercise has been performed. This comparison would answer the question, do the two groups of people differ with respect to resting pulse rates? This question is very important since it establishes at the outset of the experiment that the two groups of subjects are similar before any experimental treatments are applied. In this particular question, we have two sets of measurements of pulse rates (the variable), and each sampling unit (each person) has contributed only one measurement for the purposes of the comparison. In other words, each person is a member of only one of the experimental groups (either an exerciser or a non-exerciser). Thus, the groups of measurements are *independent*.

2. We could perform a similar comparison using the second set of pulse rate measurements (after exercise). This would answer the question, do the pulse rates of the two groups differ after one group has exercised? As with the first example, the two sets of measurements have been made on independent groups and each sampling unit has only contributed a single pulse rate measurement to the comparison.

Contrast those two examples of comparing independent groups with the following:

3. We could answer the question, does exercise affect pulse rates? We do this by comparing the pulse rates of the individuals in the exercise group after exercise with their previously recorded resting pulse rates. Here, we have a comparison where we are using *two* measurements of the same variable from each sampling unit, i.e., each person has contributed two pulse rate measurements (before and after exercise). In this case, the data are really pairs of related or dependent measurements. For example, an individual with a high resting pulse rate would probably have

a very high pulse rate after exercise, while a trained athlete may have a very low resting pulse rate and may show very little or no response to mild exercise. The outcome of this comparison would depend not only on the treatment imposed (the exercise), but also on the baseline level of resting pulse rate for each individual. This is an example of a comparison of dependent or paired data, and is sometimes referred to as a *before and after* comparison.

These two distinct types of two-sample comparisons (Examples 1 and 2, versus Example 3) are analysed in quite different ways, and it is very important to be able to differentiate between them. In doing so, the criterion used to distinguish between the two types of comparisons is whether we are measuring a variable (in this case, pulse rate) on a subject on either *one* or *two* occasions. The important factor in determining the type of question posed is the number of measurements of a specific variable per sampling unit contributing to a particular comparison.* In order to check that you have grasped how to tell the difference between these two types of comparisons, study the problems in the exercises at the end of this chapter where several paired and unpaired tests are presented.

This chapter is divided into two principal sections, which, respectively, deal with comparisons of independent groups and those involving paired data. Like one-sample comparisons, we can construct hypotheses that are one or two-tailed depending upon the question being addressed. Similarly, we have to consider the nature of the data involved in the investigation in order to enable us to choose between tests based upon the normal distribution and alternative, non-parametric tests, just as with the single-sample comparisons in Chapter 5. The methods with which we can tackle one or two-tailed comparisons, along with the various parametric and non-parametric techniques, will be detailed within each section.

6.2   Comparing two independent samples

This section is divided into two parts. The first describes the procedures used to compare samples that can be assumed to have been drawn from normally distributed populations. These techniques com-

* We shall see in Chapter 8 (Sequential Relationships) that measuring several *different* variables on the same sampling unit is an altogether different matter.

pare sample means. The second details the alternative tests that are employed when the data do not conform to a normal distribution. Like the non-parametric tests described in Chapter 5, these tests compare sample medians.

Comparing two sample means

I shall illustrate the principles of this procedure with the following example in which two groups of *Drosophila* fruit flies have been maintained at different temperatures (10°C and 15°C). The biological question that we shall ask is:

> *Do* Drosophila *maintained at these two temperatures lay different numbers of eggs?*

We can compare the mean number of eggs laid by the two groups of flies and arrive at an answer to this question. The null hypothesis that we need to test is:

H_N: There is no difference between the mean number of eggs laid by *Drosophila* maintained at 10°C and *Drosophila* maintained at 15°C.

The complementary, alternative hypothesis is:

H_A: There is a difference between the mean number of eggs laid by *Drosophila* maintained at 10°C and *Drosophila* maintained at 15°C.

Given that this test is concerned with a difference in either direction, it is two-tailed. But, before proceeding to the actual method, let us consider exactly what we are doing by comparing the two sample means.

The means of the two groups will, almost certainly, not be identical. The difference may be due to sampling error. If this is the case, then we will have no evidence to reject the null hypothesis, which states that the two samples have been drawn from a single, statistical population of *Drosophila*, and that temperature has no real impact on the number of eggs laid. This conclusion is illustrated in Figure 6.1.

Alternatively, the difference between the two samples may be indicative of two distinct statistical populations, as illustrated in Figure

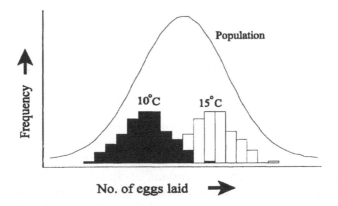

Figure 6.1. The frequency histograms of two samples of eggs counts (one at 10°C and the other at 15°C) drawn from a single population (represented by the continuous line).

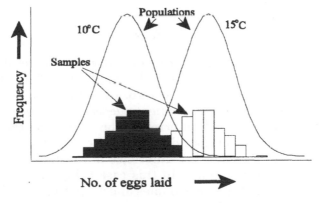

Figure 6.2. The frequency histograms of two samples of egg counts (one at 10°C and the other at 15°C) drawn from two populations (represented by the continuous lines).

6.2. These two figures can, respectively, be regarded as visual representations of the null and alternative hypotheses, and it is the task of the statistical test to distinguish between them.

We are using the sample means as representations of the means of the population from which the samples have been drawn. Moreover, the difference between the sample means is an estimate of the difference that may exist between the population means. Therefore, we can reduce our question to, is the difference between the population means equal to 0 or not? Thus, we can also restate the null and alternative hypotheses as:

$$H_N: \mu_{10} - \mu_{15} = 0$$

and

$$H_A: \mu_{10} - \mu_{15} \neq 0$$

where μ_{10} and μ_{15}, respectively, represent the population means of the number of eggs laid by *Drosophila* at 10°C and 15°C. We use sample means to estimate the difference between the population means. Following a parallel procedure to that which was used in single sample comparisons in Chapter 5, we can standardise the difference between the sample means in terms of Student's t-distribution by dividing the difference by an appropriate standard error (Equation 6.1). Like all standard errors, this standard error of the difference (*SED*) measures the precision with which the difference between the sample means ($\bar{x}_{10} - \bar{x}_{15}$) estimates the difference between the population means ($\mu_{10} - \mu_{15}$).

$$t = \frac{\bar{x}_{10} - \bar{x}_{15}}{SED} \tag{6.1}$$

This formula evaluates, in terms of t, how far the results deviate from the value suggested by the null hypothesis. Thus, if the null hypothesis is true, then t will equal zero. This statistical test is known as the *unpaired t-test* or *t-test for two independent samples*.

In Chapter 3, the standard error of the mean was calculated as the square root of the variance divided by the sample size (Equation 3.3). The standard error of the difference is calculated in a similar way, the single pooled variance in this case being the result of combining the variances of the two samples and weighting them in relation to the relative size of each sample.

$$pooled\ variance\ (s_p^2) = \frac{(n_1 - 1)s_1^2 + (n_2 - 1)s_2^2}{n_1 + n_2 - 2} \tag{6.2}$$

where n_1 and n_2 are the sample sizes, and s_1^2 and s_2^2 are the variances of the two samples. The standard error of the difference is given by Equation 6.3, where s_p^2 is known as the *pooled variance*.

$$SED = \sqrt{\frac{s_p^2}{n_1} + \frac{s_p^2}{n_2}} = s_p \sqrt{\frac{1}{n_1} + \frac{1}{n_2}} \tag{6.3}$$

No. of eggs laid	
10°C	15°C
9	9
7	10
10	10
7	11
7	10
7	12
6	10
9	9
8	12
6	10
$\bar{x}_{10} = 7.60$	$\bar{x}_{15} = 10.30$
$s_{10} = 1.35$	$s_{15} = 1.06$
$n_{10} = 10$	$n_{15} = 10$

Table 6.1. Number of eggs laid by two groups of *Drosophila* fruit flies, each group being maintained at either 10°C or 15°C.

We can now proceed to testing this null hypothesis using the unpaired t-test. There were 10 fruit flies in each sample, and the number of eggs laid by each fly was recorded, and the data are listed in Table 6.1.

The test statistic, t, is calculated using Equation 6.1.

$$t = \frac{7.60 - 10.30}{1.2136\sqrt{\dfrac{1}{10} + \dfrac{1}{10}}} = \frac{-2.7}{0.5427} = -4.975$$

The pooled variance was calculated using Equation 6.2. Confirm this value by calculating it yourself. The value of t, −4.975, is negative because $\bar{x}_{10}$ is less than $\bar{x}_{15}$. The calculation could have been performed equally well with the smaller mean being subtracted from the larger to give a positive value for t.

As with the t-test for single samples described in Chapter 5, we interpret the significance of the test statistic using t-tables in conjunction with degrees of freedom. With the two sample t-test, the number of degrees of freedom is given by Equation 6.4.

$$d.f. = (n_1 - 1) + (n_2 - 1) = n_1 + n_2 - 2 \tag{6.4}$$

In this particular example, with each sample comprising 10 measurements, we have 18 degrees of freedom. We can see from t-tables (Appendix A.2.) that, if the null hypothesis is true, then the probability of observing a value of t of this magnitude (or this difference in number of eggs laid) is less than 0.001. Therefore, we would reject the null hypothesis and accept the alternative: the mean number of eggs laid by *Drosophila* kept at the two temperatures is different. We would report the result of this test by means of a sentence of a form similar to:

An unpaired t-test was performed, showing that *Drosophila* maintained at 15°C produced significantly more eggs than similar *Drosophila* kept at 10°C ($t = 4.975$; *d.f.* = 18; P <0.001).

This sentence contains several important pieces of information, and it provides a model for all statements that report the results of statistical tests. The various components are:

- the test used: the *unpaired t-test*
- the sampling units: *Drosophila*
- the treatments: *10°C* and *15°C*
- the direction of the difference: *the mean no. of eggs laid at 15°C is greater than at 10°C*
- the quantitative details of the statistical result: *values of the test statistic, degrees of freedom, and the probability.*

By rejecting the null hypothesis and accepting the alternative, we conclude that there is a difference in the mean number of eggs laid by *Drosophila* kept at the two temperatures. However, we also have to discuss the results, and in the discussion we must comment upon the *magnitude* of the difference, as well as the fact that there is a difference, before suggesting a biological explanation for our observations. This is the reason why we must always include some form of descriptive statistics in the report (such as the means, standard deviations, and sample sizes). Of course, descriptive statistics do not always have to be presented in the form of tabular data; graphical illustrations may be more appropriate on occasion.

We saw in Chapter 3, that we can represent an estimate of a population mean as a single value or *point estimate* (the sample mean), or as an *interval estimate* (the confidence interval). We can do the same with our estimate of the difference between the means. In this example, we can present a point estimate of the size of the difference between the two treatment groups. On average, 2.7 more eggs are produced by each *Drosophila* at 15°C than by those kept at 10°C (i.e., the difference between the two sample means). Alternatively, we can calculate the confidence interval around the difference between the sample means.

The 95% confidence interval around the difference between the sample means defines the region within which we are 95% confident of finding the difference between the population means. The calculation follows the same pattern as that of the confidence interval around a sample mean (see Box 3.8.).

$$95\% \ conf. \ int. \ around \ the \ difference \ between \ the \qquad (6.5)$$
$$sample \ means = (\bar{x}_1 - \bar{x}_2) \pm (t_{0.05, \ v})(SED)$$

where v represents the degrees of freedom. In this *Drosophila* example, there are 18 degrees of freedom (giving a value of t equal to 2.1009) and the difference between the sample means is 2.7. Therefore, the 95% confidence limits will be:

$$upper \ 95\% \ conf. \ limit = 2.7 + (2.1009)(0.5427) = 3.84$$
$$lower \ 95\% \ conf. \ limit = 2.7 - (2.1009)(0.5427) = 1.56$$

This confidence interval indicates that we are 95% confident that the difference between the population means lies between 1.56 and 3.84 eggs per *Drosophila*. We can use this confidence interval to make judgements about the biological significance of the results. While the statistical conclusion is to reject the null hypothesis, the biological conclusion will depend upon the significance of the *magnitude* of the difference. It is this biological significance that is the object of the *discussion* section of a report.

The Minitab commands to perform unpaired t-tests are to be found in the same menu as the single sample t-test:

STAT > Basic Stats > 2-Sample t

The dialog box is shown in Exhibit 6.1. ready to perform the test on the *Drosophila* data.

The data can be stored in a Minitab worksheet in two basic formats. In this example, the measurements (numbers of eggs laid by each *Drosophila*) are stored in two separate columns, one for each treatment group. Analysis of the data stored in this way is achieved by selecting the option marked *Samples in different columns* in Exhibit 6.1. The

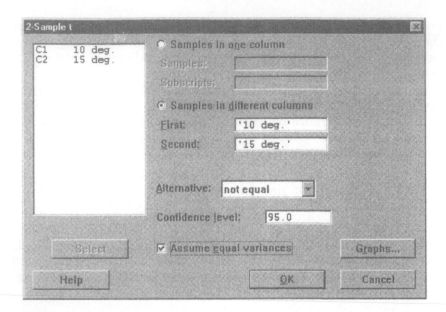

Exhibit 6.1. The Minitab dialog box to conduct an unpaired t-test between independent samples. Note that the box to assume equal variances is checked, and that a two-tailed alternative hypothesis is specified.

alternative storage option is to group all the measurements in a single column and use a second column to store codes (Minitab refers to them as *subscripts*) to indicate from which treatment group each measurement originated. We have already seen examples of this type of format in the pulse data set in which there are columns marked *ran, sex,* and *smokes*, each of which contain 1s and 2s to indicate *exercise/non-exercise, male/female,* and *smoker/non-smoker.* If the data are stored in this way, then we have to choose the *samples in one column* option. You will see in the following chapters that Minitab requires this format for data storage for several other analytical techniques.

The output of the Minitab unpaired t-test is shown in Exhibit 6.2. It includes the appropriate descriptive statistics, confidence intervals, a statement of the null and alternative hypotheses, and the results of the analysis. All the key quantitative information necessary for reporting the results of the analysis (as listed above). The subcommands, **ALTERNATIVE 0** and **POOLED,** specify that (a) the hypothesis to be tested is two-tailed and (b) that a pooled variance has been used because the variances are assumed to be equal.

In Chapter 5, we saw that a basic assumption of the single sample t-test was that the data conform to the normal distribution. That is

```
MTB > TwoSample 95.0 '10 deg.' '15 deg.';
SUBC>    Alternative 0;
SUBC>    Pooled.
```

Two Sample T-Test and Confidence Interval

```
Two sample T for 10 deg. vs 15 deg.
           N      Mean      StDev    SE Mean
10 deg.   10      7.60      1.35      0.43
15 deg.   10     10.30      1.06      0.33


95% CI for mu 10 deg. - mu 15 deg.: ( -3.84,  -1.56)
T-Test mu 10 deg. = mu 15 deg. (vs not =): T= -4.98  P=0.0001  DF=  18
Both use Pooled StDev = 1.21

MTB >
```

Exhibit 6.2. The Minitab output for the unpaired t-test. Note that the confidence limits produced by Minitab are both negative (−3.84 and −1.56) and those calculated manually and shown in the text above are both positive. The reason for this difference is that the '10 deg.' columns have been specified first in the Minitab command, and the mean number of eggs laid by those *Drosophila* was smaller than that of the *Drosophila* maintained at '15 deg.'

also the case with the unpaired t-test. If the samples do not appear to be distributed normally, but the sample sizes are large ($n > 30$), then one can use the t-test. If, on the other hand, the sample sizes are not large, then a non parametric test, the Mann-Whitney test, should be used, which is described in the section entitled *Comparing two independent samples when normality cannot be assumed* (see Section 6.2). Furthermore, there are other assumptions that are implicit in this test, namely that the variances (and, obviously, the standard deviations) of the two samples should be similar. In this example, the standard deviations are approximately equal (1.35 and 1.06), and so we have used a pooled variance in the calculation of the test statistic, t. Provided that the sample sizes are similar, this test is fairly robust to deviations to this assumption. If the samples are large or conform to a normal distribution, but the standard deviations are not similar, then we cannot use a pooled variance. Instead, we employ the following formula to calculate the standard error of the difference:

$$t = \frac{\bar{x}_1 - \bar{x}_2}{\sqrt{\dfrac{variance_1}{n_1} + \dfrac{variance_2}{n_2}}} \qquad (6.6)$$

where n_1 and n_2 denote the sizes of the two samples. The degrees of freedom are calculated according to the Equation 6.7.

$$d.f. = \frac{((s_1^2 / n_1) + (s_2^2 / n_2))^2}{\dfrac{(s_1^2 / n_1)^2}{n_1 - 1} + \dfrac{(s_2^2 / n_2)^2}{n_2 - 1}} \tag{6.7}$$

where s_1^2 and s_2^2 are the variances of the two samples. The default option in Minitab is *not* to assume that the variances are equal, and thus, by default, Minitab uses Equations 6.6 and 6.7 to calculate t and the degrees of freedom. Exhibit 6.1. shows the box to check if the variances can be assumed to be equal.

This *Drosophila* example involves testing a two-tailed hypothesis, but one can also test one-tailed hypotheses using unpaired t-tests. A clear example of this is provided by the pulse rate data. One might imagine that exercise would lead to an increase in pulse rate, and we can use this data set to test that theory. The appropriate hypotheses would be:

H_N: Pulse rates after exercise will be less than or equal to the pulse rates of people who have not exercised.

H_A: Pulse rates after exercise will be greater than the pulse rates of people who have not exercised.

This example also serves to illustrate that this test does not require that the sample sizes be equal, and, in the Minitab output below (Exhibit 6.3.), we can see that equal variances have not been assumed.

Comparing two independent samples when normality cannot be assumed

When normality cannot be assumed for data used with single sample tests (Section 5.4), there are parallel, non-parametric procedures that could be employed. Similarly, for comparisons of independent samples, there are non-parametric tests that perform a similar function as the unpaired t-test, but do not assume that the data conform to a normal distribution. The non-parametric test that I shall describe here is known as the Mann-Whitney test, the Wilcoxon-Mann-Whitney test, or the Wilcoxon Rank Sum test.

```
MTB > TwoT 95.0 'Pulse2' 'Ran';
SUBC>   Alternative 1.

Two Sample T-Test and Confidence Interval


Two sample T for Pulse2
Ran            N        Mean      StDev    SE Mean
1            35        92.5       18.9        3.2
2            57       72.32       9.95        1.3

95% CI for mu (1) - mu (2): ( 13.2,   27.2)
T-Test mu (1) = mu (2) (vs >): T= 5.83  P=0.0000  DF=  45


MTB >
```

Exhibit 6.3. The Minitab output of an unpaired t-test comparing the pulse rates of a group of people who had exercised with a group that had not exercised. This is a one-tailed test, the alternative hypothesis being that the mean pulse rate of the exercise group is greater than that of the non-exercise group. Note that the variances of the two groups have not been assumed to be equal (contrast this with the *Drosophila* example). Also, this illustrates that Minitab can accept a data set where all the measurements are stored in a single column (*Pulse2*) and codes indicating the treatment groups are stored in another column (*Ran*).

This procedure tests a null hypothesis that states that the medians of two populations are equal if the populations have a similar form. I shall describe the Mann-Whitney test using the data set that we saw in Chapter 5 that related to the retention time of a drug in the bloodstream. Those data were actually part of a larger investigation that compared three methods of administering the drug. The whole data set is listed in Table 6.2., and I shall illustrate this test by comparing Methods A and B by asking the question, do these methods of administration lead to differences in the retention time of the drug in the blood? The specific null hypothesis that will be tested is:

H_N: The median retention time (in hours) of the drug administered by Method A is the same as that when administered by Method B.

H_A: The median retention time (in hours) of the drug administered by Method A is not the same as that when administered by Method B.

The procedure for this test follows a similar pattern to those of the non-parametric tests that were introduced in Chapter 5, in that it is

Method A	Method B	Method C
6.0	7.5	4.5
7.0	8.0	5.4
8.8	8.0	6.3
9.2	8.8	6.8
11.2	9.0	6.8
12.2	10.4	6.9
12.5	11.0	8.4
14.0	11.4	8.7
17.0	12.2	8.7
19.2	12.2	8.7
20.5	12.5	8.9
22.0	13.0	10.2
22.6	13.2	10.9
23.2	13.6	11.5
24.0	14.0	11.5
24.0	14.0	12.8
24.0	14.1	14.5

Table 6.2. Retention time (in hours) for a drug
in the bloodstream. The drug was administered
by three different methods, A, B, or C. All
measurements were taken on different patients.

based upon ascribing ranks to the data. Essentially, in the Mann-
Whitney test, we combine the two samples into one data set and ascribe
a rank to each measurement, giving the lowest rank to the smallest
measurement. Therefore, in this example where Methods A and B are
compared, ranks 1 and 2 would be given to 6.0 and 7.0 from Method
A, and a rank of 3 would be given to 7.5 from Method B, and so on.
The same rules apply to allocating ranks to tied values, as were
introduced in Chapter 5. After ranking all of the measurements, we
calculate the sum of the ranks in each treatment group. The underlying
principle is that, if the null hypothesis is true, then the rank sums of
the two groups will be similar. The test statistic, T, is calculated using
the formula shown in Equation 6.8 and is essentially a measure of
how far the rank sum differs from the prediction of the null hypothesis.
In this equation, W_1 represents the rank sum of one of the samples,
and n_1 denotes the size of that sample.

$$T = W_1 - \frac{(n_1)(n_1 + 1)}{2}$$
(6.8)

Calculate the rank sums for the samples of Methods A and B (you should obtain values of 356 and 239, respectively). The test statistic is then compared against critical values of T (tabulated in Appendix A.6.). These tables are much more extensive than those that we have already encountered because they are detailed for many combinations of sample sizes. Their size and rather different format necessitates some explanation.

First, the tables only apply to comparisons in which the sample size of both groups does not exceed 20. With larger samples, problems associated with non-normality become less serious and an alternative calculation is more appropriate (this method is described at the end of this section).

Second, Equation 6.8 allows us to calculate two values of T, depending upon which sample we use. The question then arises, which value do we use to compare with tables? In the drug retention question, for Method A we have:

$$T = 356 - \frac{(17)(18)}{2} = 203$$

and for Method B:

$$T - 239 - \frac{(17)(18)}{2} = 86$$

The values in the tables are *lower* critical values. In other words, if a calculated value of T is *lower* than the critical value in tables, we reject the null hypothesis. Therefore, the answer to the question above is that we use the *smaller* of the two values of T. In the context of the drug example, we use $T = 86$ to compare against the value in tables for $n_1 = 17$, $n_2 = 17$, in the column $P = 0.05$ for a two-tailed test ($T = 88$ in Appendix A.6.). The calculated value in this example is lower than that in tables (but not less than the value in the $P = 0.02$ column), so we reject the null hypothesis and conclude that there is a difference in the duration of drug retention in the bloodstream between the two modes of administration, Method A leading to a more prolonged and active life than Method B ($T = 86$; $n_1 = 17$, $n_2 = 17$; $P < 0.05$).

In this test and in others we need to choose one value (the smaller rank sum) when we have more than one from which to select. A short-cut method is available that enables us to use one rank sum to find the other (Equation 6.9).

$$W_2 = \frac{N(N+1)}{2} - W_1 \tag{6.9}$$

where N is the total number of measurements $(n_1 + n_2)$. We can see this for the drug example:

$$W_2 = \frac{34(34+1)}{2} - 356 = 239$$

The Minitab procedure to perform the Mann-Whitney test is straightforward. The command is found in the **STAT** menu with all of the other non-parametric tests:

STAT > Nonparametrics > Mann-Whitney

The Mann-Whitney dialog box is shown in Exhibit 6.4. and the output for the comparison of drug administration between Methods A and B is illustrated in Exhibit 6.5. The Minitab output specifies the test in terms of *ETA1* and *ETA2*. These refer to the two population medians (η_1 and η_2) which, in a similar fashion to the population mean (*cf* μ), are often denoted by a Greek letter, in this case, eta or η.

If both of the samples exceed 10 measurements, then the distribution of the rank sum, W, approximates to a normal distribution. As a consequence, we can calculate a z-score for the data and compare this against normal probability tables (Appendix A.1.). The z-score is calculated using Equation 6.10.

$$z = \frac{W_1 - n_1(N+1)/2}{\sqrt{n_1 n_2 (N+1)/12}} \tag{6.10}$$

where, W_1 is the smaller of the two rank sums, n_1 and n_2 are the sample sizes of the two samples (with n_1 being the size of the sample with the smaller rank sum), and N is the total number of measurements. So, for example, if we conducted a two-tailed test and obtained a value of $z = 2.2$, we can use normal probability tables to calculate the probability of obtaining such a value:

$$(1 - 0.9861) + (1 - 0.9861) = 0.0139 + 0.0139 = 0.0278$$

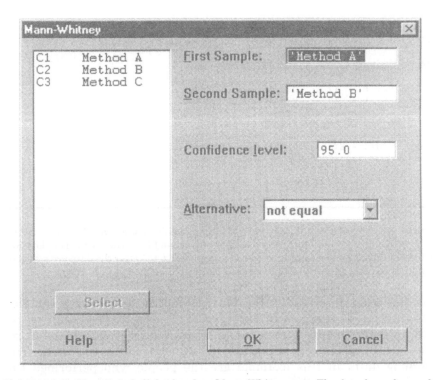

Exhibit 6.4. The Minitab dialog box for a Mann-Whitney test. The data for each sample are stored in separate columns and one or two-tailed test can be selected using the *Alternative* box.

where 0.9861 is the value in normal probability tables for a value of z less than or equal to 2.2. Alternatively, if the test had been one-tailed, then the probability would simply be 0.0139. Minitab automatically uses this normal approximation when appropriate.

The comparison of independent samples is extremely widespread in biology, and the t and Mann-Whitney tests provide the biologist with very powerful tools to compare two independent samples. In Chapter 7, I shall describe the methods by which we can compare more than two groups of measurements.

6.3   Comparison of two related (or paired) samples

At the start of this chapter, I described two main types of comparisons involving two samples: those involving *independent samples*, which

```
MTB > Mann-Whitney 95.0 'Method A' 'Method B';
SUBC>    Alternative 0.
```

Mann-Whitney Confidence Interval and Test

```
Method A   N =  17      Median =      17.000
Method B   N =  17      Median =      12.200
Point estimate for ETA1-ETA2 is       5.200
95.0 Percent CI for ETA1-ETA2 is  (-0.002,9.998)
W = 356.0
Test of ETA1 = ETA2  vs  ETA1 not = ETA2 is significant at 0.0457
The test is significant at 0.0455 (adjusted for ties)

MTB >
```

Exhibit 6.5. The Minitab output for the drug example, comparing the effect of two methods (A and B) of administering a drug on its active life in the blood. The data are detailed in Table 6.2. The significance level is given as 0.0455 after the data are adjusted for tied ranks: this is the probability value that would be quoted in a report.

have just been described in Section 6.2, and those involving samples that are related in some way. I illustrated the latter type of comparison with an example based upon the pulse rate data, in which the pulse rates of a set of subjects before and after exercise were compared. The sampling units in this example are the people who performed the exercise, and two pulse rate measurements were taken from each individual. These paired measurements were, therefore, not independent of each other. In this section, I shall describe various methods (parametric and non-parametric) that are available for comparing related samples.

Before discussing the different techniques in detail, we can consider some general principles associated with this type of comparison, and, by doing so, see the great similarity between this type of question and those concerning single-sample comparisons that were discussed in Chapter 5.

Continuing with the pulse rate example, if we want to know whether exercise has any effect on pulse rate, we can compare the pulse rates of individuals after they have exercised with their resting pulse rates before exercise. However, those individuals who have high resting pulse rates will probably have very high pulse rates afterwards, and similarly, those who have low resting pulse rates will probably have relatively low pulse rates after exercise. In order to perform a meaningful comparison, we have to take into consideration the baseline measurement for each individual to take account of the individual variability between subjects. The null hypothesis that we might want to test is:

H_N: For each individual there is no change in pulse rate after exercise relative to the resting pulse rate.

H_A: For each individual there is a change in pulse rate after exercise relative to the resting pulse rate.

Another way of expressing this null hypothesis would be to say that *on average,* the difference between the *before* and *after* pulse rates will be 0, giving hypotheses like:

H_N: The mean change in pulse rate due to exercise equals 0.

H_A: The mean change in pulse rate due to exercise does not equal 0.

Null and alternative hypotheses for paired comparisons will generally be stated in terms of *change* or *difference.* The statistical testing of such hypotheses is then reduced to two stages:

1. Find the difference between each pair of related pulse rate measurements.
2. Compare the mean of the differences against the hypothesised mean of 0.

The testing procedure thus reduces to a comparison of a single sample, *the differences,* against a hypothesised mean, *0.* In other words, this type of question is a special case of the one sample comparisons described in Chapter 5, the hypothesised mean in paired comparisons being *0* or *no change.* Not surprisingly, therefore, the test procedures used to answer to this type of question are the same as those introduced in Chapter 5: the z and t-tests, and the sign and Wilcoxon tests.

Comparison of paired samples when normality can be assumed

Recalling the discussion in Chapter 5, the z-test can be used when we have large samples ($n > 30$), and t-tests are used when we can assume that the data are normally distributed when we have 30 or fewer sampling units. When used in paired comparisons, these test procedures are known as the paired z-test and the paired t-test. In paired comparisons, it is important to note that it is the number of sampling units and not the number of measurements that determines which test we should use. Remember that we are making two measurements on each sampling unit. The only difference between the methods of this test and the method of comparing a single sample against a hypothesised value is the first stage of finding the difference between each pair of measurements.

The procedures and the formulae for the two tests are so similar and have already been introduced (*cf.* the modified version of Equation 3.14 used in Box 5.2. and Equation 5.1), thus, I shall illustrate this type of comparison with only one worked example. For Minitab examples of how these tests may be performed, I refer you to the single sample examples in Chapter 5. The example that I shall use relates to an investigation into the potency of two closely related viruses of maize (Virus A and Virus B). The researchers were aware that leaves varied in their level of susceptibility so they applied both viruses to every leaf, one virus on each half. In order to reduce the potential for bias, the viruses were allocated at random to the left or right half of each leaf. There were 15 leaves in all, and the number of lesions on each half were recorded (the data are listed in Table 6.3.). Since this experiment involves only 15 leaves, a *paired t-test* is appropriate and the calculation is shown in Box 6.1. We have seen that when using the t-distribution, we also need to calculate the degrees of freedom. Just as in the single sample comparisons, the degrees of freedom are calculated as:

$$degrees\ of\ freedom = no.\ of\ sampling\ units - 1$$
$$= no.\ of\ pairs - 1 \qquad (6.11)$$

The result of the paired t-test shows that, on average, the application of Virus B leads to more lesions than Virus A ($t = 2.99$; $d.f. = 14$; $P <0.01$). We can conclude that Virus B is more potent than Virus A, and, in a report, we should include a comment on the magnitude of the difference, in this case quoting the mean difference of 0.867 lesions per leaf.

This example has illustrated the use of the paired t-test. If the number of sampling units had been greater than 30, then we would have used the paired z-test and followed a similar pattern using normal probability tables (Appendix A.1.) to calculate the probability.

In order to conduct these paired tests using Minitab, simply find the difference between the two sets of measurements (using the **LET**

	No. of lesions	
Leaf no.	Virus A	Virus B
1	11	12
2	6	8
3	7	8
4	13	14
5	14	13
6	10	9
7	9	10
8	8	10
9	12	13
10	14	14
11	14	15
12	10	12
13	8	10
14	7	6
15	9	11

Table 6.3. The number of lesions on maize leaves in response to the application of two viruses. For each leaf, one virus was applied to each half of the leaf.

command) and then use the **1-Sample** z or **1-Sample** t commands that were described in Chapter 5 setting the *Test mean* option to 0, i.e., specifying the value of the hypothesised mean to be 0.

Comparison of paired samples when normality cannot be assumed

Paired comparisons can be carried out using two of the non-parametric tests that were described in Chapter 5: the *sign test* and the *Wilcoxon signed-ranks test* (in the context of paired comparisons, this test is known as the *Wilcoxon matched-pairs signed-ranks test*). In similar fashion to the application of single sample z and t-tests to paired comparisons, these two non-parametric tests are applied to a single sample of differences. The choice between the two tests is made on the basis of whether the measurements (i.e., the differences) are simply in the form of the direction of the difference (either positive or negative) or whether there is also a measure of the magnitude of the difference. We use the sign test in the former situation and the Wilcoxon matched-pairs signed-ranks test when we have a measure of both the magnitude and direction of the difference.

The differences in the number of lesions found on each half of 15 maize leaves.

$$-1 \quad -2 \quad -1 \quad -1 \quad \ \ 1 \quad 1 \quad -1 \quad -2 \quad -1 \quad 0$$
$$-1 \quad -2 \quad -2 \quad \ \ 1 \quad -2$$

Mean difference in number of lesions $(\bar{d}) = -0.867$
Standard deviation (s) $= \ \ 1.125$
Sample size (n) $= \ 15$

$$t = \frac{230.93 - 240}{3.95} = -2.3$$

The number of degrees of freedom = no. of sampling units − 1
$$= 15 - 1 = 14$$

Using t-tables (Appendix A.2.), we can see that the probability of obtaining this result if the null hypothesis is true is less than 0.01, so we reject H_N and conclude that *Virus B* causes more lesions than *Virus A*.

Box 6.1. Calculation of a paired t-test examining the potency of two related viruses of maize plants. The results shows that Virus B causes more lesions than Virus A.

Paired sign test

The sign test is one of the simplest non-parametric tests, and, as we saw earlier, can be applied when measurements (in this case, differences) can be classified in one of two possible ways, *plus* or *minus* (hence its name). The example in Chapter 5 described how rats were tested for left-right bias in a maze. In that example, we ascribed −1 to a left turn and +1 to a right turn, and then compared these scores against a hypothesised median of 0. We can extend the application of this procedure to paired comparisons by calculating the differences between the members of each pair and constructing null and alternative hypotheses as follows:

H_N: The median difference is 0.

H_A: The median difference is not 0.

As an example, consider the following experiment in which the effectiveness of a fungicide was assessed on rose trees. In this experiment, 10 rose bushes, which were all suffering from the same fungal infection to varying degrees, were treated with a new fungicide. The question posed is, *does this fungicide have a beneficial effect on roses?*

All the rose trees were assessed for level of infection before and after treatment on a 4-point scale: 0 = no infection; 1 = low; 2 = moderate; 3 = extensive. Although the differences in scores range more widely than just −1 or +1, for the sake of this example, we will record the difference simply as either negative, positive, or no change. The results of these assessments are listed in Table 6.4.

If the fungicide has had any beneficial effect, then we would expect the fungal score in the second assessment to be lower than in the first, and, since we are only interested in a beneficial change, we should analyse these results as a one-tailed test. The specific hypotheses will be:

H_N: The median change in fungal infection will be greater than or equal to 0.

H_A: The median change in fungal infection will be less than 0.

	Fungal score		
Rose	Before treatment	After treatment	Difference
1	2	1	−
2	3	2	−
3	2	1	−
4	2	1	−
5	2	2	0
6	2	2	0
7	3	2	−
8	2	1	−
9	3	1	−
10	2	3	+

Table 6.4. Assessments of fungal infection on rose trees before and after treatment with a new fungicide. The fungal infection is scored: 0 = no infection; 1 = low; 2 = moderate; 3 = extensive.

The data in Table 6.4. reveal that 7 differences are negative (i.e., represent a reduction in fungal infection), 1 difference is positive, and on 2 occasions there was no change. In order to evaluate the probability of this result occurring if the null hypothesis is true, we have to compare the larger number of differences (i.e., 7) with the values in sign test tables (Appendix A.4.). In the tables, n represents the number of measurements that do not coincide with the hypothesised median, which, in this case, means the 8 non-zero values. This contrasts markedly with the paired t-test where the degrees of freedom are calculated using all the pairs of data (Equation 6.11). The sign test tables are designed for two-tailed hypothesis testing, so we have to halve the probability levels at the top of the column in order to obtain one-tailed probabilities. We can see that an observed value of 7 negative scores has a one-tailed probability of less than 0.05. Therefore, we can conclude that the fungicide has had a statistically significant effect, the level of infection having decreased over the treatment period. Obviously, this result could not be viewed in isolation because the economics of using this fungicide would have to be considered before a firm recommendation of its use could be made. For example, if the new fungicide is extremely expensive, its use may not be cost effective. Furthermore, in these times of environmental awareness, the environmental impact of the fungicide would also have to be considered before commercial use.

Exhibit 5.10. illustrates the Minitab dialog box to specify a sign test. Using that as a guide, analyse the rose data to discover the exact probability of obtaining these results.[*]

The sign test is based upon the binomial distribution, which approximates to the normal distribution when sample sizes are large. I described in the section entitled *Comparing nominal data against a standard value* (see Section 5.4) how a normal approximation to the sign test could be calculated when the sample size exceeds 35. This approximation can also be used for paired comparisons when the number of non-zero differences exceeds 35. A full worked example of this calculation applied to a single sample comparison is illustrated in Box 5.4.

Wilcoxon matched-pairs signed-ranks test

This test makes use of both the magnitude and direction of the differences, and is more powerful than the sign test, but the null and

* Answer: $P = 0.0352$.

alternative hypotheses for these two tests are similar. Like all procedures used to perform paired comparisons, the first stage of the calculation is to create a single column of differences and then to follow the same method as for single-sample comparisons (as detailed in Box 5.5.). Note also that, like the sign test, when comparing the test statistic with the critical values in tables (Appendix A.5.), the sample size used is the number of non-zero values, and not the actual number of differences, and the Wilcoxon tables are two-tailed.

As an illustration of this test, we can apply it to the rose data in Table 6.4. The calculation requires us to calculate the rank sum of the positive differences and that of the negative differences, and the test statistic, T, is the larger of these two values, shown in Box 6.2. Using Minitab to perform this test (see Exhibit 5.12. for an example), we obtain an exact probability of 0.029.

The differences in the fungal scores on rose trees after treatment with a new fungicide:

$$-1 \quad -1 \quad -1 \quad -1 \quad 0 \quad 0 \quad -1 \quad -1 \quad -2 \quad +1$$

Ranks of the non-zero differences:

$$4 \quad 4 \quad 4 \quad 4 \qquad 4 \quad 4 \quad 8 \quad 4$$

The number of non-zero differences (N)   = 8
Sum of the ranks for the negative differences = 32
Sum of the ranks for the positive differences = 4
We use the larger of these two values of T to compare against Wilcoxon tables.

Comparing the larger ranks sum against Wilcoxon tables (Appendix A.5.), we can see that the one-tailed probability of obtaining this result if the null hypothesis is true is less than 0.05, so we reject H_N and conclude that the fungicide does have a statistically significant effect on the roses: the fungal scores are reduced after treatment.

Box 6.2. Calculation of a Wilcoxon matched-pair signed-ranks test examining the effectiveness of a new fungicide on rose trees. The results shows that the fungicide reduces the fungal infection.

Like the sign test, when samples are large the Wilcoxon matched-pairs signed-ranks test can be approximated by a z-test based upon the normal distribution. The methods for this approximation were detailed in the section entitled *Comparing measured data against a standard value* (see Section 5.4). This normal approximation can be used to compare paired samples when there are more than 20 pairs that have non-zero differences.

6.4   Exercises

As in previous chapters, use manual and computer methods to answer these questions.

1.   Aldosterone is a hormone produced in the adrenal cortex that is critical to the control of fluid balance in the body, particularly promoting the retention of sodium. The data below relate to rabbits inhabiting a mountainous region of Australia where sodium is very scarce, and a coastal region where there is a plentiful supply of sodium. Compare the concentrations of aldosterone (ng 100 ml^{-1} of plasma) using a t-test. Explain your results in terms of environmental adaptation.

Area	Aldosterone concentration (ng 100 ml^{-1} plasma)		
	Sample size	Mean	Standard deviation
Mountains	32	64.2	12.00
Coast	32	20.5	9.00

2.   Using the data presented on soya leaf areas presented in Exercise 9 of Chapter 2, compare the leaf areas of Variety B with Variety C. Is there any difference? Explain your choice of test.

3.   Use a Mann-Whitney test to compare the drug retention times for Methods B and C that are detailed in Table 6.2.

4.   The table below shows the frequency of ticks found on sheep that grazed two areas of moorland.

	No. of sheep	
Number of ticks	Area A	Area B
0	4	8
1	5	4
2	11	6
3	10	7
4	9	4
5	11	2
6	3	4
7	5	6
8	3	4
9	2	8
10	2	7
11+	5	10

The data set is censored (11+ ticks per sheep is the last category) and there is a suspicion of non-normality. Use a Mann-Whitney test to compare the incidence of ticks on the 2 groups of sheep. Is there any evidence to suggest that the 2 areas differ with respect to the presence of ticks?

5. In a new enclosure there are 2 feeders from which pigs can feed. The number of visits to each feeder was recorded every day for 15 days and is listed below. The number of pigs present in the enclosure varied during the recording period. Is there any evidence to suggest that the feeders are used to different extents? Should you use a paired or unpaired t-test to answer this question? Explain your choice of test.

	No. of Visits	
Day	Feeder A	Feeder B
1	25	28
2	24	24
3	29	28
4	23	26
5	32	34
6	33	34
7	21	24
8	30	32
9	26	29

| | No. of Visits | |
Day	Feeder A	Feeder B
10	22	22
11	25	24
12	27	29
13	31	30
14	24	26
15	26	28

6.   Darwin (1876) studied the effects of hybrid vigour by comparing the heights of seedlings produced from either self or cross-fertilisation by rearing pairs of seedlings in near identical conditions. The data tabulated below show the heights (inches) of the plants after a period of time. Analyse these data using a paired t-test. What conclusion can you draw about the effect of fertilisation on growth?

| | Growth (inches) | |
Pair	Self-fertilisation	Cross-fertilisation
1	17.4	23.5
2	20.4	12.0
3	20.0	21.0
4	20.0	22.0
5	18.4	19.1
6	18.6	21.5
7	18.6	22.1
8	15.3	20.4
9	16.5	18.3
10	18.0	21.6
11	16.3	23.3
12	21.0	18.0

7.   The following data indicate whether the concentration of β-endorphin (an opiate that is produced naturally in the brain as response to stress) in the blood of 19 cats increased or decreased after surgery. A plus sign indicates that the concentration increased, a minus sign indicates a decrease, and 0 indicates no change. Use a sign test to evaluate whether there is any evidence of an elevation in β-endorphin after surgery.

Change	–	+	+	+	+	–	+	+	0	+	+	+	+	+	+	–	–	–	0

8. In contests between members of a species, the winner is often the heavier individual. The general conclusion is that the winner is the individual that has the better body condition. In an investigation of contests between damselflies (*Odonata*), two sets of measurements of body condition were made: thorax weight and fat content of the thorax (both measured in mg). The following table contains the two sets of differences (mg) calculated as the winner minus the loser.

Fat (mg)	Thorax (mg)
–0.05	0.60
0.50	–0.10
1.50	–0.75
0.25	2.00
–0.75	–0.75
–0.75	0.00
0.25	1.25
0.10	0.00
–0.50	–0.75
–0.10	–0.75
0.25	–0.75
0.10	0.75
1.00	1.50
0.00	0.10
0.25	0.75
0.75	–1.50
2.00	–0.10
3.00	1.00
0.75	–1.50
1.00	–1.00
0.25	–1.25
0.50	1.00
0.75	1.00
0.10	–1.00

Conduct paired *t*-tests on these sets of differences. What can you say about the influence of body weight and fat content on the outcomes of contests?

9.   Reanalyse the damselfly data of Exercise 8 using a Wilcoxon matched-pairs signed-ranks test. Do you arrive at the same conclusions as with a paired t-test?

6.5   Summary of Minitab commands

LET
MANN
STEST
TTEST
TWOSAMPLE
TWOT
WTEST

6.6   Summary

1.   Two forms of 2 sample comparisons have been introduced for independent (unpaired) and dependent (paired) samples.
2.   The following tests have been introduced to analyse comparisons of independent samples: the unpaired t-test (parametric) and the Mann-Whitney test (non-parametric).
3.   The following tests have been introduced to analyse comparisons of dependent samples: the paired t-test (parametric), the sign test, and the Wilcoxon matched-pairs signed-ranks test (non-parametric).
4.   In the case of the Mann-Whitney, a method was introduced for comparisons with large samples ($n > 20$).

CHAPTER 7

Multiple Comparisons

7.1 Introduction

Throughout the biosciences, questions of comparison are very common, but the simple one- and two-sample comparisons described in Chapters 5 and 6 merely represent the tip of a very large iceberg. For example, how do we go about comparing the number of eggs laid by *Drosophila* at three, four, or five different temperatures (say, 10°C, 12°C, 15°C, 18°C, and 20°C)? Or, can we construct an experiment in which we can simultaneously compare the effects of several different concentrations of nitrate and phosphate on crop yield? These are certainly questions of comparison, but they are altogether more sophisticated than those described earlier. In this chapter, I shall describe various procedures that can be used to answer such questions.

The examples detailed above illustrate two fundamentally different types of comparisons. In the *Drosophila* example, a single variable (or *factor*) is under investigation, namely temperature, and we are interested in comparing the effects of five different temperatures (10°C, 12°C, 15°C, 18°C, and 20°C). These different temperatures are known as the *levels* of the factor under investigation. In the second example, the experiment examines the effects of more than one factor (the concentrations of both nitrate and phosphate). These terms, *factor* and *level*, are the keys to the organisation of this chapter since they help us to distinguish between two broad types of multiple comparison.

In Section 7.2, I will describe techniques to help analyse single factor comparisons where there are more than two levels, as in the *Drosophila* example. In the following section, methods will be introduced that enable us to perform multi-factor comparisons. The parametric tests that will be described are all forms of *the analysis of variance* (often abbreviated to ANOVA).This method enables us to perform many different types of comparisons involving either a single factor or many factors. In short, the analysis of variance is not a single test like the paired and unpaired t-tests, but is really a family of very powerful tests, and it is no exaggeration to say that they are the most important group of statistical tests used by bioscientists.

You will recall from Chapter 5 that if the assumptions of a parametric test are violated, we can use either an equivalent non-parametric test that does not rely upon assumptions, or apply a suitable transformation to the data and then use the parametric test. The final section of this chapter describes the main data transformations used in biological investigations. The transformations outlined here are applicable to data from any source (not just biological measurements) when the assumptions of a test are not valid, and, furthermore, their use is not confined solely to tests that perform comparisons as we shall see in later chapters.

Before proceeding to detail the methods used to perform these different types of comparisons, it is important to understand the reasons why we actually need this test, and to consider the basic principles underlying the analysis of variance. These principles are applicable to both single and multi-factor ANOVAs and are best illustrated by means of an example.

Let us imagine that we have maintained three groups of *Drosophila* fruit flies that originated from the same source at three different temperatures (10°C, 15°C, and 20°C) and we counted the number of eggs laid by each individual. How would we compare the outcomes of the three different treatments? Since the groups are independent, one might first be inclined to perform a series of unpaired t-tests:

$$10°C \text{ vs. } 15°C, 10°C \text{ vs. } 20°C, \text{ and } 15°C \text{ vs. } 20°C$$

Therefore, with three treatments we would need to perform three comparisons. If we designed an experiment with four treatments (say A, B, C, and D), we would need to perform six comparisons:

$$A \text{ vs. } B, A \text{ vs. } C, A \text{ vs. } D, B \text{ vs. } C, B \text{ vs. } D, \text{ and } C \text{ vs. } D$$

In fact, the number of comparisons needed for an experiment with n treatments can be predicted with the following formula:

$$no. \ of \ comparisons = \frac{n(n-1)}{2}$$

Clearly, as the number of treatments increases, the number of comparisons increases very rapidly. One might be reluctant, therefore, to use multiple t-tests simply on practical grounds. However, there is an even more important reason why this approach cannot be used.

Recall the way we used probabilities as the criteria for accepting or rejecting a null hypothesis in Chapters 5 and 6; we evaluated the probability of obtaining a set of results if the null hypothesis was true, and, if $P < 0.05$, we rejected the null hypothesis. We can never say that a particular null hypothesis is untrue, just that if it were true, the observed set of results would be unlikely. One way of regarding the probability value is that it represents the probability of wrongly rejecting the null hypothesis (i.e., committing a type I error). For example, if $P = 0.05$ and we reject H_N, then there is a 1 in 20 chance that we are wrongly rejecting a null hypothesis that is actually true. By performing many tests, we are more likely to commit type I errors, just by chance. In fact, it can be shown that if we use $P = 0.05$ as the reject/accept criterion, then the probability of committing a type I error increases rapidly if we conduct multiple tests. As the evidence presented in Table 7.1. clearly demonstrates, the use of multiple t-tests is definitely an inappropriate methodology for answering this sort of question. This fundamental problem led to the development of the analysis of variance by Sir Ronald Fisher in whose honour the test statistic (F) of the ANOVA was designated.

The underlying principle of the analysis of variance is relatively simple. We know that in all sets of measurements there will be some degree of variability because it is unlikely that all the measurements will be identical. The analysis of variance can be described very simply as a procedure that quantifies the total amount of variation within a data set. It then partitions variation into that which is due to identifiable treatment effects and that which is due to other factors (such

No. of means	No. of comparisons	Probability of committing a Type I error ($P = 0.05$)
2	1	0.05
3	3	0.14
4	6	0.26
5	10	0.40
6	15	0.54
10	45	0.90
∞	∞	1.00

Table 7.1. The probability of committing a type I error by performing multiple comparisons when using $P = 0.05$ as the criterion for rejection/acceptance.

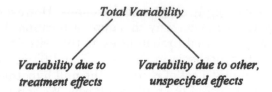

Figure 7.1. General plan of the analysis of variance: total variability being partitioned between that due to treatment effects and that which is due to other effects. The treatment effects component is split even further in investigations involving more than one factor.

as natural variability in the experimental subjects). This is shown in a diagram in Figure 7.1. This generalisation is true for all forms of the test irrespective of the level of complexity.

The type of null hypothesis that would be tested by a single factor experiment such as the *Drosophila* example would be:

H_N: The mean number of eggs laid by *Drosophila* maintained at 10°C, 15°C, and 20°C is similar.

The complementary, alternative hypothesis would be:

H_A: The mean number of eggs laid by *Drosophila* maintained at 10°C, 15°C, and 20°C are not all the same.

Note very carefully that the alternative hypothesis does not say that *all* the means would differ from each other; it implies only that *at least* one pair would not be the same. Remember, the two hypotheses have to cover all possible outcomes of a test, and so a difference between only one pair of means would be sufficient for us to reject H_N.

As a consequence, an analysis of variance will only tell us that at least one pair of treatments differ in their effect. It will not specifically reveal which pairs of treatments lead to dissimilar responses nor will it tell us how many differences exist. This is accomplished by further examination. Thus, in order to answer questions using the analysis of variance, we employ a two-stage process that answers the following questions:

(a)   Do the treatments differ in their effects in any way? and
(b)   In what way do the treatments differ?

The ANOVA enables us to answer question (a), and then, if the answer to (a) is yes, use another test to identify the detail of the differences. The following two sections describe the methods by which these two questions can be answered in the context of single factor and multi-factor investigations, respectively.

7.2 Comparisons involving many levels of a single factor

All of the comparisons described in Chapter 6 (unpaired and paired) involved a single factor and two levels. In this section, I shall describe methods that are, essentially, multi-level equivalents of unpaired and paired comparisons. Before describing in detail the techniques used to analyse such comparisons, I shall illustrate the different types of multi-level, single-factor comparisons.

The *Drosophila* example detailed above is a simple extension of a comparison of independent samples. In the two-sample (or *two-level*) case, this would be analysed using an unpaired *t*-test or Mann-Whitney test. Each *Drosophila* contributes a single data point (egg count) to the comparison, and each fly experiences only one of the treatments. This type of comparison is often referred to as a *between-subject* comparison because we are comparing measurements on different subjects or looking at differences between subjects. We could draw conclusions like:

The mean number of eggs laid by *Drosophila* maintained at 15°C is greater than that laid by similar flies kept at 10°C.

The conclusion is phrased in terms of individuals (mean numbers of eggs laid per individual). The multi-level equivalent to this type of question, where there are more than two independent treatments, will be considered in the next two sub-sections. The parametric ANOVA is described in the first sub-section and the non-parametric equivalent (the Kruskal-Wallis test) is detailed in the second sub-section.

Contrast the *Drosophila* experiment with the following example where the growth of rats in response to different food types is compared. Growth is determined by a combination of genetic and environmental (e.g., food) factors. We could devise an experiment that compared the effects of two food types by allocating one member from each of several pairs of twins to each of the two food types (all other rearing conditions being the same). Differences within pairs of twins could be attributed to the effects of nutrition, while the differences between sets of twins could be associated with familial or genetic factors. This experiment would be analysed by a paired test like those described in Chapter 6. We could extend this experiment to look at four different diets by using four siblings from each of a number of litters, each sibling being allocated at random to one of the four diets. This comparison parallels the paired comparisons of Chapter 6, with both analyses looking for differences *within* related sets of measurements (in this case, pairs of twins or within litters). Consequently, they are known as *within-subject* comparisons, the subjects being the sets of related measurements. In the example described here, the subjects are the litters from which the siblings are drawn. The parametric and nonparametric analyses for this type of comparison will be described in the sections entitled *Within-subject comparisons* and *Friedman nonparametric twoway ANOVA* (see Section 7.2), respectively.

Between-subject comparisons where normality can be assumed

As an example of between-subject comparisons, let us consider the *Drosophila* example in more detail. In Chapter 6 (Table 6.1.), the numbers of eggs laid by *Drosophila* maintained at 10°C and 15°C were compared. Those data were part of a larger data set from an experiment that also included *Drosophila* kept at 20°C. The entire data set is listed in Table 7.2.

The question that we are trying to answer in this particular investigation is whether there is any difference between the three temperatures in terms of the number of eggs laid by *Drosophila* maintained at those temperatures, or, in statistical terms, have these samples been drawn from populations with the same mean? Restating the null and alternative hypotheses:

H_N: The mean number of eggs laid by *Drosophila* maintained at 10°C, 15°C, and 20°C are similar.

H_A: The mean number of eggs laid by *Drosophila* maintained at 10°C, 15°C, and 20°C are not all the same.

The practical procedure would be to allocate flies from a common source at random to each of the treatments and then to incubate each group for a period of time at one of the specified temperatures. All

	No. of eggs laid		
	10°C	15°C	20°C
	9	9	12
	7	10	11
	10	10	12
	7	11	11
	7	10	10
	7	12	10
	6	10	9
	9	9	11
	8	12	11
	6	10	13
Total	76	103	110

Table 7.2. Number of eggs laid by three groups of *Drosophila* fruit flies, each group being maintained at either 10°C, 15°C, or 20°C.

three treatment groups would experience the experiment simultaneously so that the only known differences between the treatments would be the differences in temperature. In this example, the sizes of the samples are similar ($n = 10$ for all three temperature groups). This form of ANOVA is called the oneway analysis of variance because we are dealing with a single factor, and it can be performed with unequal sample sizes, although the power of the test to discriminate between groups is maximised when the sample sizes are equal.

The results from the three treatments are summarised in Table 7.3. This is the sort of table that would appear in a report in which the descriptive statistics are detailed; alternatively, one might include a graph such as Figure 7.2. in which the treatment means and 95% confidence intervals are plotted. This graph was produced using the

	No. of eggs laid		
	10°C	15°C	20°C
Mean	7.6	10.3	11.0
St. dev.	1.35	1.06	1.15
n	10	10	10

Table 7.3. Numbers of eggs laid (means, standard deviations, and sample sizes) by three groups of *Drosophila* fruit flies, each being maintained at either 10°C, 15°C, or 20°C.

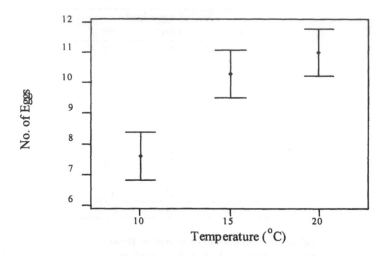

Figure 7.2. Mean (± 95% confidence intervals) number of eggs laid of *Drosophila* fruit flies maintained at 10°C, 15°C, and 20°C.

Interval Plot command that can be found by the following menu route within Minitab:

Stat > ANOVA > Interval Plot

The principle of the analysis is to calculate the total variability in the data set and then partition it into the component that is due to treatment differences and other factors (generally referred to as either the *error* or *residual* effects). As the name of the analysis suggests, the treatment and error variability are measured by their variances.

Before proceeding with the analysis itself, it is worth reconsidering the general formula for the variance (Equation 2.6, repeated below) and noting the names of the terms in this formula. These names appear repeatedly in ANOVA calculations.

$$\frac{\Sigma x^2 - \dfrac{(\Sigma x)^2}{N}}{n-1}$$

where:

Σx^2 is the sum of the square of each measurement, and it is known as the *uncorrected sum of squares*.

$\dfrac{(\Sigma x)^2}{N}$ is the square of the sum of all the measurements, divided by the total number of measurements (N). It is known as the *correction factor*.

The difference between these two terms (the numerator of the formula) is known as the *corrected sum of squares*.

$n - 1$ is known as the *degrees of freedom; n* represents the number of measurements. (In the analysis of variance, the specific measurements to which this relates may vary. For example, for the total variance, n, will be the total number of measurements, but, for the treatment variance, it will equal the number of treatments. There is a full explanation in the worked example that is detailed in Box 7.1.)

Unlike all the previous tests, several separate calculations are performed during the course of an analysis of variance, and conventionally, the results of the calculations are recorded in an ANOVA table such as the one displayed in Table 7.4. The form of the table is standard, and most computer packages (including Minitab) produce similar tables in their output of ANOVA results. The number of columns remains constant for all analyses of variance, but, as the analysis becomes more complex, additional rows are included in the table. Since the analysis involves several calculations, it is good practice when performing manual calculations to construct a blank table before commencing the calculations. The table is then ready to receive each result as it is obtained, and thus the results can be managed in an orderly fashion. The details of each stage of the analysis for this particular example are shown in Box 7.1. at the end of this section.

The test statistic, F, is the ratio of the treatment and error variances (Equation 7.1), each of which have their own associated degrees

Source of variation	DF	Corrected sum of squares (SS)	Variance	F	P
Treatment					
Error					
Total					

Table 7.4. Standard layout for a table of analysis of variance results. The *Variance* column is alternatively labelled as MS or MSE (*Mean Square* or *Mean Square Error*). The test statistic for ANOVA is denoted by F, and P represents the probability of obtaining the results if the null hypothesis is true.

of freedom. Tables of the critical values of the F-distribution are included as Appendix A.7., and their use will be illustrated in the context of the *Drosophila* example.

$$F = \frac{Variance_{treatment}}{Variance_{error}} \qquad (7.1)$$

This test statistic evaluates the degree to which the treatment variance is greater than the error variance. If the variance due to treatment effects is significantly greater than that due to other factors, we can reject the null hypothesis and conclude that temperature has influenced the numbers of eggs laid by *Drosophila*. In terms of the statistical test, therefore, the analysis of variance is always a *one-tailed* test. We will only reject the null hypothesis if F is *greater* than 1, i.e., the proportion of the total variability in the data that can be attributed to treatment effects is *greater* than that due to other, unidentified factors (the residual or error effects).

The probability of obtaining these results if the null hypothesis is true is evaluated using F-tables (Appendix A.7.). There are two sets of tables (A.7a. and A.7b.): $P = 0.05$ and $P = 0.01$. Since $P = 0.05$ and is the conventional criterion for acceptance/rejection of the null hypothesis, we initially consult the $P = 0.05$ table of critical values. If the result is statistically significant, we then proceed to use the $P = 0.01$ table to evaluate the probability as precisely as possible.

The column headings of the F tables are labelled as the *treatment degrees of freedom,* while the row headings relate to the *error degrees of freedom.* In this particular example, the treatment and error degrees of freedom are 2 and 27, respectively. Looking down the rows of error degrees of freedom, there is no entry for 27. We have to make a decision about whether to use 20 or 30 degrees of freedom, but we shall make

Source of variation	DF	SS	Variance	F	P
Treatment	2	64.467	32.234	22.61	<0.01
Error	27	38.500	1.426		
Total	29	102.967			

Table 7.5. The results table for the analysis of variance of the *Drosophila* egg-laying experiment. The probability of obtaining the results if the null hypothesis is true is less than 0.01 (it is established by using F tables). We reject the null hypothesis, and conclude that temperature has had an effect on the number of eggs laid by *Drosophila* fruit flies.

an error with whichever value we choose. As a general rule, we use the lower degrees of freedom, because, by doing so, we will err on the side of caution; we would tend to accept the null hypothesis in error rather than wrongly reject it.

Thus, in this example the required critical value of F is 3.49 (using 20 as the error d.f.). The calculated value of 22.61 is greater than that in tables, therefore we can conclude that the treatment variance is significantly greater than the error variance and we can thus reject the null hypothesis. Proceeding to the $P = 0.01$ tables, we can see that the appropriate value of F is 5.85, which again is smaller than the observed result. We can, therefore, complete the ANOVA results table (Table 7.5.) by adding 0.01 as the probability of obtaining these results if the null hypothesis is true.

We can conclude with the results of this analysis of variance that temperature does have an effect on egg-laying in *Drosophila*. However, the analysis of variance does not provide any detail of the effects. We cannot tell from the results of this analysis, for example, whether there are distinct differences in egg laying among all 3 temperatures, or whether 2 temperatures have similar effects to each other but have very different effects compared to the third. In order to discover the details of these temperature effects, we need to utilise additional tests that compare the treatment means. There are several tests that can be used to fulfill this task (for review, see Zar, 1996); but here, I shall describe one of those tests, Tukey's wholly significant difference test (or Tukey's test), which is considered to be reasonably robust under most circumstances. All tests of this sort are grouped under the general heading of *multiple comparison tests*.

The principles of this test closely parallel those of the unpaired *t*-test in that the difference between two means is divided by a standard error to give the test statistic, q (Equation 7.2).

$$q = \frac{\bar{x}_1 - \bar{x}_2}{SE} \tag{7.2}$$

where $\bar{x}_1$ and $\bar{x}_2$ are the two treatment means compared and SE is the standard error, which is calculated according to Equation 7.3.

$$SE = \sqrt{\frac{MS_{error}}{2}\left(\frac{1}{n_1} + \frac{1}{n_2}\right)} \tag{7.3}$$

where MS_{error} is the error variance from the analysis of variance results table, and n_1 and n_2 represent the sizes of the two samples being compared. When the sizes of the samples compared are the same, this equation simplifies to:

$$SE = \sqrt{\frac{MS_{error}}{n}} \qquad (7.4)$$

where n is the size of each sample. (In the *Drosophila* example, where the sample sizes are equal, n would be 10.)

This calculation has to be applied to each pair of treatment means in an experiment. Thus, in the *Drosophila* example, we will need to perform this calculation 3 times. One might ask how this procedure differs from a series of unpaired t-tests that were earlier rejected as invalid. The difference between the Tukey test and the unpaired t-test lies in the way that the standard error is calculated. The Tukey test differs from multiple t-tests because the standard error used in the Tukey test is based upon the variance of *all* the treatments in the experiment, not just of the particular pair being compared.

For this example, the following Tukey tests have to be performed:

10°C vs. 15°C

$$q = \frac{7.6 - 10.3}{\sqrt{\dfrac{1.426}{10}}} = \frac{-2.7}{0.267} = -10.11$$

10°C vs. 20°C

$$q = \frac{7.6 - 11}{\sqrt{\dfrac{1.426}{10}}} = \frac{-3.4}{0.267} = -12.73$$

15°C vs. 20°C

$$q = \frac{10.3 - 11}{\sqrt{\dfrac{1.426}{10}}} = \frac{-0.7}{0.267} = -2.62$$

The calculated values of q (ignoring negative signs) have to be compared against critical values in q, (also known as Studentized range tables; listed in Appendix A.8.). In this table, the rows refer to the error degrees of freedom (taken directly from the results table of the ANOVA), and the columns relate to the number of levels of this factor (in this case, 3). Any calculated value that is greater than the appropriate critical value in the tables indicates that 2 treatments differ significantly from each other. In this particular example, the critical value is 3.58 (again using 20 as the error degrees of freedom), and we can see that the number of eggs laid by *Drosophila* at 10°C is lower than at 15°C and 20°C, but there does not appear to be any difference in effect between the latter 2 temperatures.

When reporting an experiment such as this, we must state our conclusions in a similar style to that outlined in Section 6.2 in relation to the unpaired *t*-test. We might conclude this analysis thus:

A oneway analysis of variance was performed (include a reference to software package or textbook) that demonstrated a significant effect of temperature on egg-laying in *Drosophila* ($F = 22.61$; d.f. $= 2,27$; $P < 0.01$). Subsequent comparisons of treatment means using Tukey's wholly significant difference test (include a reference) showed that *Drosophila* maintained at 10°C laid significantly fewer eggs than those maintained at 15°C and 20°C. There was no evidence of any difference in effect between 15°C and 20°C.

Although we have used the Studentized range tables to determine which specific pairs of treatment groups differ from each other, it is not necessary to quote the q value and probability level in the summary of the results. It is sufficient to report the result of the analysis of variance in detail in order to show that the treatments have had an effect.

The reader's understanding of this conclusion will have been enhanced by the presence of descriptive statistics displayed either in the form of a table (Table 7.3.) or a graph (Figure 7.2.) earlier in the *results* section of the report. The confidence intervals displayed in Figure 7.2., illustrate very clearly that *Drosophila* laid fewer eggs at 10°C than at the higher temperatures: the 95% confidence intervals of the 10°C group do not overlap with those of the other 2 treatment groups. Remember also, that the statistical test only tells us that there is *statistical* difference. We also need to assess the *biological* significance of the difference by examining the mean egg counts of the treatment groups, explaining how such a difference arose, and discussing the consequences of the difference. This biological significance should be the focus for the *discussion* section of the report.

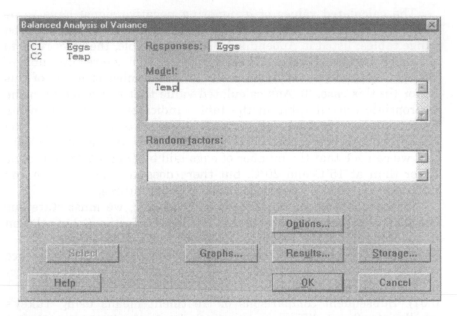

Exhibit 7.1. The Minitab dialog box to perform an analysis of variance. The measurements are stored in *C1* (*Eggs*) and the levels of the experimental factor are stored in *C2* (*Temp*). A summary of treatment means for each level of *Temp*, can be obtained through the *Results* button in this dialog box (see Exhibit 7.2.).

This analysis can be performed using Minitab, the dialog box being accessed through the **STAT** menu:

STAT> ANOVA> Balanced ANOVA

This Minitab command requires that the measurement data (the number of eggs in this example) are stored in a single column, and codes for the levels of the factor should be stored in another column. For this example, I have used 10, 15, and 20 as the codes for the levels (treatments); Minitab allows the codes to be either text or integers. In the dialog box, Minitab uses the terms *response* and *model* to refer to the *measurements* and *factor*, respectively, as shown in Exhibit 7.1.

The results of the analysis of variance produced by Minitab are shown in Exhibit 7.3. In addition to the standard ANOVA results table similar to that shown in Table 7.5., Minitab also summarises the levels of the treatment factors, indicating that there are 3 levels, which have been coded as 10, 15, and 20. Below the ANOVA results is the table of means and sample sizes for each treatment group.

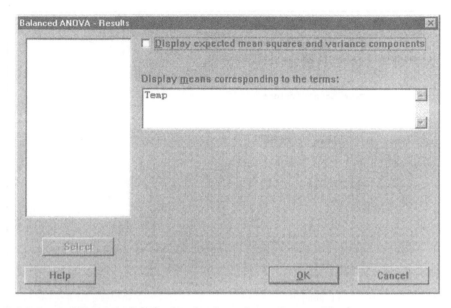

Exhibit 7.2. The Minitab dialog box that is used to produce a table of treatment means from an analysis of variance.

Although, Minitab does not provide a function to perform the Tukey tests in the balanced ANOVA command, it does provide a function within the oneway command (which is only relevant for this specific form of the analysis of variance). Since the balanced ANOVA command is generally applicable for many analyses of variance and is, therefore, of far greater use, I have not described any other analysis of variance commands.

Just as with the comparisons of one and two samples that were described in Chapters 5 and 6, the analysis of variance makes assumptions about the data to which it is applied. The same three assumptions that applied to the t-test also apply to the analysis of variance. The variances of the treatment groups are similar, the variances are independent of the means, and the data conform to a normal distribution. An additional assumption also applies to the analysis of variance, that the treatment effects must be additive (there is a detailed explanation of this assumption in Section 7.4). If the measurements do not satisfy these assumptions, then one of the simple data transformations that are detailed in Section 7.4 could be applied to the data. Alternatively, one can use a non-parametric test that is the direct equivalent to the

```
MTB > ANOVA Eggs = Temp;
SUBC>   Means Temp.

Analysis of Variance (Balanced Designs)


Factor       Type Levels Values
Temp         fixed    3    10    15    20


Analysis of Variance for Eggs

Source       DF         SS         MS       F       P
Temp          2     64.467     32.233   22.61   0.000
Error        27     38.500      1.426
Total        29    102.967




Means

Temp         N        Eggs
10          10       7.600
15          10      10.300
20          10      11.000

MTB >
```

Exhibit 7.3. Minitab output from an analysis of variance of the *Drosophila* egg-laying example.

oneway analysis of variance. It is called the Kruskal-Wallis test, and it is described in the next section.

Box 7.1. shows the details of the calculation of the analysis of variance for the *Drosophila* example.

The Kruskal-Wallis non-parametric oneway ANOVA

The non-parametric test that is the equivalent to the oneway analysis of variance is called the Kruskal-Wallis test. It is used in a similar way to the parametric ANOVA. It tests very similar null hypotheses, and, if a significant treatment effect is identified, we must employ an additional multiple comparison test (like that of Tukey) in order to discover the detail of the treatment differences.

Just as with the non-parametric tests described in earlier chapters, the Kruskal-Wallis test compares the *medians* of the different treatment populations, rather than the means. Furthermore, the circumstances in

1. Calculate the *Correction Factor (CF)*:
where N is the total no. of *Drosophila* in the experiment.

$$Correction \ factor \frac{(\Sigma x)^2}{N} = \frac{(Total \ no. \ of \ eggs \ laid)^2}{Total \ no. \ of \ Drosophila \ in \ the \ experiment}$$

$$= \frac{(9+7+10+\cdots+11+11+13)^2}{30}$$

$$= \frac{289^2}{30} = 2784.033$$

2. Calculate the *Total uncorrected sum of squares*:

$$\Sigma x^2 = 9^2 + 7^2 + 10^2 + ... + 11^2 + 11^2 + 13^2 = 2887$$

3. Calculate the Total corrected sum of squares (SS_{total}):
= Total uncorrected sum of squares − Correction factor
= 2887 − 2784.033 = 102.967

4. Calculate the uncorrected sums of squares for the treatment effects using the total no. of eggs laid at each temperature (see Table 7.2.):

$$= \frac{Sum \ of \ squares \ of \ treatment \ totals \ (T)}{number \ of \ replicates \ per \ treatment}$$

$$= \frac{\Sigma T^2}{n} = \frac{T_{10}^2}{n_{10}} + \frac{T_{15}^2}{n_{15}} + \frac{T_{20}^2}{n_{20}}$$

In this example, the size of the samples are similar ($n = 10$ for all three temperature groups). The oneway analysis of variance can be performed with unequal sample sizes, although the power of the test to discriminate between groups is maximised when the sample sizes are equal. So, in investigations where the sample sizes are not equal, the sum of squares of each treatment group (T^2) will be divided by the number of measurements contributing to that group.

Continued

Box 7.1. Method of calculating the oneway analysis of variance for the *Drosophila* egg-laying example (data from Table 7.2.).

4. (*continued*)

$$= \frac{76^2 + 103^2 + 110^2}{10}$$

$$= \frac{28485}{10} = 2848.5$$

5. Calculate the corrected sums of squares for the treatment effects ($SS_{treatments}$) by subtracting the correction factor from the uncorrected treatment sum of squares:

$$SS_{treatment} = \textit{Uncorrected sum of squares of treatment totals} - CF$$

$$= 2848.5 - 2784.033 = 64.467$$

6. The corrected sum of squares for the *Residual* or *Error* effects is the difference between those of the total and treatment effects, respectively; and is found by subtraction:

$$SS_{error} = SS_{total} - SS_{treatment}$$

$$= 102.967 - 64.467 = 38.5$$

7. The *Total degrees of freedom* are given by:
   No. of independent measurements (N) $-$ 1
   $=$ No. of Drosophila $-$ 1 $=$ 30 $-$ 1 $=$ 29
8. The *Treatment degrees of freedom* are given by:
   No. of treatments $-$ 1 $=$ 3 $-$ 1 $=$ 2
9. The *Error* or *Residual degrees of freedom* are found by subtraction:
   Total d.f. $-$ Treatment d.f. $=$ 29 - 2 $=$ 27
10. The *treatment variance* is:

$$= \frac{SS_{treatment}}{d.f._{treatment}} = \frac{64.467}{2} = 32.234$$

11. The *error variance* is:

$$= \frac{SS_{error}}{d.f._{error}} = \frac{38.5}{27} = 1.426$$

Continued

Box 7.1. *Continued.*

---

12. The test statistic, F, is the ratio of the treatment and error variances (Equation 7.1):

$$F = \frac{Variance_{treatment}}{Variance_{error}} = \frac{32.235}{1.426} = 22.61$$

13. The probability of obtaining these results if the null hypothesis is true is evaluated using F-tables (Appendix A.7.): $P < 0.01$. Therefore, we can conclude that temperature has had an effect on egg-laying in *Drosophila*.

Box 7.1. *Continued.*

which this test is used instead of the parametric ANOVA are similar to those that required the use of the Mann-Whitney and Wilcoxon tests described in Chapter 6. The data may be censored in some way; or they do not conform to a normal distribution; or the variances are unequal; or they are not independent of the mean.

As an illustration of the use of the Kruskal-Wallis test, consider the following example. Imagine that we wished to compare the learning capabilities of laboratory rats that have been reared under three different regimes by recording the number of trials (or amount of time) that rats took to learn a task. The reason why this particular example might require a non-parametric test is because some rats might never learn the task, thus giving rise to censored data. We can then test the following null hypothesis:

H_N: The median number of trials taken to learn the task does not differ between rearing conditions.

With the alternative hypothesis being:

H_A: The median number of trials taken to learn the task is not the same for all rearing conditions.

In this experiment, the three rearing treatments after weaning were: (a) reared in isolation; (b) reared with siblings, with minimal handling by humans; and (c) reared with siblings and subjected to systematic, gentle handling by humans. The results for the learning experiment are detailed in Table 7.6. with the number of trials taken

	Reared in Isolation (a)	Reared with siblings, but minimal handling (b)	Reared with siblings, systematic handling (c)
	12+ (14.5)	9 (8)	8 (5.5)
	10 (10.5)	11 (12.5)	7 (3.5)
	11 (12.5)	10 (10.5)	9 (8)
	12+(14.5)	8 (5.5)	6 (1.5)
	9 (8)	7 (3.5)	6 (1.5)
Rank sum (R)	60	40	20

Table 7.6. The number of trials required to learn a task for 5 rats in each rearing treatment. Numbers in parentheses indicate the rank within the whole experiment. The mean rank is allocated to scores that are tied. The sum of the ranks (R) for each treatment is also included; it is required in the calculation of the Kruskal-Wallis test statistic.

to learn the task shown. Like the Mann-Whitney and Wilcoxon tests described in Chapter 6, the calculation of the Kruskal-Wallis test is based upon the ranks of the measurements, and thus the data have been ranked and are shown in parentheses. When measurements are tied, ranks are averaged in the same way as with the Mann-Whitney and Wilcoxon tests.

The test statistic, H, is calculated using Equation 7.5.

$$H = \frac{12}{N(N+1)} \sum_{i=1}^{k} \frac{R_i^2}{n_i} - 3(N+1) \tag{7.5}$$

where,

n_i = the number of observations in treatment group i

N = the total number of observations

R_i = the rank sum of treatment i

k = the number of treatment groups

Critical values of H are given in Appendix A.9. for experiments with 3 treatments and sample sizes as large as 5 (such as the rat learning example). If an investigation comprises more than 3 treatments, or includes sample sizes greater than 5, the distribution of H can be approximated by the chi-squared (χ^2) distribution with $k - 1$ degrees of freedom. Critical values of the chi-squared distribution are

No. of trials taken to learn task	6	7	8	9	10	11	12+
No. of ties (no. of rats with same score)	2	2	2	3	2	2	2

Table 7.7. The groups of tied scores in the rat example: there are 7 groups of tied scores, with 3 rats taking 9 trials to learn the task and 2 rats taking each of 6, 7, 8, 10, 11, and 12+ trials.

provided in Appendix A.10. The calculation of the Kruskal-Wallis test statistic for the rat learning experiment is shown in Box 7.2.

Where a data set includes one or more tied measurements (as in the rat learning experiment), an adjustment has to be made to the value of H. The magnitude of the correction factor depends upon the number of sets of tied values. In the rat data listed in Table 7.6., we can see that there are seven sets of tied measurements.

The correction factor can be calculated thus:

$$C = 1 - \frac{\Sigma t}{N^3 - N} \tag{7.6}$$

where, Σt is given by:

$$\Sigma t = \sum_{i=1}^{m} (t_i^3 - t_i) \tag{7.7}$$

Here m denotes the number of sets of tied measurements (in the rat example, this would be seven), and t_i is the actual rank of the tied score.

Therefore, for each group of tied measurements, we would calculate $t_i^3 - t_i$ and add them to obtain Σt.

Finally, we can apply the correction factor, C, to H in order to obtain the corrected value of the test statistic, which is denoted by H_C:

$$H_C = \frac{H}{C} \tag{7.8}$$

Once a Kruskal-Wallis test has demonstrated a treatment effect, one has to employ a non-parametric equivalent test to the Tukey test to determine the detail of the differences between the treatments. This test compares the mean ranks of each treatment group. As with the Tukey test, one has to compare each pair of treatments, and the results

1. Applying Equation 7.5 to the data in Table 7.6. to calculate the test statistic, H:

$$H = \frac{12}{15(16)}\left[\frac{60^2}{5} + \frac{40^2}{5} + \frac{20^2}{5}\right] - 3(16)$$

$$H = 0.05[720 + 320 + 80] - 48$$

$$H = 8$$

2. Given that there are tied measurements, H must be corrected using Equation 7.7

$$\Sigma t = 6 + 6 + 6 + 24 + 6 + 6 + 6 = 60$$

3. According to Equation 7.6,

$$C = 1 - \frac{60}{15^3 - 15} \approx 0.9821$$

4. Therefore, the corrected value of H_C is:

$$H_C = \frac{8}{0.9821} = 8.145$$

5. Comparing this calculated value with the critical values for the Kruskal-Wallis statistic (Appendix A.9.), we can conclude that there is a significant treatment effect, $P < 0.01$.

Box 7.2. Calculation of the Kruskal-Wallis test statistic (with correction for ties), H_C, for the rat learning experiment.

of this test are interpreted using critical values of the test statistic, Q, which are in the table for Appendix A.11. (based upon the function provided by Zelen and Severo, 1964). Thus, with this rat learning investigation, we would have to perform a total of three comparisons. The formula for this comparison is given in Equation 7.9, the form of which is very similar to that of the Tukey test. A difference between descriptors (in this case the mean rank) of the two treatment groups is divided by a standard error.

$$Q = \frac{\overline{R}_A - \overline{R}_B}{SE} \qquad (7.9)$$

where,

Q is the test statistic

$\overline{R}_A$ and $\overline{R}_B$ denote the mean ranks of the two treatments being compared

SE is the standard error, the calculation of which is given by:

$$SE = \sqrt{\left(\frac{N(N+1)}{12} - \frac{\Sigma t}{12(N-1)} \right)\left(\frac{1}{n_A} + \frac{1}{n_B} \right)} \qquad (7.10)$$

Therefore, in the case of the rat example with $k = 3$ treatment groups, the critical value would be 2.394. The three comparisons necessary for this example are shown in Box 7.3.

The dialog box used to direct Minitab to perform a Kruskal-Wallis test is found by following the menu route:

Stat > Nonparametrics > Kruskal-Wallis

The dialog box is shown in Exhibit 7.4. Minitab requires the data to be stored in the same way as for a parametric ANOVA with the measurements in a single column (labeled *Response* in the dialog box) with a second column holding the codes for the treatments (labeled *Factor*). In the rat learning experiment, the columns have been named *Trials* and *Rearing*, respectively. The response variables must be numeric, but the factor codes may be numeric or text values. In the example shown, rearing has been coded *a, b,* or *c.* The results of the analysis are shown in Exhibit 7.5. Two values for H are provided, one without any correction for ties ($H = 8.00$) and the other with the corrected value ($H_C = 8.15$). The z scores indicate the differences of each of the treatment medians from the overall median, and thus, in the context of the rat experiment, are not of immediate use. The calculations of Q must be performed manually (there is no Minitab function) using the method outlined above.

1. Compare treatment (a) reared in isolation against treatment (b) reared with siblings, but with minimal handling by humans:

$$Q = \frac{12-8}{\sqrt{\left(\dfrac{15(16)}{12} - \dfrac{60}{12(14)}\right)\left(\dfrac{1}{5}+\dfrac{1}{5}\right)}} = \frac{6}{\sqrt{(20-0.357)0.4}} = 2.141$$

2. Compare treatment (a) reared in isolation against treatment (c) reared with siblings and subjected to systematic, gentle handling by humans.

$$Q = \frac{12-4}{2.803} = 5.197$$

3. Compare treatment (b) reared with siblings, but with minimal handling by humans, with treatment (c) reared with siblings and subjected to systematic, gentle handling by humans.

$$Q = \frac{8-4}{2.803} = 1.427$$

The critical value in Q tables for three treatment groups is 2.394. Therefore, we can conclude that rats raised in isolation take significantly more attempts to learn a task than those raised in a social group and exposed to systematic, gentle handling. No other differences between treatments were observed.

Box 7.3. Calculation of the non-parametric equivalent to the Tukey test for the rat learning example.

Within-subject comparisons

The previous two sub-sections have described the parametric oneway analysis of variance and its non-parametric equivalent, the Kruskal-Wallis test. These two analyses can be applied to experimental designs that are known as *fully randomized designs*. Such designs are suitable

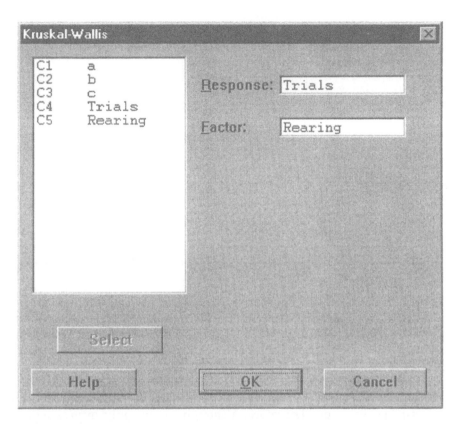

Exhibit 7.4. The Minitab dialog box for the Kruskal-Wallis non-parametric analysis of variance.

for situations where the experimental environment is uniform. Each experimental subject has the same experience other than the difference between the levels of the experimental factor under investigation. Furthermore, in such designs each subject provides a single measurement and is independent of all the other sampling units. As mentioned in Section 7.2, we may wish to perform investigations in which the data are not independent, extending the paired t-test to incorporate more than two measurements. The example that I used to illustrate this type of investigation was a comparison of growth by rats fed on 1 of 4 different foods, and, in order to reduce genetic effects, I suggested that 4 siblings from each of 8 litters were allocated to each treatment group. The data from this experiment are shown in Table 7.8.

```
MTB > Kruskal-Wallis 'Trials' 'Rearing'.
```

Kruskal-Wallis Test

```
Kruskal-Wallis Test on Trials

Rearing      N     Median    Ave Rank        Z
a            5     11.000      12.0        2.45
b            5      9.000       8.0        0.00
c            5      7.000       4.0       -2.45
Overall     15                  8.0

H = 8.00  DF = 2  P = 0.018
H = 8.15  DF = 2  P = 0.017 (adjusted for ties)

MTB >
```

Exhibit 7.5. Minitab output of the Kruskal-Wallis analysis of rat learning data shown in Table 7.6.

The objective of this experiment is to determine whether variations in weight gain can be attributed to differences in diets. In this respect, there is no difference between this experiment and, say, the *Drosophila* egg-laying investigation; we have one factor at several levels in both. However, in this rat experiment, we are not dealing with a uniform group of subjects. The rats *within* a litter will be more similar than those from different litters, the latter difference reflecting different genetic backgrounds. Therefore, although we are only interested in a

Litter No.	Food Type				
	A	B	C	D	Total
1	35.10	39.24	57.55	56.61	188.50
2	35.53	40.54	58.63	56.97	191.67
3	37.60	42.66	59.74	58.68	198.68
4	35.75	40.90	58.67	57.25	192.57
5	36.40	41.69	58.86	57.57	194.52
6	35.09	40.62	57.38	56.43	189.52
7	36.09	41.08	59.39	58.30	194.86
8	36.45	42.10	59.40	58.12	196.07
Total	288.01	328.83	469.62	459.93	
Mean	36.00	41.10	58.70	57.49	
Std. dev.	0.83	1.06	0.86	0.82	

Table 7.8. Weight increase (g) of rats on 4 different diets over an experimental period of 42 days.

single experimental factor, *food type*, we have an additional, identifiable source of variation, namely *litter*. Following a similar argument to that proposed in Chapter 6 with regard to the paired *t*-test, we must take these between-litter differences into account because, if we fail to do so, they may mask real effects caused by differences between the diets.

The basic principle of the analysis of variance is that it partitions the total variability between treatment (identifiable) and unspecified (residual or error) effects. In this rat-feeding experiment, we require the analysis of variance to partition the variability between three components: food types; between-litter differences; and the remaining residual or error effects. It is worth briefly reconsidering how the test statistic, *F*, is calculated. By doing so, we can understand the basic principles upon which the more sophisticated forms of the analysis of variance are based. We saw in Equation 7.1 that the test statistic is calculated by dividing the treatment variance by that of the error:

$$F = \frac{Variance_{treatment}}{Variance_{error}}$$

Every time we include an additional identifiable source of variation to the analysis, we reduce the proportion of the total variation that is unattributed and included in the error variance. As a consequence, the value of *F* will increase because the magnitude of the treatment variance will increase relative to the error variance. This is the arithmetic reason why failing to take account of the between-litter effects may allow real food-type effects to be masked — because they would remain as part of the error variance thereby reducing the value of *F*.

As a general rule, therefore, any modification to an experimental design or analysis that decreases the value of the error variance will improve the chance of the test identifying treatment effects. Conversely, any increase in the error variance will reduce the effectiveness of the test. The sample size of each treatment group is, therefore, crucial since this will greatly influence the error degrees of freedom, and thus the error variance.

In this rat-feeding experiment, we wish to establish whether the different food types yield different weight gains. Although it is reasonable to anticipate that there will be differences between litters, we are not really concerned with quantifying them, we simply wish to ensure that they do not obscure any real differences due to the experimental food treatments applied. This type of problem is found throughout biology, and, as a consequence of its original development in horticul-

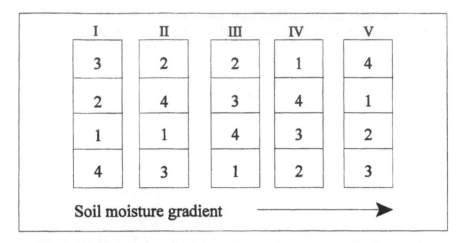

Figure 7.3. Schematic diagram of a field experiment in a randomized block design. Five blocks (I, II, III, IV, and V) in which each of 4 treatments have been randomly allocated to the sections within each block. The long axes of the blocks are perpendicular to the soil moisture gradient.

tural experiments, this particular experimental design has become known as a *randomized block design.* The blocking element of this design has become so important in biological experimentation that it is worth considering the underlying principles before using the technique to analyse the rat-feeding experiment.

Figure 7.3. illustrates a horticultural experiment where four seed varieties are used in a field in which there is a known, but unquantified, soil moisture gradient. The experimental environment is, therefore, not uniform and may mask any treatment effects that might be present. It is a direct parallel to the rat-feeding trial where differences between litters play the same role as the soil moisture gradient. If we divide the field into several, relatively thin blocks whose long axes run perpendicular to the soil moisture gradient, we can assume that the variability due to soil moisture within a block will be relatively small compared to that between blocks. If we then split each block into 4 sections and randomly allocate each seed variety to one of the plots within each block, then the blocks in this experiment would parallel exactly the litters in the rat-feeding experiment, and the sections within the blocks would parallel to the individual rats. The randomisation process in allocating the treatments to the sections within the blocks is very important in order to avoid any other unknown systematic variability. For example, if we allocated a particular seed variety

Source of variation	DF	SS	Variance	F	P
Treatment	3	3163.800	1054.570	10043.524	<0.01
Litters (Blocks)	7	20.190	2.884	27.467	
Error	21	2.210	0.105		
Total	31	3186.190			

Table 7.9. The results table for the analysis of variance of the rat feeding experiment. The probability of obtaining the results if the null hypothesis is true is less than 0.01 (it is established by using F tables). We reject the null hypothesis, and conclude that diet has had an effect on the weight gain of rats.

to the top section in every block, it might coincide with an old field drain thereby confounding the performance of the variety with a characteristic of the environment (c.f. the section entitled *Stratified random sampling* (see Section 4.2), where the advantages of stratified sampling are discussed).

The final ANOVA results table is detailed in Table 7.9., where we can see evidence of differences in weight gain between the diets. The detailed analysis of the rat feeding experiment is shown in Box 7.4.

In exactly the same way as with the oneway analysis of variance, we must now perform Tukey tests to discover the detail of the differences between the four treatment groups. The calculation of these Tukey tests follows the same method as before, using Equations 7.2, 7.3, and 7.4. The results are summarised in Table 7.10. (check that you obtain the same answers).

We can see from these results that there are differences in weight gain between rats on all the diets. We report the results of an experiment that has been designed as a randomized block in exactly the same way as that of a fully randomized design (oneway ANOVA)

Comparison	q	P
diet A vs. diet B	44.52	<0.05
diet A vs. diet C	198.14	<0.05
diet A vs. diet D	187.58	<0.05
diet B vs. diet C	153.62	<0.05
diet B vs. diet D	143.06	<0.05
diet C vs. diet D	10.56	<0.05

Table 7.10. Summary of Tukey tests for the rat-feeding experiment. The standard error used in all calculations is 0.1146.

Diet	Mean weight gain (g)	s.e.	n
A	36.001[a]	0.294	8
B	41.104[a]	0.374	8
C	58.690[a]	0.299	8
D	57.491[a]	0.290	8

Table 7.11. Weight gained (g) by rats maintained on four different diets over a 42-day experimental period. Means with similar superscripts differ at $P < 0.05$.

because the two designs answer the same questions and the analyses are testing the same type of null hypothesis. The blocking technique is simply a device to overcome methodological difficulties, and so we do not need to refer to any block effect in the results. One way of reporting the results is to summarise the data from the treatment groups in a table (such as Table 7.11.). Notice in the table that superscripts have been used to indicate significant differences (based upon the results of the Tukey tests) between each of the treatment groups; this is a common technique for reporting differences.

The table of descriptive statistics would be followed by a paragraph summarising the results of the analysis itself:

> Weight gain in rats fed on four different diets (A, B, C, and D) was investigated using a randomized block design (insert a reference). Analysis of variance showed significant differences due to diet effects ($F = 10043.53$; d.f. = 3,7; $P < 0.01$), and Tukey's wholly significant difference tests subsequently revealed significant differences between all 4 treatments.

In order to emphasise the usefulness of the blocking technique, re-analyse these data using a oneway analysis of variance (i.e., ignoring the litters) and conduct the appropriate Tukey tests. What differences can you identify between treatments? In what way do the results differ between the two designs?*

The analysis of this experiment using Minitab uses the same commands as used earlier, but for a randomized block design, both the *Food* and *Litter* variables are entered in the *Model* box (as illustrated in Exhibit 7.6.). The results of the analysis are shown in Exhibit 7.7. Note

---

* Re-analysing these data using a fully randomized design reveals no evidence of any difference in weight gained between rats on diets C and D, where the blocking enabled a difference between these two treatments to be detected.

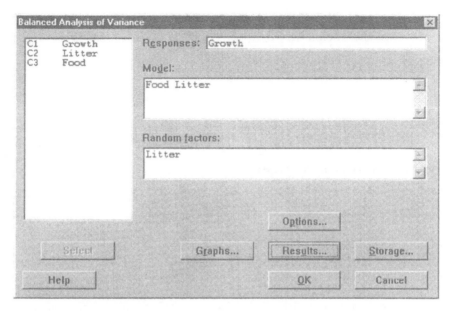

Exhibit 7.6. The Minitab dialog box used to specify the randomized block design analysis for the rat growth experiment. Note that the blocking factor, *Litter*, is also specified as a random factor.

that the factor, *Litter*, is also entered in the box labeled *Random factors*. In any experiment in which there is blocking, the blocks are randomly selected units that are simply acting as examples; we do not have any specific interest in each block. For example, we have no desire to discover anything in particular about each litter in the rat-feeding experiment, and each block of land in the example illustrated in Figure 7.3. is simply a portion of land used in the experiment. As such, blocks are referred to as *random factors,* while the principal factors under investigation are known as *fixed factors.* In terms of the calculation of the analysis of variance, the distinction between fixed and random factors only becomes important when there are two or more fixed factors under examination. We shall see examples of this sort of investigation in Section 7.3.

The calculations for the rat-feeding experiment are detailed in Box 7.4. Take particular note of the how the uncorrected and corrected sums of squares for the litter (block) effect are calculated in Stages 6 and 7; the pattern is the same as the calculations for the main diet effects.

In the next section, I shall describe how to analyse randomized block designs using non-parametric techniques. However, before

```
MTB > ANOVA 'Growth' = Food Litter;
SUBC>    Random 'Litter';
SUBC>    Means Food.
```

Analysis of Variance (Balanced Designs)

```
Factor        Type Levels Values
Food          fixed    4    A    B    C    D
Litter        random   8    1    2    3    4    5    6    7    8


Analysis of Variance for Growth

Source        DF        SS         MS       F        P
Food           3    3163.80    1054.60 1.0E+04  0.000
Litter         7      20.19       2.88   27.46  0.000
Error         21       2.21       0.11
Total         31    3186.19



Means

Food     N    Growth
A        8    36.001
B        8    41.104
C        8    58.690
D        8    57.491

MTB >
```

Exhibit 7.7. The Minitab output for the analysis of the rat growth example.

proceeding, it is important to mention another application of the randomized block design that is particularly important in many areas of the biosciences, particularly in biomedical, behavioural, and psychological research, and very often where longitudinal studies are conducted. This application is known as the *repeated measures* or *within-subjects* design. Although the computational outcome is similar to the randomized block design (and so I have not included a worked example here), it has been given a different name because of a specific difference in the experimental methodology. In the randomized block design, one allocates a separate subject to each treatment in every block. In a repeated measures design, each subject receives *every* treatment; the subject is essentially taking the place of the block.

Imagine, for example, that the effects of 4 drugs on blood pressure were being investigated using 8 mice. Following a method that directly

1. Calculate the *Correction Factor (CF):*
    where N is the total no. of rats in the experiment.

    $$Correction \ Factor \frac{(\Sigma x)^2}{N} = \frac{(Total \ weight \ gained)^2}{Total \ no. \ of \ rats}$$

    $$= \frac{(35.1 + 35.53 + 37.6 + ... + 56.43 + 58.3 + 58.12)^2}{32}$$

    $$= \frac{1546.3^2}{32} = 74719.1489$$

2. Calculate the *Total uncorrected sum of squares:*

    $\Sigma x^2 = 35.1^2 + 35.53^2 + 37.6^2 + ... + 56.43^2 + 58.3^2 + 58.12^2 = 77905.3399$

3. Calculate the Total corrected sum of squares (SS_{total}):
    = Total uncorrected sum of squares − Correction factor
    = 77905.3399 − 74719.1489 = 3186.1910

4. Calculate the uncorrected sums of squares for the treatment effects using the total weight gained on each food type (see Table 7.8.):

    $$= \frac{\Sigma T^2}{n} = \frac{288.01^2}{8} + \frac{328.83^2}{8} + \frac{469.62^2}{8} + \frac{459.93^2}{8}$$

    $$= 7782.9455$$

5. Calculate the corrected sums of squares for the treatment effects ($SS_{treatments}$) by subtracting the correction factor from the uncorrected treatment sum of squares:

    $SS_{treatment}$ = *Uncorrected sum of squares of treatment totals − CF*

    = 77882.9455 − 74719.1489 = 3163.7966

    Continued

Box 7.4. Method of calculating the rat weight gain experiment (data from Table 7.8.).

6. Calculate the uncorrected sums of squares for the litter effects using the total weight (ΣB) gained in each litter. This is calculated in a similar way to the uncorrected sum of squares for the treatments:

$$= \frac{188.5^2}{4} + \frac{191.67^2}{4} + \ldots + \frac{194.86^2}{4} + \frac{196.07^2}{4}$$

$$= 74739.3374$$

7. Calculate the corrected sums of squares for the treatment effects ($SS_{litters}$) by subtracting the correction factor from the uncorrected treatment sum of squares:

$SS_{litters}$ = *Uncorrected sum of squares of treatment totals – CF*

$$= 74739.3374 - 74719.1489 = 20.1885$$

8. The corrected sum of squares for the *Residual* or *Error* effects is the difference between those of the total and treatment effects, respectively; and is found by subtraction:

$$SS_{error} = SS_{total} - SS_{treatment} - SS_{litters}$$

$$= 3186.91 - 31.63.7966 - 20.1885 = 2.2059$$

9. The *Total degrees of freedom* are given by:
   No. of independent measurements (N) – 1 = No. of rats – 1
   = 32 – 1 = 31
10. The *Treatment degrees of freedom* are given by:
    No. of treatments – 1 = 4 – 1 = 3
11. The *Litter degrees of freedom* are given by:
    No. of litters – 1 = 8 – 1 = 7
12. The *Error* or *Residual degrees of freedom* are found by subtraction:
    Total d.f. – Treatment d.f. – Litter d.f. = 32 – 3 – 8 = 21

Continued

Box 7.4. *Continued.* Randomized block analysis of the rat feeding experiment (data in Table 7.8.) showing that the different diets give rise to differences in weight gain.

13. The *treatment variance* is:

$$= \frac{SS_{treatment}}{d.f._{treatment}} = \frac{3163.7966}{3} = 1054.57$$

14. The *litter variance* is:

$$= \frac{SS_{litter}}{d.f._{litter}} = \frac{20.1885}{7} = 2.884$$

15. The *error variance* is:

$$= \frac{SS_{error}}{d.f._{error}} = \frac{2.2059}{21} = 0.105$$

16. The test statistic, F, is the ratio of the treatment and error variances (Equation 7.1):

$$F = \frac{Variance_{treatment}}{Variance_{error}} = \frac{1054.57}{0.105} = 10043.52$$

17. The probability of obtaining these results if the null hypothesis is true is evaluated using F-tables (Appendix A.7.): $P < 0.01$. We can conclude that significant differences in weight have arisen due to the different feeding treatments.

Box 7.4. *Continued.*

parallels that of the randomized block design, we would randomize the *sequence* with which each mouse would experience the 4 drugs. The analysis, just like all other analyses of variance, partitions the total variability among the different components, and this is illustrated in Figure 7.4. The total variation can be split into 2 main components: the between-subject variation and the within-subject variation. In addition, the within-subject component is split further into a treatment element (the drug effects) and a residual or error component. In effect, the between-subject variation represents the consistent differences

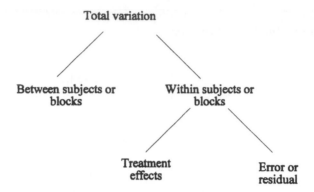

Figure 7.4. Partitioning of variation within a repeated measures or randomized block design. Since each subject (or block) experiences every treatment, the within-subject variation is partitioned into the treatment effects and the error or residual effects.

between subjects (e.g., Mouse A has consistently higher blood pressure than Mouse B, irrespective of treatment: the *block* effect). The error term reflects individual differences associated with specific treatments, or, in other words, it can be viewed as the interaction between the individual subject and the treatments.

Having outlined the principles of the repeated measures design, we need to identify why it is so important. First of all, if every subject experiences all the treatments, we have the potential to reduce the number of subjects, which must be good on ethical grounds. Second, the uncontrolled variability found in between-subject designs (due to the random allocation of subjects to treatments) is eliminated. Therefore, it is a very powerful design. If it is a very powerful design that is ethically sound, why do we not use it all the time? Inevitably, there are some difficulties with the design and they can be divided into two main types: *practice* (sometimes referred to as *general practice*) and *carryover effects*. The first of these is a general effect, while the latter is very specific.

Practice effects are those effects that are not related to a specific treatment, but are the result of the operation of the whole experiment. For example, in an investigation of responses to different stimuli, a subject may become fatigued or bored as the experiment progresses. This can be overcome by including rest periods of an appropriate duration between treatments. Another example of this type of effect is when the subject is required to perform tasks during the course of the experiment. There is a danger that, as the experiment progresses, the subject becomes more proficient through learning and practice. If this happens, the response on later treatments will be affected. This is a particular danger in operant studies, but it can be overcome by training the subject before the experimental period, so that the individual is totally proficient and will not improve in performance during the testing period. Of course, the randomisation process will reduce the impact of this sort of effect because some subjects will experience a treatment early in the experiment, while others will encounter it later. In terms of repeated measures experiments, the randomisation process is known as *counterbalancing*.

In contrast to the general practice effects, carryover effects are very specific. Since each subject experiences all treatments sequentially, there is the danger that the effect of one treatment influences the response of the subject to the next one; it is the specific effect of one treatment on the next. There are ways of overcoming this sort of problem. For example, in pharmacological experiments, a period of time can be inserted between treatments to allow for residues of one drug to be eliminated from the individual before the start of the next drug. In psychological experiments in which attitudes are under examination, the carryover effects between one treatment and the next would probably be quite significant; it is hard to imagine that the experimenter could shift a subject's attitude in one direction and then turn it to another. In such circumstances, the repeated measures design is probably not the most appropriate, and a between-subject design should be used. This example re-emphasises the point made in Chapter 1 in relation to planning the investigation thoroughly before starting any experimentation.

Friedman non-parametric twoway ANOVA

Data from a single factor randomized block design that do not satisfy the assumptions of the parametric analysis can be analysed using the Friedman non-parametric randomized block test. In the same way that the non-parametric tests were described earlier (e.g., the Kruskal-Wallis test), the Friedman test is based upon ranked data, thereby avoiding any reliance on particular statistical distributions. The Friedman test and its parametric equivalent also test similar null hypotheses, the only difference being that null hypotheses tested by Friedman analyses are framed in terms of *medians* rather than *means*.

Consider the following evaluation of four systemic insecticides of tomato plants as an example of the use of the Friedman test. This experiment was carried out in a glasshouse, and, in order to accom-

	Insecticide			
Block	A	B	C	D
1	mild (1.5)	severe (4)	moderate (3)	mild (1.5)
2	moderate (3)	moderate (3)	moderate (3)	mild (1)
3	mild (1.5)	severe (4)	mild (1.5)	moderate (3)
4	none (1.5)	moderate (4)	mild (3)	none (1.5)
5	mild (2)	severe (4)	moderate (3)	none (1)
6	moderate (2.5)	severe (4)	moderate (2.5)	mild (1)
7	none (1)	moderate (4)	mild (2.5)	mild (2.5)
8	mild (1.5)	severe (4)	moderate (3)	mild (1.5)
9	mild (1.5)	severe (4)	moderate (3)	mild (1.5)
10	none (1)	moderate (3)	moderate (3)	moderate (3)
Rank sum (R_i)	17	38	27.5	17.5

Table 7.12. The scores for leaf damage on tomato plants treated with 4 different systemic insecticides. The damage was scored on a 4-point scale. The numbers in parentheses are the ranked scores *within* each block. The rank sums for each treatment (R_i) are also included.

modate inevitable variations in sunlight in parts of the glasshouse, it was laid out as randomized block design with 10 blocks. The measured variable was the amount of leaf damage after a predetermined period of time, the damage being measured on an ordinal scale of *none, mild, moderate,* and *severe.* The data are detailed in Table 7.12.

The Friedman test statistic is calculated using Equation 7. 11, and the significance of the result is evaluated using χ^2 tables with $a - 1$ degrees of freedom (where a is the number of treatment groups; χ^2 tables are provided in Appendix A.10.).

$$\chi^2 = \frac{12}{ab(a+1)} \sum_{i=1}^{a} R_i^2 - 3b(a+1) \qquad (7.11)$$

where, a is the number of treatment groups, and b is the number of blocks, and R_i is the rank sum for the ith treatment.

As with the Kruskal-Wallis test, tied scores can be taken into consideration using a correction factor, C, that is evaluated thus:

$$C = 1 - \frac{\Sigma t}{b(a^3 - a)} \qquad (7.12)$$

where Σt is calculated using Equation 7.7 (the same equation as for the Kruskal-Wallis test). The corrected value of χ^2 is given by:

$$\chi_c^2 = \frac{\chi^2}{C} \qquad (7.13)$$

The full calculation of the Friedman analysis of variance for the insecticide comparison (including the corrections for ties) is shown in Box 7.5. As with the other analyses of variance, a separate comparison of the treatment groups must be performed if treatment effects are detected. This is achieved using Equation 7.14 for each pair of treatments, the test statistic being q (the same as for the Tukey test with critical values in Appendix A.8.) with infinite degrees of freedom. The results of these analyses are summarised in Table 7.13.

$$q = \frac{R_1 - R_2}{SE} \qquad (7.14)$$

where R_1 and R_2 denote the rank sums for the two treatment groups being compared and SE is the standard error, which is given by

$$SE = \sqrt{\frac{ba(a+1)}{12}} \qquad (7.15)$$

The results of the Friedman test and the subsequent comparison of treatments enable us to make the following formal concluding statement.

A comparison of four systemic insecticides was carried out in order to evaluate their performance in reducing leaf damage to tomato plants. A randomized block design was used, and a Friedman nonparametric twoway analysis of variance (*reference*) showed that there were significant differences between insecticides ($\chi^2 = 20.86$; d.f. = 3; $P < 0.001$). Subsequent comparisons using a nonparametric equivalent of the Tukey test revealed that tomato plants treated with Insecticides A and D sustained lower levels of leaf damage than those treated with Insecticide B.

1. Calculate the uncorrected χ^2 value using Equation 7.11.

$$\chi^2 = \frac{12}{40(5)}(17^2 + 38^2 + 27.5^2 + 17.5^2) - 3(10)(5)$$

$$= 0.06(289 + 1444 + 756.25 + 306.25) - 150$$

$$\chi^2 = 167.73 - 150 = 17.73$$

2. Using Equation 7.7

$$\Sigma t = 90$$

3. The correction factor, C, is:

$$C = 1 - \frac{90}{10(64 - 4)} = 0.85$$

4. The corrected value of the Friedman test statistic is:

$$\chi_c^2 = \frac{\chi^2}{C} = \frac{17.73}{0.85} = 20.86$$

With three degrees of freedom (no. of treatment groups − 1), this result indicates that there is a significant difference between insecticides ($\chi^2 = 20.86$; d.f. = 3; $P < 0.001$).

Box 7.5. Calculation of the Friedman nonparametric twoway analysis of variance of the insecticide data in Table 7.12. The result is interpreted using χ^2 tables (Appendix A.10.).

The Minitab method of conducting the Friedman test is illustrated in Exhibits 7.8. and 7.9. The dialog box shown in Exhibit 7.8. is found by the following menu route:

STATS > Nonparametrics > Friedman

The data are stored in the same way as the other analyses of variance: all of the measurements in a single column, and two additional

Comparison	q	P
Insecticide A vs. Insecticide B	5.14	<0.05
Insecticide A vs. Insecticide C	2.57	n.s.
Insecticide A vs. Insecticide D	0.12	n.s.
Insecticide B vs. Insecticide C	2.57	n.s.
Insecticide B vs. Insecticide D	5.02	<0.05
Insecticide C vs. Insecticide D	2.45	n.s.

Table 7.13. Summary of comparisons between treatments for the insecticide experiment. The standard error used in all calculations is 4.083. The critical value of q in the studentized range tables against which these calculated values are compared is 3.63 (four treatment groups and infinite degrees of freedom). From these comparisons, we can conclude that tomato plants treated with insecticides A and D sustain significantly less leaf damage than plants with insecticide B.

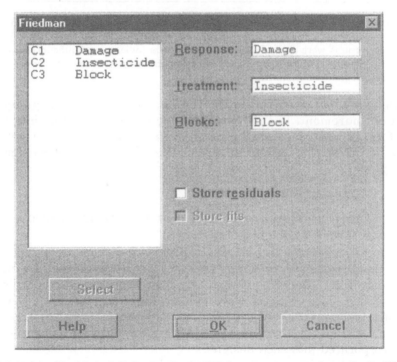

Exhibit 7.8. The Minitab dialog box for the Friedman nonparametric randomized block analysis. The codes for the treatment groups are stored in Column 2, and those for the blocks in C3.

```
MTB > Friedman 'Damage' 'Insecticide' 'Block'.

Friedman Test


Friedman test for Damage by Insectic blocked by Block

S = 17.73   DF = 3   P = 0.000
S = 20.86   DF = 3   P = 0.000 (adjusted for ties)

                        Est      Sum of
Insectic        N     Median     Ranks
A              10      1.500      17.0
B              10      4.000      38.0
C              10      3.000      27.5
D              10      1.500      17.5

Grand median   =      2.500

MTB >
```

Exhibit 7.9. The Minitab output from the Friedman nonparametric randomized block analysis using the insecticide example (data in Table 7.12.). The codes used for the treatment groups were A, B, C, and D. The results indicate a significant treatment effect, the details of which can be obtained using the nonparametric equivalent of the Tukey test (Equations 7.14 and 7.15).

columns holding codes indicating the block and treatment group for each measurement. In this particular example, where the data take the form of text-based scores (*none, mild, moderate,* and *severe*), we have to convert the data to numeric values for Minitab (for example, 0, 1, 2, and 3).

7.3   Comparisons involving more than one factor

In the introduction to this chapter, I divided questions involving multiple comparisons into two broad groups: those involving a single factor, and those where we focus on more than one experimental variable. In this section, I describe how we can evaluate the effects of several factors simultaneously. These designs are often described as *factorial designs*. I shall also outline how blocking or repeated measures can be incorporated into such designs.

Factorial design experiments not only provide an opportunity to evaluate the effects of several factors, but they also enable us to determine whether there is any interaction between the effects of the various factors. The concept of interaction effects is best illustrated by means of an example. Consider the following experiment in which researchers compared the growth responses of two closely-related plant species to light and dark. Plants on the light treatment experienced the normal daily light/dark cycle, while those on the dark treatment were kept in darkness all the time. The growth response was measured in terms of root length. Ten plants of each species were allocated at random to each of the dark and light treatments. The data that were obtained are listed in Table 7.14., and the mean root growth for each treatment group are plotted in Figure 7.5. In that figure, we see very clearly that root growth in Species A is much greater in the light than in the dark, but, for Species B, the converse appears to be the case; the two species respond differently to light and dark. This is an example of an interaction between two factors (species and light/dark), and, graphically, they are characterised by lines that are not parallel with each other. In many instances, the existence of an interaction between factors is of greater interest to a researcher than significant effects of the individual factors independent of each other. This figure also emphasises the importance of describing the data prior to reporting the results of analysis: this graph leaves the reader no doubt as to the difference between the two species in their responses to light and dark; it was produced by the Minitab command, **Interactions plot**, that is accessed via the menu route:

Stat > ANOVA > Interactions plot

The calculation of the analysis of variance for the factorial is very similar to those of the fully randomized and randomized block, the only additional elements being concerned with the interaction. A full calculation is detailed in Box 7.6. The table of results for the ANOVA is shown in Table 7.15., and we can see that the interaction between the species and the light/dark factors is statistically significant, as are the effects of the two factors individually (the probability values are all less than 0.05).

Just as with the analyses of variance described earlier in this chapter, the detail of the effects of the interaction between factors and the factors acting individually are determined using Tukey tests. Since interactions are generally of great interest to researchers, it is customary to calculate those first. Use Equations 7.2 and 7.4 to compare the means of the four treatment groups within each species and within each light/dark treatment. Remember that we use the number of

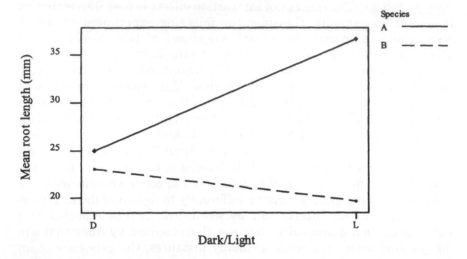

Figure 7.5. Mean root length (mm) of two plant species under light (L) and dark (D) regimes. The two species respond differently: Species A producing more roots under the light than in the dark and Species B having greater root growth in the dark than the light.

	Species A		Species B	
	Dark	Light	Dark	Light
	25.69	34.80	18.45	22.98
	22.56	32.66	20.11	16.51
	24.52	34.50	23.43	21.06
	23.60	39.55	24.24	15.62
	23.64	37.30	26.95	17.23
	26.68	38.63	24.14	21.91
	28.58	37.06	22.49	19.65
	21.60	33.81	24.40	21.81
	26.38	39.10	23.53	21.01
	25.87	39.18	22.52	19.39
Mean	24.91	36.66	23.03	19.72
St. Dev.	2.11	2.52	2.37	2.51
Total	249.12	366.59	230.26	197.17

Table 7.14. Root length (mm) of two closely-related plant species grown under normal light conditions (Light) and in darkness (Dark).

Source of variation	Degrees of freedom	Sums of squares	Mean square	F	P
Species	1	886.23	886.23	155.87	<0.01
Light/Dark	1	177.99	177.99	31.30	<0.01
Species* Light/Dark	1	566.71	566.82	99.67	<0.01
Error	36	204.69	5.69		
Total	39	1835.63			

Table 7.15. Results of the analysis of variance for the factorial design experiment of growth response of two species to different light/dark treatments. There is a significant interaction between species and light/dark treatment.

treatment means to choose the appropriate column of the Studentized range tables. Within a factorial experiment, this may vary from one factor to another depending upon the number of levels. In this example, it will be 4 for the interaction Tukey tests, and 2 for both of the main factor effects. However, since this particular experiment has only two levels of each of the main factors, Tukey tests are not necessary to determine the detail of the main factor effects; we only need to inspect the means. What conclusions can you draw? The results are summarised in Table 7.16.

Comparison	q	P
Species A Light vs. Species A Dark	15.58	<0.05
Species B Light vs. Species B Dark	4.39	<0.05
Species A Light vs. Species B Light	22.47	<0.05
Species A Dark vs. Species B Dark	2.49	n.s.

Table 7.16. Summary of Tukey tests between treatments for the root growth experiment. The standard error used in all calculations is 0.754. The critical value of q in the studentized range tables against which these calculated values are compared is 3.85 (four treatment groups and 30 degrees of freedom [30 being the nearest below the error d.f. which is 36]).

Four conclusions can be drawn from the analysis of the interaction effects, and they could reported thus:

The root growths of two plant species were compared under light and dark conditions in a two-factor factorial design experiment (*reference*). A significant interaction between species and light/dark treatment was revealed ($F = 99.67$; d.f. = 1.36; $P < 0.01$). Subsequent Tukey tests showed that, under dark conditions, Species A and Species B did not differ in terms of root growth. However, under light conditions: (a) the root growth of Species B was significantly lower than Species A; (b) the root growth of Species B was significantly lower than it was under dark conditions; and (c) the root growth of Species A was significantly higher than it was under dark conditions (all $P < 0.05$).

The quantity of conclusions in this relatively simple factorial experiment is considerably greater than from the single factor designs of Section 7.2, and thus it is very important to help the reader grasp the essence of the results. In this particular example, Figure 7.5. provides a very clear description of the results that can be backed up by a table summarising the means for the different treatment combinations (Table 7.17.).

	Species A	Species B	Overall
Dark	24.91[a]	23.03[c]	23.97
Light	36.66[a,b]	19.72[b,c]	28.19
Overall	30.78	21.37	

Table 7.17. Summary of mean root length (mm). Interaction means within a column or row that have similar superscripts differ significantly at $P < 0.05$.

Factorial designs, with the additional capability to evaluate interactions between factors, are considerably more sophisticated analytical tools than single factor analyses of variance. However, this added level of complexity brings with it several potential difficulties that must be considered very carefully during the planning stages of an investigation.

The first point to bear in mind is that, in order to evaluate an interaction effect, there has to be replication at each treatment combination for which an interaction is required. If there is no replication, then only main factor effects can be evaluated. If we wish to evaluate interaction effects, then as factors and/or levels are added, the number of subjects will increase rapidly, and so will the cost. Consider the following example. If the absolute bare minimum number of replicates is two per treatment combination (though this will rarely be sufficient), then eight subjects will be needed for a two-factor factorial with two levels in each factor. However, by adding a third level to each factor, the number of subjects increases to 18. Additionally, in the case of animal experiments, the ethical implications of increasing the number of subjects will also have to be considered very carefully.

A second major consideration with elaborate factorial designs is the increased complexity of the interactions. In the root growth example, there are only two factors producing a single interaction (Species * Light/Dark). If a design includes three factors (e.g., A, B, and C), there will be four interaction terms: A * B, A * C, B * C, and A * B * C and, as the number of potential interactions increases, several difficulties may arise:

(i)   How can we explain the interactions (some may be very complex)?
(ii) Some interactions may be spurious or irrelevant to the objectives.
(iii) How can the clarity of the report be achieved even though the results are complex?

The key to all three of these potential difficulties lies in defining the project objectives very clearly in the planning stages and then, in the report, discussing only the results that relate to those objectives. If the objectives are refined as much as possible, the experiment can be designed with the minimum level complexity. By restricting the report to the detail that is essential to meeting the objectives, you will not divert the reader from the message that you are trying to communicate.

We can analyse factorial designs in Minitab using the same commands and dialog boxes as for the earlier examples. Exhibit 7.10. illustrates this process for the root growth example. Both of the factor columns have been specified in the *Model* box, and, in order to calculate the interaction effect, a vertical bar has been inserted between these two columns (an exclamation mark would achieve the same outcome), without it no interaction would be computed. This character is also included in the *Results* dialog box that specifies the means to be displayed (an example of this dialog was shown in Exhibit 7.2.). We can specify in a long-hand form all the effects that we wish to be computed by entering the following in the Model box:

*Species 'Dark/Light' Species * 'Dark/Light'*

In factorial analyses with many factors, the long-hand version will become very cumbersome, and the shorter version will be the preferred form in factorial designs where all the factors are fixed. However, we

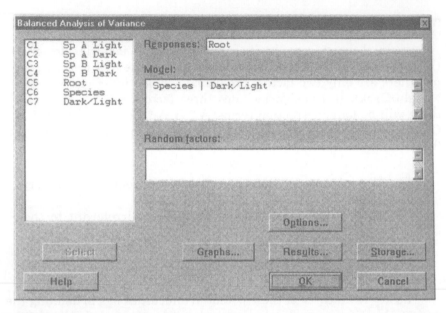

Exhibit 7.10. The balanced ANOVA dialog box to specify a factorial analysis for the root growth example. Note the vertical bar separating the two factors in the *Model* that instructs Minitab to calculate the effects of both factors and also the interaction between them.

will see in the final example of this section (where we have a random factor) that sometimes we need to use the long-hand version.

The output of this analysis is shown in Exhibit 7.11. As before, the Tukey tests have to be performed manually. Remember that there will be different standard errors for the interaction and the main factor computations because of the sample size differences ($n = 20$ for Species and Dark/Light, and $n = 10$ for the interaction). The full calculation of the analysis of variance for the root growth example is detailed in Box 7.6.

I conclude this section with another example of a factorial design in which I have included the additional level of complexity of blocking. By including this example, my intention is to provide a template from which it will be possible to construct even more elaborate designs. Although this specific example includes blocks, it could equally well have been a repeated measures design. The computation is the same; *subjects* would simply take the place of *blocks*.

In this example, soil nitrogen concentration has been measured at two different growth stages of wheat under two different fertiliser

```
MTB > ANOVA 'Root' = Species |  'Dark/Light';
SUBC>   Means Species| 'Dark/Light'.
```

Analysis of Variance (Balanced Designs)

```
Factor      Type Levels Values
Species    fixed    2    A     B
Dark/Lig   fixed    2    D     L


Analysis of Variance for Root

Source              DF         SS         MS        F       P
Species              1     886.23     886.23   155.87   0.000
Dark/Lig             1     178.00     178.00    31.32   0.000
Species*Dark/Lig     1     566.71     566.71    99.67   0.000
Error               36     204.69       5.69
Total               39    1835.63


Means

Species     N       Root
A          20     30.786
B          20     21.371

Dark/Lig    N       Root
D          20     23.969
L          20     28.188

Species Dark/Lig     N       Root
A          D        10     24.912
A          L        10     36.659
B          D        10     23.026
B          L        10     19.717

MTB >
```

Exhibit 7.11. Results of the Minitab analysis of the root growth experiment (data in Table 7.14.). The Dark/Light factor label is abbreviated to *Dark/Lig* because Minitab allows only 8 characters for a name in the output.

treatments, and the experiment was replicated in four blocks. The two fixed factors are, therefore, *growth stage* and *fertiliser*, with *blocks* being the random factor. A typical block would appear thus:

| Fertiliser 2 | Fertiliser 1 | Fertiliser 1 | Fertiliser 2 |
| Growth Stage 2 | Growth Stage 2 | Growth Stage 1 | Growth Stage 1 |

The data from this experiment are listed in Table 7.18.

1. Calculate the *Correction Factor (CF)*:
where N is the total no. of plants in the experiment.

$$Correction\ Factor \frac{(\Sigma x)^2}{N} = \frac{(Total\ root\ length)^2}{Total\ no.\ of\ plants}$$

$$= \frac{(25.69 + 22.56 + 24.52 + \ldots + 21.81 + 21.01 + 19.39)^2}{40}$$

$$= \frac{1043.14^2}{40} = 27023.5265$$

2. Calculate the Total corrected sum of squares (SS_{Total}):

$$\Sigma x^2 - CF = 25.69^2 + 22.56^2 + 24.52^2 + \ldots + 21.81^2 + 21.01^2 + 19.39^2 - CF$$

$$= 29039.1578 - 27203.5265 = 1835.63$$

3. Calculate the Species corrected sum of squares ($SS_{Species}$):

$$\Sigma \frac{(Species\ total)^2}{n} - CF = \frac{615.71^2}{20} + \frac{427.43^2}{20} - CF$$

$$= 28089.7605 - 27203.5265 = 886.234$$

4. Calculate the corrected sums of squares for the light/dark treatment effects ($SS_{L/D}$):

$$\Sigma \frac{(Light/\ Dark\ total)^2}{n} = CF = \frac{563.76^2}{20} + \frac{479.38^2}{20} - CF$$

$$= 27381.5261 - 27203.5265 = 177.9996$$

Continued

Box 7.6. Calculation of the factorial ANOVA for the root growth experiment (data in Table 7.14.). Note that, in stage 5 where the corrected sum of squares for the interaction is calculated, not only is the correction factor subtracted from the SS_{Total}, but also the corrected sums of squares for both of the factors as well. If this is not done, the effect of *Species* and *Light/Dark* will be included both in the individual factors and also in the interaction term.

5. Calculate the corrected sums of squares for the interaction effects ($SS_{interaction}$) by subtracting the correction factor, $SS_{Species}$, and $SS_{L/D}$ from the uncorrected interaction sum of squares. We must subtract the $SS_{Species}$ and the $SS_{L/D}$ in order to avoid including the individual factor effects in the interaction sum of squares.

$$\Sigma \frac{(Interaction\ total)^2}{n} - CF - SS_{Species} - SS_{L/D}$$

$$= \frac{249.12^2}{10} + \frac{366.59^2}{10} + \frac{230.26^2}{10} + \frac{197.17^2}{10} - CF - SS_{Species} - SS_{L/D}$$

$$= 28834.468 - 27203.5265 - 886.234 - 177.9996 = 566.7078$$

6. The corrected sum of squares for the *Residual* or *Error* effects is the difference between those of the total and the various treatment effects, respectively, and is found by subtraction:

$$SS_{error} = SS_{Total} - SS_{Species} - SS_{L/D} - SS_{Interaction}$$

$$= 1835.63 - 886.234 - 177.9996 - 566.7078 = 204.6886$$

7. The *Total degrees of freedom* are given by:
No. of independent measurements (N) $- 1 =$
No. of plants $- 1 = 40 - 1 = 39$

8. The *Species degrees of freedom* are given by:
No. of Species $- 1 = 2 - 1 = 1$

9. The *degrees of freedom* associated with the Light/Dark treatment are given by:
No. of Light/Dark treatments $- 1 = 2 - 1 = 1$

10. The *interaction degrees of freedom* are given by:
(Species d.f.)(Light/Dark d.f.) $= (1)(1) = 1$

11. The *Error* or *Residual degrees of freedom* are found by subtraction:
$=$ Total d.f. $-$ Species d.f. $-$ Light/Dark d.f. $-$ Interaction d.f.
$= 39 - 1 - 1 - 1 = 36$

Continued

Box 7.6. *Continued.* Calculation of the factorial ANOVA for the root growth experiment (data in Table 7.14.).

12. The *Species variance* is:

$$= \frac{SS_{Species}}{d.f._{Species}} = \frac{886.0975}{1} = 886.234$$

13. The *Light/Dark variance* is:

$$= \frac{SS_{Light/Dark}}{d.f._{Light/Dark}} = \frac{177.9948}{1} = 177.9996$$

14. The *Light/Dark variance* is:

$$= \frac{SS_{Interaction}}{d.f._{Interaction}} = \frac{566.7078}{1} = 566.7078$$

15. The *error variance* is:

$$= \frac{SS_{error}}{d.f._{error}} = \frac{204.6886}{36} = 5.6858$$

16. The test statistic, *F*, for *Species* is:

$$F_{Species} = \frac{Variance_{Species}}{Variance_{error}} = \frac{886.234}{5.6858} = 155.87$$

17. The test statistic, *F*, for *Dark/Light* is:

$$F_{Dark/Light} = \frac{Variance_{Dark/Light}}{Variance_{error}} = \frac{177.9996}{5.6858} = 31.30$$

Continued

Box 7.6. *Continued.* Calculation of the factorial ANOVA for the root growth experiment (data in Table 7.14.).

18. The test statistic, F, for *Interaction* is:

$$F_{Interaction} = \frac{Variance_{Interaction}}{Variance_{error}} = \frac{566.7078}{5.6858} = 99.67$$

19 Using F-tables (Appendix A.7.), we can see that *Species*, *Dark/Light*, and the interaction between the two factors are all significant at $P < 0.01$.

Box 7.6. *Continued.* Calculation of the factorial ANOVA for the root growth experiment (data in Table 7.14.).

The computation of the sums of squares, degrees of freedom, and variances for all three factors (*growth stage, fertiliser,* and *block*) and the associated interactions for the analysis of this experiment, follows exactly the same pattern as the previous factorial example. The difference in the calculation lies in the way that the various F-ratios are computed, specifically we use different error variances for each value of F. This notion may seem strange at first, but, if we recall the way that the total variation was partitioned in the single factor randomized block or repeated measures design, the reasoning will become clear.

In the single-factor randomized block or repeated measures design, the total variation was partitioned into the between-block effects and the within-block effects as illustrated in Figure 7.4. In this two-factor experiment, it is the within-block component that we have to examine further in order to evaluate the effects of the two-

| Block | Fertiliser 1 | | Fertiliser 2 | |
	Growth Stage 1	Growth Stage 2	Growth Stage 1	Growth Stage 2
I	28.00	26.00	38.00	32.00
II	30.00	27.00	40.00	34.00
III	26.00	24.00	39.00	31.00
IV	29.00	30.00	42.00	39.00
Mean	28.25	26.75	39.75	34.00
St. dev.	1.71	2.50	1.71	3.56
Total	113	107	159	136

Table 7.18. Nitrogen concentration (units per sample) in a two-factor factorial design with blocking. Each of the two fixed factors has two levels.

fixed factors. Therefore, this component is partitioned even further into the effects of the two factors and the associated interactions. In terms of the objectives of the experiment, we are interested in evaluating only three effects: *fertiliser, growth stage,* and the interaction between *fertiliser and growth stage.* Recalling the discussion of the error term of the randomized block and repeated measures designs in the section entitled *Within-subject comparisons* (see Section 7.2), we saw there that it could be viewed as the interaction between the fixed factor and the block or subject. The same principle is applied here in the multi-factor design. The error variances that we use to calculate the three F-ratios of interest are the variances of the interactions between the factor of interest and the block, namely:

$$F_{Fertiliser} = \frac{Variance_{Fertiliser}}{Variance_{Fertiliser*Block}}$$

$$F_{Growth\,stage} = \frac{Variance_{Growth\,stage}}{Variance_{Growth\,stage*Block}}$$

$$F_{Fertiliser*Growth\,stage} = \frac{Variance_{Fertiliser*Growth\,stage}}{Variance_{Fertiliser*Growth\,stage*Block}}$$

The error term used in the last F-ratio is, in fact, the residual or error variance of the whole analysis and it is labelled *Error* in Minitab output of this type of analysis (Exhibit 7.13.). The specification of the analysis in Minitab is shown in Exhibit 7.12. The various components in the *Model* box in Exhibit 7.12. are:

Growth | Fertiliser specifies the main factor effects

*Growth * Fertiliser Block* interaction specifies the between-block effect

*Growth * Block* and *Fertiliser * Block* denote the *Growth * Block* and *Fertiliser * Block* interactions that will be used as error terms.

Compare it with Exhibits 7.6. and 7.10. for the randomized blocks and factorials, respectively.

Since the calculations of all but the F-ratios in this analysis follow the same pattern as the earlier factorial design, I have not included a full calculation here. Only the detailed calculations of the F-ratios are shown in the expressions below:

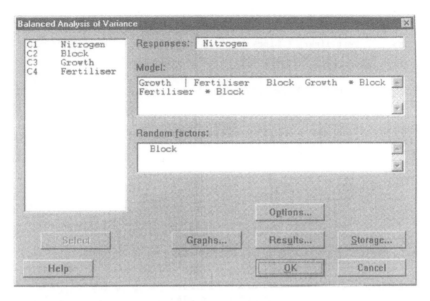

Exhibit 7.12. The Minitab specification of the soil nitrogen experiment (a two-factor factorial with blocking), the data for this experiment are detailed in Table 7.18.

$$F_{Growth\ stage} = \frac{Variance_{Growth\ stage}}{Variance_{Growth\ stage*\ Block}} = \frac{52.563}{3.229} = 16.28$$

$$F_{Fertiliser} = \frac{Variance_{Fertiliser}}{Variance_{Fertiliser*\ Block}} = \frac{351.563}{1.896} = 185.44$$

$$F_{Growth\ stage*\ Fertiliser} = \frac{Variance_{Growth\ stage*\ Fertiliser}}{Variance_{Growth\ stage*\ Fertiliser*\ Block}} = \frac{18.062}{0.396} = 45.63$$

The analysis of variance shows significant interaction and main effects and we must compare the means using Tukey tests. The same formulae are used as before (Equations 7.2 and 7.4), but we must take care to use the appropriate *error variance* in each set of calculations of the standard error. In this particular example, with both main factors having only two levels, Tukey tests are only required for the interaction comparisons, and the standard error would be calculated thus:

$$SE = \sqrt{\frac{Variance_{Growth*\ Fertiliser*\ Block}}{n}} = \sqrt{\frac{0.396}{4}}$$

```
MTB > ANOVA 'Nitrogen' = Growth  | Fertiliser   Block   Growth  * Block &
CONT>        Fertiliser  * Block;
SUBC>    Random    'Block';
SUBC>    Means   Growth |Fertiliser.
```

Analysis of Variance (Balanced Designs)

```
Factor       Type Levels Values
Growth       fixed    2    1    2
Fertilis     fixed    2    1    2
Block        random   4    1    2    3    4
```

```
Analysis of Variance for Nitrogen

Source              DF        SS        MS        F      P
Growth               1    52.563    52.563    16.28  0.027
Fertilis             1   351.563   351.563   185.44  0.001
Growth*Fertilis      1    18.062    18.062    45.63  0.007
Block                3    57.688    19.229     4.07  0.088 x
Growth*Block         3     9.688     3.229     8.16  0.059
Fertilis*Block       3     5.687     1.896     4.79  0.115
Error                3     1.187     0.396
Total               15   496.438

x Not an exact F-test.
```

```
Means

Growth    N   Nitrogen
1         8    34.000
2         8    30.375

Fertilis    N   Nitrogen
1           8    27.500
2           8    36.875

Growth Fertilis    N   Nitrogen
1        1         4    28.250
1        2         4    39.750
2        1         4    26.750
2        2         4    34.000

MTB >
```

Exhibit 7.13. The Minitab output for the soil nitrogen example (two-factor factorial with blocks). The only F-ratios (and associated P values) of interest are those of Growth stages and Fertiliser and their Interaction; all the others are irrelevant and should be ignored.

Comparison	q	P
Fert. 1 Growth 1 vs. Fert. 1 Growth 2	4.77	*n.s.*
Fert. 1 Growth 1 vs. Fert. 2 Growth 1	36.55	$P<0.05$
Fert. 1 Growth 2 vs. Fert. 2 Growth 2	23.04	$P<0.05$
Fert. 2 Growth 1 vs. Fert. 2 Growth 2	18.28	$P<0.05$

Table 7.19. Summary of the Tukey tests for the interaction effects in the soil nitrogen example. The critical value of q obtained from studentized range tables is 6.83 (4 treatment means and 3 error d.f.).

A summary of the results of the Tukey tests is given in Table 7.19. Having analysed this experiment, we must report the results. Although the design has been more elaborate than that of the earlier root growth example, these two investigations have tested very similar null hypotheses. The only difference has been methodological; the soil nitrogen experiment has incorporated blocking into its design. Put simply in terms of the experimental objectives, both experiments have explored the impacts of two factors on a single response. The conclusions of the two investigations can, therefore, take a similar form and the descriptive techniques that we employ to communicate our findings will be very similar. Figure 7.6. illustrates the contrasting effects of the two fertilisers on soil nitrogen concentration at the two growth stages of wheat. This can be supported by a table similar to Table 7.17. and a concluding paragraph of a form like:

The response of soil nitrogen to two different fertilisers and at two different growth stages of wheat was investigated in a two-factor factorial design experiment (*include a reference*). A significant interaction between fertiliser and growth stage was revealed ($F = 45.63$; d.f. = 1, 3; $P < 0.01$). Subsequent Tukey tests showed that, with Fertiliser 1, there was no difference in soil nitrogen between the 2 growth stages. However, with Fertiliser 2, there was a significantly lower concentration of nitrogen at growth stage 2 than stage 1; and at both growth stages, Fertiliser 2 produced significantly greater levels of nitrogen than Fertiliser 1 (all $P < 0.05$).

The discriminatory power of the analysis of variance is at its greatest when sample sizes are equal. In fact, apart from the fully randomized (oneway) ANOVA, all analyses of variance should have equal sample sizes. Inevitably, however, this is not always possible; accidents

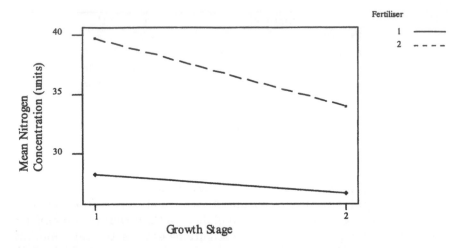

Figure 7.6. Mean soil nitrogen concentration under two fertilisers and at different growth stages of wheat. Results show that there is a greater soil nitrogen concentration at growth stage 1 than stage 2 with Fertiliser 2, and more nitrogen with Fertiliser 2 than Fertiliser 1 at both growth stages.

can result in "unbalanced" designs. Methods have had to be developed to cope with such eventualities. It would be remiss of me to omit mention of these methods, but the computational detail would occupy more space than is available. However, Minitab offers a convenient method to perform analyses of variance with missing data that does not require much space to explain. The technique is called the *generalised linear model* (often abbreviated to GLM), and it can be found through the same menu route as the analysis of variance commands:

STAT > ANOVA > General linear model

Conveniently, the dialog boxes take the same forms as those described throughout this chapter. The calculations adjust the corrected sums of squares of the factors that have missing data. The most profound effects are felt by the error degrees of freedom; one degree of freedom is lost for each datum that is missing. So, in the soil nitrogen experiment, with only three error degrees of freedom, the loss of one datum could be very serious indeed. Once again, the importance of good planning before starting experimental work cannot be stressed too greatly. It is common practice in medical research, for example, to increase the number of control patients in order to compensate for the potential loss of patients in the experimental groups.

By now you will no doubt have realised that I have been able to provide only a glimpse of the diverse family of analytical tools used in the analysis of variance. My objective has been to focus on the biological question, not the tool. In the planning stages of any investigation, you must ask yourself whether you need to study one or more factors, whether blocking would be appropriate, or whether you need to make multiple (repeated) measures on each subject. It is beyond the scope of this book to describe any more designs, but I hope that I have given a flavour of the techniques that are available and also the building blocks that should enable you to construct and analyse more elaborate experimental designs. If you need to delve more deeply, there are several good monographs on experimental design (e.g., Cox, 1958; Crowder and Hand, 1990).

7.4   Data transformations

In this and earlier chapters, I have described non-parametric methods that can be used to analyse data when the assumptions of the parametric tests have not been satisfied. There is another alternative way of coping with such difficulties; we can transform our data so that it does satisfy the assumptions. In this section, I will outline the principal methods that are used in biological investigations and also explain why and how we can use them.

First of all, let us consider in a little more detail the assumptions that are sometimes violated. There are 4 main assumptions associated with most of the tests described so far:

(a)   the data should be drawn from a population that conforms to a normal distribution

(b)   the variances of the different treatment groups should be similar

(c)   variances should be independent of the mean

(d)   the treatment effects should be additive

The meaning of the first three of these assumptions is fairly self-evident, but the fourth may not be immediately obvious. Additivity can be best explained by means of an example. Imagine that a treatment is applied to a number of subjects and we measure their weights before and after the treatment is applied. We would say that the treatment effect is additive if all the subjects experience the same magnitude of change, say an average increase of 500 g, irrespective of any differences in starting weights between the subjects. If, on the other hand, the subjects experience weight gains of, say, 10% of their body weights, this would be a *multiplicative* effect; their absolute weight gains would be dependent upon their starting weights.

This particular assumption provides a good starting point for a discussion of data transformations because it is violated so often. In all areas of biology, from microbiology through physiology to ecology, we measure growth in a variety of forms: growth of populations, body size, or tissue. In the early stages of growth of a population, individual or tissue, the increase in size over a given period of time depends upon the original size, e.g., two individuals can give rise to more offspring than a single individual, and the increase is clearly multiplicative rather than additive. It is customary, therefore, to use the logarithms of measurements rather than the raw data because logarithms convert multiplicative relationships to additive ones. We can see this effect in action in the table below (Table 7.20.), where two pairs of measurements and the differences between them are listed along with the logarithms of the same measurements. The second measurement in each pair is double the first. The differences between the members of each pair are not identical, but, in contrast, the differences between the logarithmic data are similar; the logarithmic transformation has had the desired effect of creating an additive relationship.

Having transformed the data, we can apply the statistical techniques that have the assumption of additivity, but what does that mean in terms of the conclusions that we can draw? The data that are analysed are not the measurements that we gathered. Are we cheating in some way? The simple answer to that question is that we are not cheating, but simply using a different scale of measurement. Remember in the discussion of data types in Chapter 1, we compared temperatures on Celsius and Fahrenheit scales using a simple arithmetic

Raw measurement			Logarithmic data		
1st	2nd	Difference	1st	2nd	Difference
15	30	15	1.1761	1.4771	0.3010
30	60	30	1.4771	1.7781	0.3010

Table 7.20. Illustration of the effect of transforming data that are multiplicative. Each pair of raw data double between the 1st and 2nd measurement, and the differences are, consequently, not identical. The logarithmically transformed data are additive, with the differences being identical.

transformation to convert between the two scales; there is no real difference in principle between that conversion and the logarithmic transformation that we have applied here.

When we discover violations of any of the other assumptions, we can apply different transformations to the data. On occasions, we may find that more than one assumption is not satisfied. This is because some of these assumptions are inter-related. For example, if data conform to a Poisson distribution, they clearly do not satisfy the normality assumption, but, additionally, they will not satisfy assumption (c) because, recalling the description of the Poisson distribution in Chapter 3, the mean will equal the variance. Similarly, if the variances of treatment groups in a data set are proportional to the means, and the means are dissimilar, then assumptions (b) and (c) would both be violated. This inter-relationship is the reason why we can often cure more than one problem by applying a single transformation to troublesome data. In addition to the logarithmic transformation, there are two others that are commonly used: *the square root* and the *arcsine* transformation.

The square root transformation is used where group variances are proportional to their means. This will be the case for data that follow a Poisson distribution, such as counts of randomly occurring objects or events. The arcsine transformation (named as such because it uses the trigonometric arcsine function) has been found to be useful when data are in the form of proportions or percentages that often follow a binomial distribution at low or high values (i.e, <30% or >70%). In this particular transformation, the converted data are calculated by finding the arcsine of the square root of the raw percentage or proportion.

Now that we have identified the key transformations, two important questions must be addressed. How do we actually implement the transformations? And how do we report the results of analyses carried out on transformed data in a way that will be meaningful to the reader?

Implementation of each of the transformations is relatively straightforward, and this is detailed in Table 7.21. along with the appropriate Minitab commands. For the logarithmic transformation, it is important to add *1* to each raw measurement to avoid problems that would arise with zero values. For similar reasons, we add *0.5* to raw data before applying the square root transformation. In the case of the arcsine transformation, there is not a problem with zero values, but, in order to transform data using Minitab, the data must fall in the range *0* to *1*. Clearly, therefore, data in the form of percentages should be divided by *100*.

Transformation	Arithmetic conversion	Minitab command
Logarithmic	$T = \log(X+1)$	**Let C2 = LOGTEN(C1+1)**
Square root	$T = \sqrt{(X+0.5)}$	**Let C2 = SQRT(C1+0.5)**
Arcsine	$T = \arcsine\sqrt{X}$	**Let C2 = ASIN(SQRT(C1))**

Table 7.21. The main data transformations and the Minitab commands to achieve them. X denotes the raw data, and T represents the resulting transformed data. With regard to the Minitab commands, the raw data are stored in column $C1$ and the transformed data will be stored in column $C2$. For the arcsine transformation, the raw data must be within the range 0 to 1.

Reporting analyses based upon transformed data in a meaningful way provides a challenge that has to be met. However correct or sophisticated the analyses that we employ, they remain worthless until the results are communicated effectively. Providing the reader with a table of means that are logarithms or arcsines of the square roots of the raw data is hardly going to provide a clear insight into the biological system under investigation. The way forward is to transform raw data for the purposes of analysis, and then quote the raw data or the detransformed descriptive statistics (such as means, standard errors, and confidence intervals) in tables summarising the results. The appropriate formulae for detransforming data are shown in Table 7.22. In the legend of a table that relates to transformed data, it is essential to specify that the analysis was carried out on transformed data.

There are two final points to be made before leaving transformations. First, parametric techniques are generally more powerful than their nonparametric equivalents. Thus, if a suitable transformation is available to apply to data that do not satisfy an essential assumption, then it is better to use the transformation than to employ the nonparametric method. Second, the use of data transformations is not confined to questions of comparison. We will see in the next section that transformations play a very significant role in questions concern-

Transformation	Arithmetic back conversion	Minitab command
Logarithmic	$X = 10^T - 1$	**LET C3 = (10**C2) – 1**
Square root	$X = T^2 - 0.5$	**LET C3 = (C2**2) – 0.5**
Arcsine	$X = (\sin T)^2$	**LET C3 = (SINE(C2))**2**

Table 7.22. The arithmetic and Minitab methods for detransforming data using the main data transformations. In the arithmetic conversions, T denotes the transformed data, and X represents the data once they have been untransformed.

ing sequential relationships. This will be exemplified in investigations of changes in growth over time.

7.5   Exercises

1. The following data have been gathered from an investigation that was designed as a fully randomised experiment to compare the serum concentrations of IgM antibodies (mg 100 ml^{-1}) under four experimental treatments (A, B, C, and D). Analyse the data and draw appropriate conclusions. Is there any difference between the treatments? Examine the variances carefully before performing the analysis.

A	B	C	D
60.8	68.7	102.6	87.9
56.8	67.7	106.4	84.2
62.2	74.0	100.2	83.1
58.6	67.2	97.1	85.7
61.7	69.8		90.3

2. Using the censored data in Table 6.2. that relate to the retention time of a drug in the bloodstream, perform a Kruskal-Wallis test to determine if there is any difference in retention time between the three methods with which the drug is administered.

3. A randomized block design was used to compare the growth (g of dry matter) of 4 varieties of lettuce (A, B, C, and D) that were grown from seeds in a glasshouse experiment (one can assume that the effects were additive). Analyse the resulting data and draw appropriate conclusions. Did the researchers gain any advantage from using this design rather than a fully randomized design?

Block	A	B	C	D
I	6.655	7.744	7.986	8.591
II	6.897	7.260	7.623	8.228
III	5.566	6.171	6.776	8.107
IV	3.872	4.356	5.687	6.413
V	3.267	4.114	4.840	4.719

4. Re-analyse the data from Exercise 3 using the Friedman nonparametric randomized block analysis. Is there any difference between the results of the two analyses?

5. In a two-factor factorial experiment to investigate the effect of spacing between plants of three varieties of strawberry, the following quantities of strawberries (kg) were obtained. What conclusions can you draw about the effects of spacing on the three varieties?

	Spacing								
	1			2			3		
	Variety			Variety			Variety		
Replicate	A	B	C	A	B	C	A	B	C
1	10.4	8.1	10.4	10.8	13.9	11.6	14.8	12.5	12.2
2	10.3	9.8	11.4	9.6	12.6	10.4	10.5	12.7	11.7
3	9.0	8.8	10.1	10.7	12.2	11.4	11.4	13.9	11.0
4	9.2	8.6	10.5	10.8	13.0	11.4	11.2	13.3	11.9
5	9.0	9.1	10.5	9.8	13.0	12.3	13.0	12.6	12.0

6. Re-analyse the data of Exercise 5 regarding the replicates as blocks. Are there any differences in the conclusions that you can draw?

7.6   Summary of Minitab commands

ANOVA
ASIN
FRIEDMAN
INTERACTIONS PLOT
INTERVAL PLOT
KRUSKAL
LET
LOGTEN
SINE
SQRT

7.7 Summary

1. The analysis of variance (ANOVA) has been introduced that partitions the variability in a data set into different sources: those that can be identified and those that cannot (the error or residual variation).
2. The underlying principle of this family of tests is that it evaluates the relative contribution to the total variation of each source of variability. Those sources that contribute more than the residual component are said to have exerted a significant effect.
3. The ANOVA can be used to compare the effects of many levels of one or more factors.
4. Blocking is methodological technique that enables the researcher to accommodate variability in the experimental environment.
5. Repeated measures on the same subject can be analysed using similar techniques as those applied to blocks.
6. Once a significant effect has been identified by the ANOVA, other tests (like the Tukey test and its nonparametric equivalents) are performed to determine the detail of the differences between treatment groups.
7. Various nonparametric methods have been described that offer alternatives to the parametric analyses of variance.
8. Data transformations were described that can be applied to data that do not satisfy the main assumptions of the analysis of variance. They can also be applied to data that are to be analysed by other statistical methods.

PART III

Sequential Relationships

CHAPTER 8

Non-Causal and Causal Relationships

8.1 Introduction

In the discussion of problem types in Chapter 1, I used the term *sequential relationships* to describe those investigations where we measure more than one characteristic of subject material simultaneously over a range or sequence of values. The objective of such investigations into sequential relationships is to determine whether there is any relationship or *trend* between the various measured variables that holds true over their measured range. In order to illustrate what we actually mean by such a relationship, let us consider two examples.

The first example concerns the relative abundance of two plant species (A and B) in a large area of woodland; many areas were sampled and, at each site, the abundance of each species was recorded. Using the data gathered, we can ask questions like:

> Is a high abundance of Species A generally associated with high or low numbers of Species B?

The sampling units of this investigation are the sampling sites, and the attributes that we have measured are the abundances of the two species. In essence, this question is trying to establish whether a relationship exists between the abundances of the two species over a range of population sizes. If a relationship is found, then one might proceed further to identify the factors underlying the relationship, e.g., both species might prefer shady conditions.

The second example is drawn from a physiological investigation where researchers were exploring the relationship between blood pressure and the concentration of a vasopressive drug (i.e., a drug that raises blood pressure). A large number of mice were injected with varying concentrations of the drug and their resulting blood pressures were measured. The question posed in this investigation would be:

Is there a relationship between blood pressure and concentration of the drug? As we change the dose of the drug, will there be a predictable change in blood pressure?

The sampling units in this example are the individual mice, and the characteristics of the mice that are being measured are the doses given and the blood pressure responses.

Let us now compare these two questions, noting both the similarities and the differences between them. In addition, let us also consider how these questions differ from questions of comparison that were discussed in Part II.

All the measured characteristics in the two examples are continuous variables: plant abundance, drug concentration, and blood pressure. We could imagine other questions of a similar nature that might involve measurements on ordinal, interval, or ratio scales, but not nominal variables that relate to qualitative characteristics. In these questions, the attributes have been measured over a range of values. It is this characteristic that enables the data from both of these investigations to be plotted on graphs; we are essentially examining trends of some sort in both cases.

Although the two questions share some similarities; in other respects they differ markedly. The first does not have an experimental variable that has been imposed by the researcher. Two characteristics have simply been measured in a random fashion, the sample sites selected at random and the population sizes recorded. In the second example, by contrast, the different doses have been fixed by the investigators. The first question, therefore, is more of an exploration, a search for the existence of a relationship without any implicit thoughts of cause and effect. The second example attempts to characterise the extent to which the blood pressure is dependent upon the dose.

An alternative way of considering the two questions is in terms of the relative symmetry of the two measured attributes. In the plant ecology example, we could interchange Species A and Species B without altering the sense of the question. However, that is not possible in the physiological example; it would be nonsensical to consider blood pressure actually *causing* a change in drug dosage. The methods that we employ to analyse these two distinct types of relationships reflect this difference. Relationships where there is no inference of cause and effect are analysed using *correlation* methods, while those investigations into cause and effect require the use of *regression analyses*.

Before proceeding to describe these two groups of methods in detail, it is important to draw a distinction between the questions to which they are applied, those that were discussed in Part II, and also those that will be introduced in the following chapter that deal with questions of association.

Questions that can be addressed by correlation techniques are quite clearly not like any of those considered in Part II (Chapters 5, 6, and 7) because, in correlation questions, there is no obvious response variable dependent upon a causal or independent factor. If the two variables that we wished to explore were qualitative characteristics (say, gender and blood group) rather than ordinal, interval, or ratio, what sort of techniques would be appropriate? Certainly, plotting two nominal variables on the axes of a graph would not be meaningful even if it were possible; they are more appropriately displayed in tables (like those shown in Table 1.1.) with the rows and columns representing the different categories of the two variables. Questions involving nominal variables that are displayed in tables are the subject of detailed consideration in Part IV as *questions of association* and *goodness of fit*.

While the difference between the exploratory questions that do not infer causality and the questions of comparison outlined in Part II is quite plain to see, you may quite rightly ask how the dose-response example described above differs in principle from, say, the investigation into the effect of rearing method on learning in rats in Chapter 7. Rearing method is a causal factor that would seem to parallel very closely the dose, with the learning performance as the measured response. The difference between the two investigations is that the rearing method is a *qualitative* characteristic, whereas the concentration of drug is measured on a ratio scale. We cannot plot rearing method (or any other nominal data) on an axis of a graph in which we can draw a line through the points. The general rule, therefore, is that whenever we are exploring the dependence of one variable on another and the experimental or *independent* variable is nominal, we must use an appropriate analysis of variance technique.

An additional perspective to this question is provided by the *Drosophila* egg laying example that was included in Chapter 7. In that example, the independent variable, temperature, was measured on a quantitative, interval scale. Therefore, given that the temperature can be plotted on a graph, why should an analysis of variance be used rather than one of the regression techniques that will be explained later in this chapter? The answer to this question is that, in principle, the relationship between eggs laid and temperature should be viewed as a sequential relationship, but, in the specific example described in Chapter 7, the temperature was applied at only three levels (10°C, 15°C, and 20°C), which do not really provide sufficient points to plot on a graph and establish a relationship (three points could be equally indicative of a straight line or a curve). If there had been five or more levels, then perhaps the question could have been analysed using the regression methods that will be outlined in Section 8.3. This example serves to illustrate very clearly the universal maxim that no system of classification is totally absent of grey areas.

I have chosen to separate problems to be analysed by correlation and regression analyses from the other problem types because, although they share some similarities with questions of comparison and association, they are very different in other ways, and so would not sit easily within either of the other sections.

Three reasons have led me to describe the two techniques in a single chapter. First, correlation and regression problems are closely related in terms of the problems that they are used to investigate. Second, the computation and reporting methods of the two analyses are very similar. Third, because of the similarity between the computational methods, it is very easy to interchange the two techniques erroneously. Too many students have difficulties deciding whether a problem requires a correlation or regression. By considering them in the same chapter, I hope to demonstrate clearly where each technique should be used correctly, and, in so doing, reduce the amount of confusion and the number of errors in their use.

The structure of this chapter reflects this distinction between the two techniques. In Section 8.2, I will describe correlation methods to examine sequential relationships in which no causal effect is inferred. Regression methods to analyse cause and effect relationships will be introduced in Section 8.3.

8.2   Correlation

Introduction

The diversity of biological questions that can be addressed using correlation techniques is extensive. For example, physiologists might correlate metabolic rate with body temperature; anatomists examine the relationships between the lengths of groups of bones or dimensions of other body structures (such as bill length and breadth) to see if similar patterns exist in many individuals; and behavioural scientists have found correlations between dominance (as measured by ranks) and body weight in many different animal species. These examples demonstrate that questions involving correlations may involve different

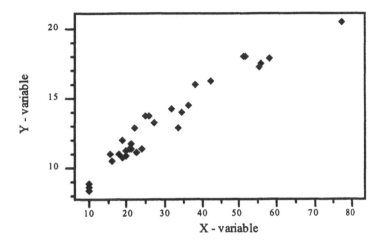

Figure 8.1. A positive correlation: low values of the X-variable are associated with low values of the Y-variable; and high values of the X-variable are associated with high values of the Y-variable.

scales of measurement: temperature (interval), lengths and weights (ratio), and ranks (ordinal). Just as with questions of comparison, distinct correlation techniques have been developed to handle the different data types. Accordingly, this section has been sub-divided into a section that describes methods that can be applied to data that conform to a normal distribution (*Correlation where normality can be assumed*), and, for those data where the normality assumption does not hold true, data transformations can be used to manipulate measurements (described in Chapter 7) or nonparametric techniques can be used (see *Correlation where normality cannot be assumed*).

The introduction to this chapter emphasised the representation of sequential relationships in a graphical format. This begs a very important question. Graphically, what does a correlation actually look like? The following series of graphs (Figures 8.1. to 8.5.) illustrate the key features of correlations. The first three figures show, respectively, relationships described by positive, negative, and no correlation.

One aspect of correlations that is a frequent source of confusion is what is actually meant by a strong correlation. The distinction is illustrated in Figures 8.4. and 8.5. in which the ellipses represent the areas of the graphs within which all the points are located. The short axis of the ellipse in Figure 8.5. is shorter than that of Figure 8.4.: the ellipse is thinner. This is indicative of a stronger correlation. To confirm this, imagine the ellipse that would surround the points in Figure 8.3.

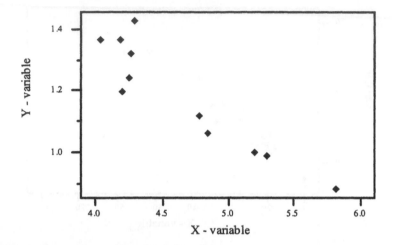

Figure 8.2. A negative correlation: low values of the X-variable are associated with high values of the Y-variable and vice versa.

where there is no correlation. The ellipse would be close to being a circle. In graphical terms, the strength of a correlation can be characterised as the narrowness of the ellipse. In the last two sub-sections, the computational and Minitab methods to evaluate correlations will be described along with examples of how we can interpret and report the results of those analyses.

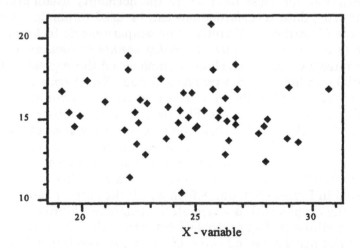

Figure 8.3. No correlation: there is no consistent relationship between values of the X- and Y-variables.

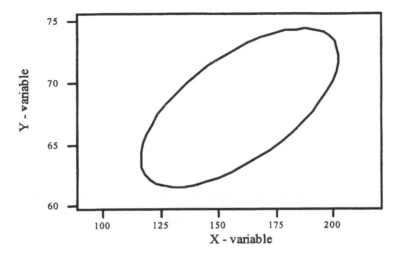

Figure 8.4. A positive correlation in which the ellipse defines the area in which all the points are located.

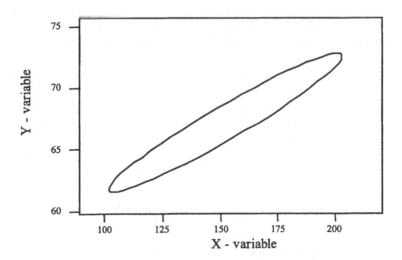

Figure 8.5. A positive correlation in which the ellipse indicates the area in which all the points are located. This ellipse is thinner than that in Figure 8.4. demonstrating that this is a stronger correlation than that illustrated in Figure 8.4.

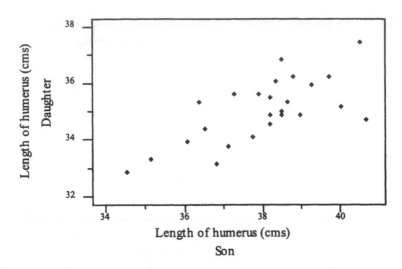

Figure 8.6. Length of humerus (cm) in sons and daughters of the same parents.

Correlation where normality can be assumed

I shall illustrate correlation techniques that can be applied to data that conform to a normal distribution by means of an example drawn from anatomy. In this example, investigators measured the lengths of the humerus bones in the upper arms of 25 pairs of sons and daughters of the same parents. The sampling units in this investigation are the sets of parents. The data for this investigation are listed in Table 8.1. They are also plotted as a scatter diagram in Figure 8.6., and we can see that there appears to be a positive correlation between the two sets of measurements.

The correlation between the two sets of measurements is described in terms of the degree to which they vary together relative to their total variation independently of each other. As this description suggests, the test statistic (denoted by r) is evaluated by a fraction:

$$r = \frac{variation\ of\ X\ and\ Y\ together}{variation\ of\ X\ and\ Y\ independently}$$

The independent variation is measured by the variances of X and Y, and we measure their joint variation by means of the *covariance*, which is calculated thus:

$$covariance = \frac{\Sigma xy - \frac{(\Sigma x)(\Sigma y)}{n}}{n-1} \qquad (8.1)$$

Contrast this equation with that of the variance (below): in the expression for the covariance, xy has replaced x^2 and $(\Sigma x)(\Sigma y)$ has replaced $(\Sigma x)^2$.

$$variance = \frac{\Sigma x^2 - \frac{(\Sigma x)^2}{n}}{n-1}$$

The formula for the parametric correlation coefficient, which is known as the *Pearson product-moment correlation coefficient*, is given by Equation 8.2.

$$r = \frac{covariance(x, y)}{\sqrt{(variance(x))(variance(y))}} \qquad (8.2)$$

In Chapter 2, we saw that the variance can only take positive values. The covariance, by contrast, can be positive or negative. Furthermore, the covariance can never be greater than the denominator of Equation 8.2. Consequently, the value of r can vary within the range -1 to $+1$. The negative extreme represents a perfect negative correlation, a value of $+1$ indicates a perfect positive correlation, and a value of zero would arise when there is no correlation between the two variables under examination. Therefore, in the example of the length of the humerus bones, we use the correlation coefficient to test a null hypothesis like:

H_N: There is no correlation between the lengths of the humerus bones in brothers and sisters.

With the complementary alternative hypothesis being:

H_A: There is a correlation between the lengths of the humerus bones in brothers and sisters.

These hypotheses are two-tailed; we would reject the null hypothesis if there was either a positive or negative correlation.

Son	Daughter	Son	Daughter
23.62	27.28	24.23	29.72
22.71	30.63	23.01	28.49
22.55	28.19	20.88	24.54
23.32	28.65	23.62	27.89
21.95	26.06	23.32	26.37
23.93	29.26	22.10	27.74
22.86	28.96	21.34	25.15
24.23	28.80	23.47	28.19
23.17	30.02	21.79	27.13
22.86	28.50	21.18	26.82
24.08	28.35	25.45	30.48
22.40	26.52	24.84	28.50
22.86	28.19		

Table 8.1. Length of humerus (cm) of 25 pairs of sons and daughters.

The calculation of the correlation coefficient is detailed at the end of this section in Box 8.1., where the value of r is given as +0.693. The statistical significance of this result is determined using the table of critical values for the correlation coefficient that are included in Appendix A.12. The degrees of freedom associated with the test statistic, r, are given by the expression below (Equation 8.3) where the subjects are the sets of parents (the sampling units):

$$d.f. \text{ for correlation coefficient} = no. \text{ of subjects} - 2 \qquad (8.3)$$

So, in this anatomical example, we could summarise the results as:

A Pearson product-moment correlation coefficient (*reference*) demonstrated a significant positive correlation between the lengths humerus bones in pairs of brothers and sisters (r = 0.693; d.f. = 23; P < 0.001).

Since this statistical technique is exploratory and does not make any reference in its calculation to a causal relationship, one must be very careful in the use of the term *significant* in the context of correlation. Student reports and the wider scientific literature are littered with discussions of totally spurious correlations, so think very carefully about what you write in connection with correlation. A statistically significant correlation result implies only that two variables vary together; there is no implication that one depends upon the other, and, like all statistical results, they should be considered in the discussion section of a report where the biological significance should be evaluated in the context of the biological problem and the existing literature.

The dialog box in Minitab where the correlation functions are defined is found by the following menu route:

STAT > Basic Statistics > Correlation

The dialog box is illustrated in Exhibit 8.1.

The Minitab output for the correlation is shown in Exhibit 8.2. Note that the number of subjects is not included in the output. Consequently, for the purposes of reporting, a count of the number of

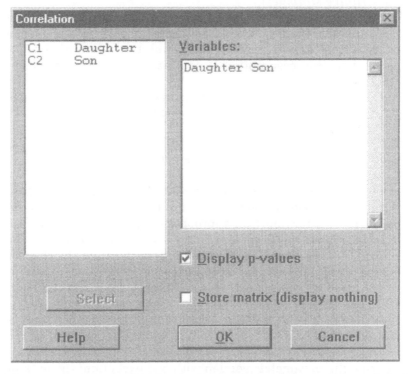

Exhibit 8.1. The Minitab dialog box for defining the correlation coefficient calculations for the humerus length example (data in Table 8.1.). (The *Store matrix* option is used to store correlation coefficients ready for multivariate statistical analysis, which are not covered in this book.)

```
MTB > Correlation 'Daughter' 'Son'.

Correlations (Pearson)


Correlation of Daughter and Son = 0.693, P-Value = 0.000

MTB >
```

Exhibit 8.2. The Minitab output of the correlation calculation for the humerus example showing that there is a statistically significant correlation between lengths of the humerus of brothers and sisters sharing the same parents.

sampling units would have to be obtained using another command (e.g., *descriptive statistics*).

The assumption implicit in the proper application of the Pearson product-moment correlation coefficient is that both sets of measurements have been drawn from normal populations. The methods that are used when this assumption is not satisfied are described in the next section.

The calculations of the correlation coefficient for the humerus length example are shown below in Box 8.1.

Correlation where normality cannot be assumed

When we are required to correlate two sets of measurements that have been drawn from populations that do not conform to a normal distribution, we can either apply an appropriate data transformation (such as those described in Chapter 7), or we can use a non-parametric technique. The one that I shall describe here is called the Spearman correlation coefficient and its test statistic is denoted by r_s (the subscript s is used to distinguish this coefficient from that of Pearson described in the previous section).

Spearman's rank correlation coefficient is based upon the same principles as that of Pearson's parametric correlation, and, as a consequence, the calculation of this non-parametric coefficient is exactly the same as the parametric version with one exception: the data must be converted into ranks. Since the two calculations are identical, I have not included a worked example. However, I would suggest that you rank the humerus length data in Table 8.1. and then calculate the

1. Calculate the covariance using Equation 8.1.:

$$\frac{1610.26 - \dfrac{(575.77)(700.44)}{25}}{25-1} = 1.19$$

2. Calculate the variance of the humerus lengths for the sons:

$$\frac{19680.75 - \dfrac{(700.44)^2}{25}}{25-1} = 2.34$$

3. Calculate the variance of the humerus lengths for the daughters:

$$\frac{13290.75 - \dfrac{(575.77)^2}{25}}{25-1} = 1.26$$

4. Calculate the correlation coefficient, r according to Equation 8.2.:

$$r = \frac{1.19}{\sqrt{(2.34)(1.26)}} = 0.693$$

5. With $n - 2$ degrees of freedom (i.e., 23), the probability of obtaining this result if the null hypothesis is true is less than 0.001 (using critical values for r in Appendix A.12.).

Box 8.1. Calculation of the Pearson product-moment correlation coefficient for the humerus length data (Table 8.1.).

Spearman coefficient* in order to familiarise yourself with the method. Minitab includes a ranking command that can be used to convert the measurements into a ranked form before using the correlation command illustrated in Exhibit 8.1. The rank command is found by the menu route:

MANIP > Rank

8.3   Regression

Introduction

In the introduction to this chapter, I drew the distinction between those sequential relationships in which cause and effect could be inferred, and those in which it could not. The latter group, along with the techniques for their analysis, were described in Section 8.2. The analysis of those relationships in which one variable is dependent on one or more other variables is the subject of this section.

There is an enormous variety of sequential relationships in which one or more factors may elicit a response in another. Just as with questions that can be addressed by analysis of variance techniques, it is impossible to describe in detail every possible type of problem and its solution within the confines of a single book. Therefore, this section provides an introduction to causal sequential relationships and the regression techniques that can be used for their analysis. However, even though there are many different types of questions that require analysis using regression methods, several principles are common to all.

I highlighted in Section 8.1 the asymmetry of regression questions by means of an example concerning the effect of a drug on blood pressure. Although we could imagine the drug dosage affecting blood pressure, the reverse would not be possible; blood pressure changes would not directly cause a change in the dose of the drug. This asymmetry in the relationship is reflected in the conventions concerning the plotting of graphs that describe causal relationships. The causal factor is *always* plotted on the x-axis (horizontal), and the response variable is always plotted on the y-axis (vertical). This asymmetry is also a feature of the calculations involved in regression analyses. Therefore, it is extremely important in any investigation to recognise the cause and effect variables correctly. While this may sound very

* The calculation gives a value of $r_s = 0.620$.

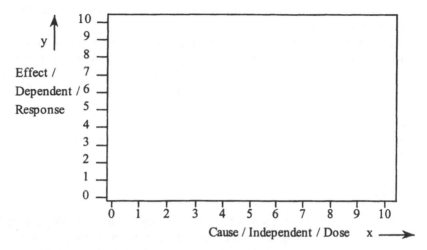

Figure 8.7. The x- and y-axes of a graph with the various pairs of terms that are often used to describe the variables that have been plotted.

obvious, I once encountered a student whose graphs and analysis seemed to suggest that the amount of water drunk by goats somehow caused changes in ambient temperature. In Figure 8.7. you will see the various terms that are often used to denote the cause and effect variables.

In general, the questions answered by regression analyses are of a form similar to:

> Does variable X have an effect on variable Y and, if so, how can it be characterised?

If we can answer this question, we will be in a position to apply that description in a predictive manner. For example, in the context of the effect of a drug on blood pressure, we could ask the further question:

> What will be the response in terms of blood pressure change if we increase the dose of drug X by a specified number of units?

In short, we are trying to characterise the relationship in such a way that we can *predict* a response. The way that this is achieved is by the analysis technique constructing a line that best fits the data, and, since every line can be described by an equation, determining the equation that describes the line of best fit. The central principles of regression analysis are the establishment that a relationship exists and the characterisation of such a relationship.

I have divided this discussion of regression techniques into two sections. In the following section, entitled *Simple linear relationships*, I shall describe the techniques that are employed to characterise simple relationships that can be described by straight lines: simple linear relationships. More complex relationships will be examined in the next section, entitled *More complex relationships*. These include situations where the relationship is not linear or where more than one independent (causal) factor is involved.

Simple linear relationships

Simple linear relationships are found in many areas of biology, and I shall describe an example in this section in which the effect of drug dosage on blood pressure will be examined in detail. Before proceeding to examine this example, we need to understand how an equation can describe a line, because this basic principle underpins the whole technique of regression analysis.

A line on a graph has two basic properties: its gradient (e.g., the change in blood pressure caused by a unit increase in drug dosage) and the location of the line with respect to the axes of the graph. In the case of a straight line (the simplest form of line), these two properties are relatively easy to specify. Since the line is straight, the gradient will be a constant value (unlike the gradient of a curve which will vary). The location of the line can be fixed by identifying the point at which it crosses the y-axis. These two properties are illustrated in Figure 8.8.

We can express these two properties in terms of the x and y variables in the form of an equation:

$$y = a + bx \qquad (8.4)$$

where

a = the intercept

b = the gradient (change in the dependent variable as a consequence of an increase of one unit of the independent variable)

x = the independent variable (e.g., drug dose)

y = the dependent variable (e.g., blood pressure response)

This particular equation describes (or *models*) a linear relationship. In the context of a simple linear relationship, the function performed

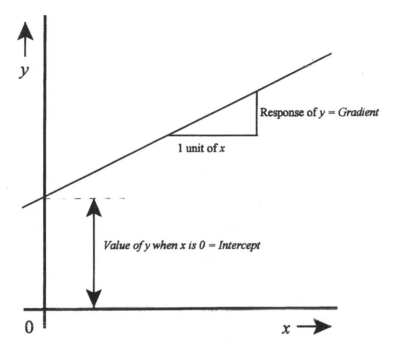

Figure 8.8. A straight line graph is characterised by its gradient and intercept.

by a regression analysis is to determine the values of the intercept and gradient (*a* and *b*); the equation resulting from a regression analysis is often referred to as a *regression equation*. Once we have established that a particular model accurately describes a relationship, we can use it to predict the way that the *y*-variable will respond to changes in the *x*-variable by substituting values of *x* into the equation and calculating the expected response in *y*. However, before we can use the regression predictively, we must answer two fundamental questions:

1. Is the model a close enough fit to be useful? Although we instruct regression analysis to fit the line of best fit through the data, there may be no relationship between the variables, or the data may be very variable. In such circumstances, the fit may be the best, but it might still not be very good, and any predictions based upon such a model would be inadequate.
2. Is the model that we have constructed the correct one? For example, if we have fitted a straight line through the data,

we must confirm that this type of line, rather than a curve, is the most appropriate.

The procedures that we employ to validate that a regression model is both appropriate and of adequate quality are, therefore, extremely important. Regression analysis is not simply a matter of performing a series of calculations in order to obtain a test statistic; we will see that the task of validating a model takes about the same amount of effort as developing the model itself.

The prospect of constructing and validating mathematical models to describe our data may strike a note of fear in the minds of many, but, in fact, the tasks are easily accomplished using a computer. Moreover, most of the validation procedures simply require us to identify patterns in graphs. Since the whole process involves many different stages (both arithmetical and graphical), I have focussed entirely on the Minitab procedures, rather than including fully worked calculations. In so doing, I hope that attention can be directed toward the analysis of biological questions using the regression process, rather than the computation detail. The best way to understand these procedures is by seeing them in action. So let us consider a study in which the effect of drug dosage on diastolic blood pressure was investigated.

In this experiment, many subjects were given different doses of a drug, and the data for two subjects are listed in Table 8.2. and plotted in Figure 8.9. At first sight, there would seem to be a steady increase in blood pressure as the dosage increases. The analysis will enable us to determine whether or not our eyes are being deceived.

How do we actually set about producing an equation like Equation 8.4 that describes the relationship illustrated in Figure 8.9? And how do we answer the last two key questions regarding the model?

The first of those questions can be sub-divided into two further questions, one concerning the gradient and the other relating to the intercept:

a.   Does the gradient differ from 0? In other words, if drug dose changes, is there any response in blood pressure?

b.   Does the intercept differ from 0? If there is no drug, can we expect blood pressure to differ from 0? Of course, in the context of this particular investigation, we would definitely expect a positive pressure, but in other contexts the answer to this question may not be so obvious.

Blood pressure (mm Hg)	Dose (units)	Blood pressure (mm Hg)	Dose (units)
91	1	111	8
92	1	114	9
95	2	115	9
94	2	118	10
97	3	118	10
96	3	120	11
99	4	120	11
101	4	124	12
102	5	123	12
103	5	127	13
106	6	126	13
107	6	130	14
109	7	129	14
109	7	132	15
112	8	132	15

Table 8.2. Diastolic blood pressure responses to changes in drug dosage.

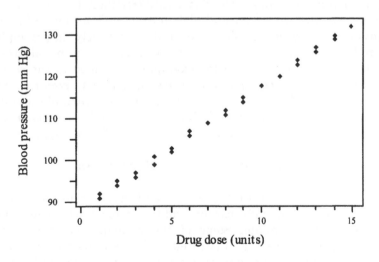

Figure 8.9. The effect of drug dose on blood pressure (data in Table 8.2.).

These two questions can be translated into two testable null hypotheses and their complementary alternative hypotheses:

H_N: The gradient does not differ from zero.

H_A: The gradient does differ from zero.

H_N: The intercept does not differ from zero.

H_A: The intercept does differ from zero.

Regression analysis enables us to test these null hypotheses and will also inform us of the magnitude of the gradient and intercept which can then be used predictively. The way that regression analysis determines the line that best fits the data is illustrated in Figure 8.10. In that diagram, we can see six observations with a line fitted through them. The line has been fitted by minimising the vertical distances between each point and the line (a process called least-squares regression). The differences between the observed values and those predicted by the model (i.e., points on the line) are known as *residuals*, and we shall see that they play a fundamental role in the process of confirming that the correct model has been fitted (i.e., they help us to answer our previous second question regarding the constructed model). Furthermore, like all the other statistical tests, regression analysis makes certain assumptions about the data. In particular, the residuals are expected to conform to a normal distribution with a mean of zero.

Having considered the objectives and the underlying principles of this analysis, let us see it in action as we analyse the blood pressure data. Just as with the correlation examples described in the sections, *Correlation where normality can be assumed* and *Correlation where normality cannot be assumed*, the data for each variable should be stored in separate columns within Minitab. The main regression dialog box is shown in Exhibit 8.3. It is accessed by the following menu route:

STAT > Regression > Regression

The main regression dialog box is relatively simple, but the four buttons at the bottom right corner lead to further boxes where important selections are made. Some of the principal choices that are selected in these boxes are illustrated in the exhibits that follow (8.4. to 8.6.). They enable us to generate all the data we require to produce a regression equation, validate that it is correct, and be able to judge its usefulness.

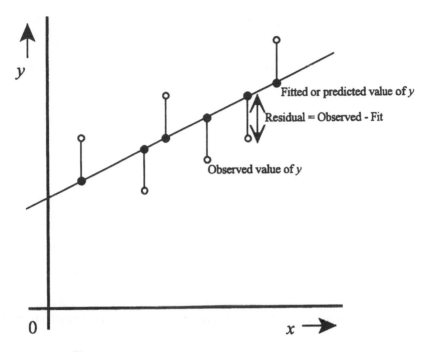

Figure 8.10. The line of best fit constructed by the regression analysis minimises the distance between the observed measurements of the y-variable (o) and the values predicted by the regression line (●). The *residual* value is the difference between the observed and fitted values. The residual values are central to the validation process.

The *Graphs* button leads to a dialog box (illustrated in Exhibit 8.4.) in which one can select various graphs that can be plotted. By examining these graphs, we can validate the regression line (by confirming that we have fitted the correct shape of line to the data) and also check that the normality assumption is satisfied by the residuals. There are two main elements to the choices available in this dialog box. First, we choose which form of the residuals to use, and, in Exhibit 8.4., I have selected <u>S</u>tandardized rather than the other two options because they facilitate very simple interpretations of the graphs that we use to validate the regression relationship. The standardized residuals are calculated by dividing each residual value by the standard deviation of all the residuals, which, if the regression model is good, should provide a set of standardized residuals that are normally distributed around a mean of 0 and with a standard deviation of one. The standardized residuals can be used under most circumstances. The way we use these graphs will become apparent when we examine the

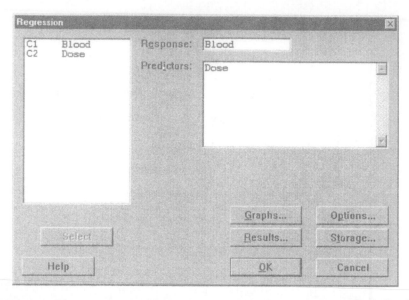

Exhibit 8.3. The main Minitab dialog box to specify the regression analysis for the blood pressure example (data in Table 8.2.). Further details of the specification are given in the text.

graphical output. Second, the choice of graphs to plot is very important. In this example, I have chosen three of the four available graphs. The last one, *Residuals versus order,* would help to reveal problems with the regression relationship that relate to the sequence in which the measurements were made. For example, if the measurements had been made over time, the regression line might be a very good fit for the early observations, but not fit very well later in the sequence.

The *Results* button leads to options relating to the regression output that Minitab will produce. The default option is illustrated in Exhibit 8.5., and, for most purposes, this will be perfectly adequate.

The most important choice available in the *Options* dialog box enables us to choose whether the intercept should be calculated by the regression process (the default option, as illustrated in Exhibit 8.6.) or whether the line should be forced through the origin (i.e., the intercept will be fixed as 0). We might take this latter option if we have theoretical or independent information that indicates that the line should pass through the origin; in such circumstances we would deselect the *Fit intercept* option. The predictive capability of regression can be exploited by specifying values of the independent variable (or columns in which several values are stored) in the box labelled *Pre-*

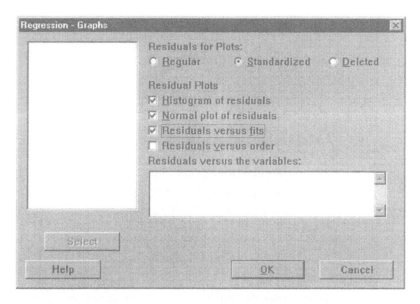

Exhibit 8.4. Minitab dialog box to select the graphs that will enable the regression relationship to be validated and the assumption of normality of the residuals to be tested.

diction intervals ... Inserting values in this box will cause Minitab to estimate the response of the dependent variable and also generate 95% prediction intervals that provide an indication of the precision of the estimation. We will see this option in operation later in this section.

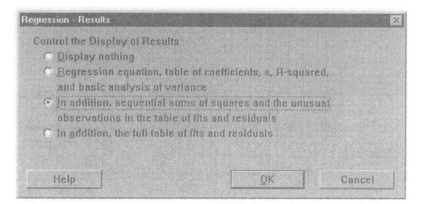

Exhibit 8.5. Dialog box to specify the detail of the regression results to be produced by Minitab; the default option that is illustrated is adequate for most circumstances.

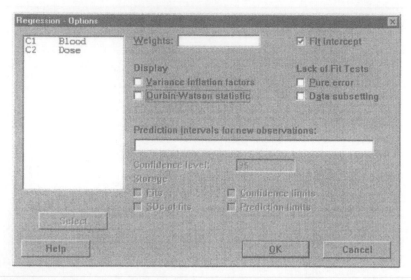

Exhibit 8.6. Regression options that can be selected. In particular, one can choose to force the line of best fit through the origin, or, as the default setting shown here, have Minitab estimate the intercept from the observed data. One can also specify additional values of the independent variable in the box labeled *Prediction intervals ...* for which estimated responses and 95% prediction intervals will be calculated.

The *Storage* button leads to a dialog box in which one can choose to store certain variables (such as the fits and residuals); however, with the graph plotting options that are available (Exhibit 8.4.), storage is not really necessary under many circumstances.

A further option that Minitab provides is to produce a graph in which the line of best fit is plotted. This dialog box is shown in Exhibit 8.7. and the graph that is produced is illustrated in Figure 8.11. The dialog box illustrated in Exhibit 8.7. requires the user to specify the response and predictor variables, and also to choose the type of line that should be fitted. There are three choices: linear (a straight line), quadratic, and cubic. The latter two options are curved and will be described in more detail in the section that follows (*More complex relationships*). This option is found via the menu route:

STAT > Regression > Fitted Line Plot

As a first stage in the interpretation of the outcomes of the regression process, one should make a simple, visual inspection of the fit of the line. Does it actually look like a reasonable fit? As with all analyses,

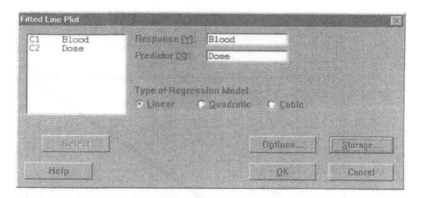

Exhibit 8.7. The Minitab dialog box to produce a fitted line through a data set.

if we have an idea of the answer from a picture, then interpretation of the quantitative analysis will be made much easier than blindly feeling our way without any assistance. In the context of this example of blood pressure and drug dosage, Figure 8.11. seems to indicate a very close fit indeed, but, in order to make an objective assessment, we must consider the results of the regression analysis.

Our examination of the analytical results should commence with a confirmation that the residuals are distributed normally. This is accomplished by inspecting two of the graphs selected in the dialog box illustrated in Exhibit 8.4.; the histogram of the residuals (Figure 8.12.) and the normal probability plot of the residuals (Figure 8.13.). In the first of these we would expect a typical bell-shaped histogram with a mean of zero. As we can see, Figure 8.12. shows a histogram that does not quite conform to the conventional shape. Although the centre of the histogram seems to be at zero, there is a dip at that point. However, the number of measurements in this data set is not great, and the dip is only four below the peak value. Other than the dip, the general appearance of the histogram does seem to be bell-shaped. Therefore, I would conclude that this dip is more a consequence of the small data set rather than a critical problem with the data. I would not say that there was any substantive evidence of non-normality in this histogram.

The second graph that assists us in checking the normality assumption is shown in Figure 8.13. Put simply, we look for a straight line to indicate that the residuals conform to a normal distribution. Here, the graph indicates that the residuals from the blood pressure data do satisfy the normality assumption. If there are any deviations from a straight line, there may be a problem with normality. However, since

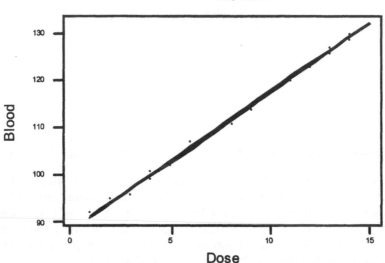

Regression Plot

Y = 88.3476 + 2.92321X
R-Sq = 99.8 %

Figure 8.11. Straight line fitted through the blood pressure/drug dose data shown in Table 8.2.

we would expect that most residuals will be concentrated in the centre of the range (i.e., in the middle of the graph), the few points at the extremes may appear to wander off the straight line. This should not be too much of a problem, but we shall see when we discuss the interpretation and use of the results of regression analysis that we should be cautious in the way that we use the predictive capabilities of regression at the extremes of the range of measurements.

After concluding that the residuals are distributed normally, we must check that the line that we have fitted is correct, i.e., that we are justified in fitting a straight line in this particular example. This is achieved by examining Figure 8.14., where the residuals are plotted against the fitted values. This scatter diagram can help us to identify a large number of potential problems, but, even though it is a very powerful tool, it is *very simple* to use: our task is simply to look for patterns. If we see a pattern of any sort, there may be a problem with the quality of the fit of the regression line to the data.

The three main types of patterns to note are:

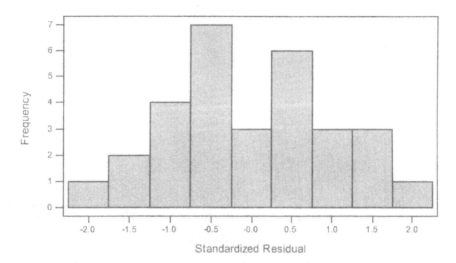

Figure 8.12. Histogram of the residuals, indicating that they conform to a normal distribution. The dip in the center is probably a consequence of the small size of the data set rather than indicating a problem with the data.

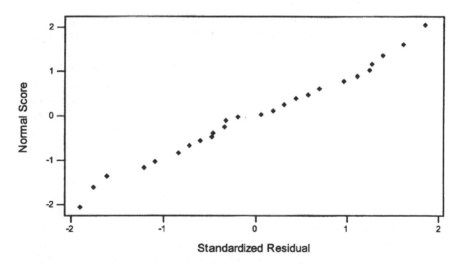

Figure 8.13. A plot of the normal scores of the standardized residuals of the blood pressure data (Table 8.2.). If the residuals conform to a normal distribution, the graph should be very close to a straight line, as they are here.

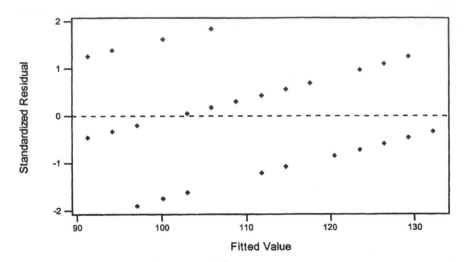

Figure 8.14. Plot of standardized residuals against fitted values for the blood pressure data. There is no systematic pattern and the residuals fall within the range −2 to +2.

1. Residuals should show a random scatter between −2 and +2, which is basically a confidence interval on either side of the mean of 0. This range is a consequence of using standardized residuals and would vary from study to study if the raw residuals were used. If there is a large proportion of points outside this range, this may be indicative of a very poor fit of the line to the data. Perhaps there is no real relationship between the dependent and independent variables. This will be confirmed by inspecting the quantitative output of the regression analysis (Exhibit 8.8.). In this particular example, there is no problem. The data are all within the defined range with no systematic pattern to the scatter.

2. If points are clustered more tightly at one of the range of fitted values than the other, this is indicative of the line fitting the data better at one end than the other. This could mean that the relationship is only linear over part of the measured range of the independent variable. For example, in population biology, there may be linear growth in the population over a period of time, but, after a while, growth might decline to a steady state.

3. If the pattern is U-shaped or an inverted U, this indicates that a straight line has been fitted to a curvilinear rela-

tionship. The greater usefulness of the standardized residuals compared to the raw residuals is of particular
importance in this context. If we erroneously fit a straight
line through a data set that is following the slightest curvilinear relationship (perhaps so slightly that it is not
perceptible to the naked eye), the standardized residuals
plotted against the fitted values will show a dramatic and
unmistakable U-shaped pattern (this is illustrated in Figure 8.16. later in this chapter).

Having confirmed that we have fitted the appropriate line through
the measurements, we are now in a position to examine the results of
the quantitative analysis of the relationship. The Minitab output for
the blood pressure investigation is shown in Exhibit 8.8. There are
several components in this output. In addition to the regression equation itself, the various elements inform us whether the model is a close
enough fit to be useful, and they also provide us with the actual
magnitudes of the gradient and intercept.

The first component of the output is the regression equation that
estimates the intercept to be 88.3 and the gradient to be 2.92. If, after
inspecting the other elements, we decide that the line fits the data
closely enough, we can make statements about the relationship
between blood pressure and drug dose based upon this equation.

The second part of the output provides statistical evidence of
whether the intercept and gradient differ from 0. If the gradient does
differ significantly from 0, then we can conclude that blood pressure
is dependent upon drug dosage. The statistical tests being used here
are t-tests to compare a single value (the intercept or gradient) against
a hypothesised value of 0. We can see that both the intercept and
gradient are restated more precisely (under the *Coef* heading) along
with their respective standard deviations, t-values and probabilities.
These results indicate that both the intercept and gradient differ
significantly from 0, confirming the visual evidence shown earlier in
Figure 8.11. If the gradient did not differ from 0, this would indicate
that the independent variable has had no effect on the dependent
variable; there would be no evidence of a relationship.

At the bottom of the output, there is an analysis of variance table
that restates the same information as that provided by the t-test
relating to the gradient, namely that there is a significant dose effect
on blood pressure. As a statistical aside, this output illustrates the
relationship between the t- and F-distributions; the value of t (112.62)
is equal to the square root of F (12682.44).

```
MTB > Name c3 = 'FITS1' c4 = 'SRES1'
MTB > Regress 'Blood' 1 'Dose';
SUBC>    Fits 'FITS1';
SUBC>    SResiduals 'SRES1';
SUBC>   GHistogram;
SUBC>   GNormalplot;
SUBC>   GFits;
SUBC>   GOrder;
SUBC>   RType 2;
SUBC>    Constant;
SUBC>    Brief 2.
```

Regression Analysis

```
The regression equation is
Blood = 88.3 + 2.92 Dose

Predictor       Coef        StDev          T          P
Constant     88.3476       0.2360     374.34      0.000
Dose         2.92321       0.02596    112.62      0.000


S = 0.6143      R-Sq = 99.8%      R-Sq(adj) = 99.8%

Analysis of Variance

Source          DF          SS         MS          F          P
Regression       1       4785.3     4785.3   12682.44      0.000
Residual Error  28         10.6        0.4
Total           29       4795.9


MTB >
```

Exhibit 8.8. Minitab output of the regression analysis of blood pressure response to drug dosage (data in Table 8.2.).

Between the details of the magnitude of the intercept and gradient and the analysis of variance is a line of further information, the most important of which is that labelled as *R-sq*. This is the *coefficient of determination,* which is a measure of the proportion of variation in blood pressure explained by the independent variable (drug dose in this example). It is, in fact, the square of the correlation coefficient (hence the label R^2), and is generally expressed in the form of a percentage. Clearly, in this example, the effect of drug dosage is the predominant factor affecting the blood pressure of the subjects in the experiment because 99.8% of the variation in blood pressure has been attributed to the effect of the independent variable.

Regression analysis is clearly not as straightforward as the analyses described in previous chapters; there are many separate stages to the process if it is performed thoroughly. Thankfully, the arrival of good quality statistical packages means that we are now able to be sufficiently thorough with minimal effort. However, having reached the end of the analysis, how will we report such a diverse collection of results? What should we include?

We must provide sufficient information to convey the biological meaning, but we must ensure that we do not overload the reader with statistical output that masks the biology. Remember, the principal objective of the report is to inform the reader of the biological signif-icance of the outcomes, not to provide extensive evidence of our capa-bility to perform regression analysis. The essential details to be trans-mitted are the regression equation (if it is statistically significant) and an indication of the quality of the fit. This can be achieved by a paragraph such as:

A linear regression analysis was performed of blood pressure on drug dose (*reference to the statistics package*) that indicated a significant positive effect of dose ($F = 12682.44$; d.f. = 1.28; $P < 0.001$; $R^2 = 99.8\%$) as described by Equation 8.5 and shown in Figure 8.11.

$$blood\ pressure = 88.3 + 2.92\ dose \qquad (8.5)$$

Although it is very frustrating to reduce a great deal of intellectual effort into such a brief summary, one should take pride in being able to convey the message of this relationship so succinctly. But what are this equation and the associated results actually saying biologically? The evidence of these results indicates that, in the absence of the drug, the base diastolic blood pressure will be 88.3 mm Hg and, with every additional unit of drug that is administered, we can expect an increase in blood pressure of 2.92 mm Hg.

As I stated earlier, a major opportunity provided by regression analysis is the predictive capability that comes with the regression equation; we can substitute a value for the independent variable and calculate the expected response in the dependent variable. For exam-ple, if we administered a dose of 6.5 units of the drug, we would expect a blood pressure of 107.28 mm Hg (calculated by substituting 6.5 into the regression equation*). Such calculations are straightforward oper-ations, but they only provide *estimates* of the response, not exact values. For example, if we repeatedly applied 6.5 units of the drug,

*$88.3 + (2.92)(6.5) = 107.28$

```
Predicted Values

    Fit  StDev Fit         95.0% CI           95.0% PI
107.349      0.119   ( 107.105, 107.592)  ( 106.067, 108.630)
```

Exhibit 8.9. The extra information that would be included in regression output if the response to 6.5 units of drug had been requested in the *Options* dialog box (see Exhibit 8.6.). The 95% prediction intervals would be used to gauge the precision of the estimate.

we would not expect a response of exactly 107.28 mm Hg on every occasion, but a mean response of that value. The precision of that estimate depends upon the quality with which the regression equation fits the data. Ideally, it would be more useful if we were able to provide the estimated mean response along with a confidence interval so that anyone using the estimate might have some guide to its quality. Minitab provides the opportunity to estimate responses with 95% prediction intervals through the *Options* dialog box shown in Exhibit 8.6. One can enter actual values of the independent variable or a column in which several values are stored into the box labelled *Prediction intervals for new observations*. The output of the regression analysis would be supplemented by that shown in Exhibit 8.9. The estimated response to a dose of 6.5 units is 107.349 mm Hg with a 95% prediction interval of 106.067 − 108.630.

The predictive qualities of regression equations are the bases for modelling many biological systems. Although they provide a very useful and powerful set of tools, they also have their limitations, and one has to interpret and use their results with great care. For example, we might record the heights (lengths) of young babies from birth to, say, six months of age and use these data to construct a regression equation. If we then use the equation to estimate the height at 18 years of age, the predictions would belong more to Brobdingnag, the land of giants in *Gulliver's Travels*, than reality! Regression equations should only be used to predict responses over the range of measurements of the independent variable that was used to develop the model. Extrapolation of the line of best fit beyond the measured range of the predictor variables is very dangerous indeed.

In this section, we have seen how to develop predictive models of linear relationships using simple linear regression analysis. The principles used in developing these simple models apply equally well to more complex models. In the next section, we shall see how to address relationships that are either not linear or that involve more than one independent variable.

More complex relationships

Biological relationships may be relatively simple, as we saw in both the previous section and in various examples in earlier chapters, but some are more complex. For example, in Section 7.4, we saw that the early growth of tissues, body size, and populations is generally related to time in a multiplicative, rather than, additive manner. We also saw that this could be corrected by applying a logarithmic transformation. Since this transformation is used so often in biology, it is worth devoting a little time to seeing how it actually works by considering a simple example of population growth. In this example, the size of a population at a given time, t, is represented by N. The population size is related to: the initial population size, N_0 (at time $t = 0$); a species specific constant known as the intrinsic rate of natural increase, which is denoted by r; and time. This relationship is described mathematically by Equation 8.6 in which e is the exponential function: a constant that is the base of natural logarithms.

$$N = N_0 e^{rt} \qquad (8.6)$$

The fact that N is the result of the product of N_0 and e^{rt} gives rise to the multiplicative nature of the relationship. If we take the natural logarithms (denoted by the symbol ln) of both sides of that equation (i.e., apply a logarithmic transformation to the data), we obtain the equation below, which describes a straight line of lnN on the y-axis and t on the x-axis with an intercept of lnN_0 and a gradient of r:

$$lnN = lnN_0 + rt \qquad (8.7)$$

The natural logarithm of e^{rt} becomes rt because natural logarithms are based on e.

In a similar fashion, the relative growth rates of tissues, bones, or organs in a body are often described by a power function like:

$$y = ax^b \qquad (8.8)$$

where, y and x are the weights of the two tissues and a and b are constants.

This is often called an allometric growth function and, like the growth relationship described by Equation 8.6, this function is multiplicative and can be linearised by applying a logarithmic transformation, giving

$$\log(y) = \log(a) + b \log(x) \tag{8.9}$$

that describes the straight line graph of $\log(y)$ against $\log(x)$ with an intercept of $\log(a)$ and a gradient of b.

Multiplicative relationships, such as those described by Equations 8.6 and 8.8, are curvilinear, but have been transformed into straight line relationships that can be analysed by the simple regression analyses described in the last section entitled *Simple linear relationships*. The areas of population biology, microbiology (particularly bacterial growth), and developmental biology are full of examples of such relationships.

Another significant example of a transformation being used to linearise a relationship is found in the area of enzyme kinetics. It is known that the rate of conversion of a substrate by an enzyme is often described by the following equation (known as the Michaelis-Menten equation):

$$V = \frac{V_{max}S}{K_M + S}$$

where,

V is the velocity of the reaction and V_{max} is the maximum velocity

S is the concentration of the substrate being converted by the enzyme

K_M is the Michaelis constant which is specific to a particular enzyme-substrate reaction.

The relationship described by this equation is curvilinear, but we can transform it into a straight line by the performing the following manipulation, and then, by simple linear regression analysis, evaluating the constants K_M and V_{max}:

$$V = \frac{V_{max}S}{K_M + S}$$

$$and, by inversion \quad \frac{1}{V} = \frac{K_M + S}{V_{max}S} = \frac{K_M}{V_{max}S} + \frac{S}{V_{max}S}$$

$$= \frac{K_M}{V_{max}S} + \frac{1}{V_{max}}$$

$$\therefore \frac{1}{V} = \frac{K_M}{V_{max}} \cdot \frac{1}{S} + \frac{1}{V_{max}}$$

If we plot $\frac{1}{V}$ on the y-axis against $\frac{1}{S}$ on the x-axis, there will be a straight line with a gradient of $\frac{K_M}{V_{max}}$ and an intercept of $\frac{1}{V_{max}}$. The constants, K_M and V_{max} can be evaluated from the output of the regression analysis. This particular plot is known as a Lineweaver-Burk plot.

The key point to note in these examples is the general form of the equation of a straight line:

$$y = a + bx$$

with two variables (x and y) and two constants (a and b). In any investigation, if theory predicts a relationship that can be manipulated into this form, then we can apply simple linear regression analysis to the transformed data. Therefore, biologists should endeavour to master simple algebra in order to be able to manipulate equations for this purpose. In all these examples of transforming data to obtain a linear relationship, the resultant data can be expressed in their original form after analysis. For example, rather than quoting the logarithmic value of an intercept, we can back transform the values just as I described in Section 7.4.

The problem presented by curvilinear relationships is relatively straightforward to solve when we know the nature of the relationship (as in the growth and enzyme examples cited above) because we know what transformation to apply. However, how do we cope with curvilinear relationships when there is no external information to direct us

Independent	Dependent
1.5903	2.456
2.1248	4.382
3.2127	8.295
3.3282	9.755
7.6283	57.192
7.8817	62.118
8.7492	76.741
15.4024	238.531
18.7772	352.771
19.7366	391.790
19.9103	397.831
22.3058	495.822
23.5668	555.135
24.9146	621.105
29.9603	898.359

Table 8.3. Artificial data to illustrate
the development of a regression model
for a curvilinear relationship.

to a particular model? The simple answer is that we have to find a
suitable model by a process of trial and error. Although this may seem
a rather unstructured or chaotic way of conducting scientific research,
it is often not too difficult to find a solution fairly quickly. There are
a number of factors that assist us in the process. First, many relationships can be identified by using one of a small number of transformations (such as logarithmic, square, and reciprocal). Second, the graph-
based validation process described in the preceding section provides a
very rapid test of any potential model. We simply try out the various
transformations and check the quality of the fit until we find a model
that satisfies the criteria set out in the last section, entitled *Simple
linear relationships*. As an example, consider the data that I have
constructed specifically to illustrate this process (Table 8.3.); they are
also plotted in Figure 8.15. and clearly show a curvilinear relationship.

Having fitted a straight line to the data, the *standardised residuals
vs. fits* graph (Figure 8.16.) is U-shaped indicating very clearly that a
straight line has been fitted to a curvilinear relationship. Note that
the gradient of the residuals plot is far steeper than that of the plot
of the original raw data, demonstrating the effectiveness of the residual plot to indicate that an inappropriate model has been fitted to the
data.

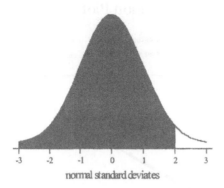

Figure 8.15. Plot of the data shown in Table 8.3. It is clearly a curve, not a straight line, and the residual vs. fits plot (shown in Figure 8.16.) confirms that a straight line fit is inappropriate.

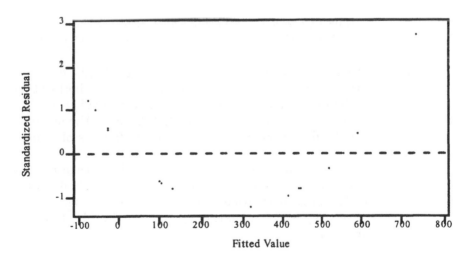

Figure 8.16. The standardised residual vs. fits plot for the data shown in Table 8.3. There is a clear pattern to the graph, residuals being positive then negative and then positive again, indicating that a straight line has been fitted to a curvilinear relationship. Note also the steepness of this curve, which is far more extreme than the graph of the raw data (Figure 8.15.), demonstrating the effectiveness of this plot in identifying an inappropriate fit.

Regression Plot

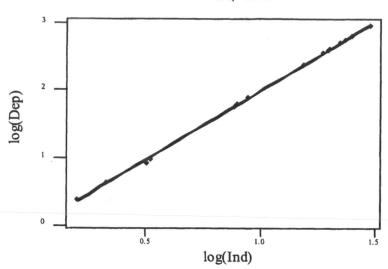

Y = -4.9E-02 + 2.03706X
R-Sq = 99.9 %

Figure 8.17. Plot of the log-transformed data with a fitted regression line. The line is a very close fit and examination of the residuals would show that this is quite acceptable.

As a first attempt to find a suitable model to describe the data, let us use a logarithmic transformation. We can quickly see whether this is a possible model by using the Minitab fitted line plot (Figure 8.17.; the dialog box for this function is shown in Exhibit 8.7.).The straight line seems to fit the data very well and the R^2 of 99.9% is obviously very good.

Under most circumstances, we would plot the residuals against fits and examine the statistical output according to the method that I described in the last section, entitled *Simple linear relationships*. But, if we examine the regression equation* generated by Minitab and shown in Figure 8.17.

$$Y = -4.9E - 02 + 2.03706X$$

* Minitab uses Y and X in this equation to denote the variables that have been plotted on the y- and x-axes, irrespective of any transformations that may have been applied to the data. So, in this example, the Y and X in the equation actually denote log (Dependent) and log (Independent), respectively.

which, because we have log transformed the data, we'll see it represents a relationship between the *Independent* and *Dependent* variables of:

$$Dependent = 10^{-0.049}Independent^{2.0376}$$

$$= 0.8933Independent^{2.03706}$$

The power to which the *Independent* variable has been raised is very close to 2, indicating that the true relationship may be more like:

$$Y = aX^2 \tag{8.10}$$

If so, we could try an alternative transformation to logarithms by regressing y on x^2. Try this and look at the R^2. Is it an improvement over the logarithmic transformation?* Do the residual plots indicate that the fit is appropriate? What is the regression equation?

In this particular example, we have found a model that fits the data well using only one independent variable (the square of the x-variable) as have all the regression examples described so far. However, we saw in the discussion of factorial designs in Chapter 7 that many biological systems involve more than one factor. For example, the growth a plant will depend upon many factors, including the concentrations of nitrogen, phosphorus, and potassium in the soil. If we could apply regression techniques to describe such a relationship mathematically, we would be in a position to determine the best combination of nitrogen, phosphorus, and potassium in the soil for maximum growth.

Fortunately, it is possible and fairly straightforward to extend the regression process to relationships involving more than one independent variable (although this task was effectively impossible before the arrival of computer-based statistics packages). However, accompanying the ability to accomplish extremely complex computations so readily is the potential to make errors with equal ease. A detailed treatment of building complex models is beyond the scope of this book, so I shall confine myself to describing a relatively simple example that illustrates the basic principles. If you need to explore this topic further, it is covered in greater depth in some statistics texts such as that by

---

* The R^2 for the logarithmic plot is 99.9% and that for the y on x^2 is 100%, which indicates that this model is an improvement over the logarithmic model that we initially used. The equation is *Dependent* = –0.4266 + 1.00151 *Independent*^{2}.

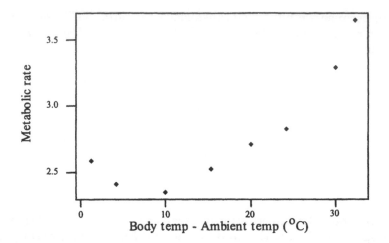

Figure 8.18. The relationship between metabolic rate (measured in arbitrary units) and the difference between body and ambient temperature (°C).

Zar (1996), and the computation instructions in the manuals of packages such as Minitab.

By means of illustration, we shall consider the relationship between the metabolic rates of mammals and the temperature of their environment relative to their body temperature. Biologically, we would expect a relationship because warm-blooded animals attempt to maintain a constant body temperature that is largely independent of the temperature of the environment around them. However, in order to achieve this level of independence, they have to operate various control mechanisms, one of which is to modify their metabolisms. For example, if the ambient temperature falls, an individual can increase metabolic rate to counteract the cooling effect of the environment. The results of the investigation are shown graphically in Figure 8.18., where metabolic rate is plotted against the difference between body and ambient temperatures (the data are listed in Table 8.4.).

Since Figure 8.18. indicates a clear curvilinear relationship, there is little point in trying to fit a straight line through the data. We could try regressing *metabolic rate* on the square of the temperature difference, but the residual plot shown in Figure 8.19. indicates that the fit is not appropriate because the residuals form a U-shape: positive at each end and negative in the middle range of fitted values. A similar outcome results from a logarithmic transformation.

Although simple relationships involving only x^2 of the form shown in Equation 8.10 are not unusual, we may quite often encounter a

Body temp – Ambient temp	Metabolic rate (arbitrary units)
1.1765	2.58824
4.1176	2.41176
10.0000	2.35294
15.2941	2.52941
20.0000	2.70588
24.1176	2.82353
30.0000	3.29412
32.3529	3.64706

Table 8.4. Data from an investigation of the relationship between metabolic rate of mammals and environmental temperature. The temperature measurements shown are the differences between body and ambient temperature.

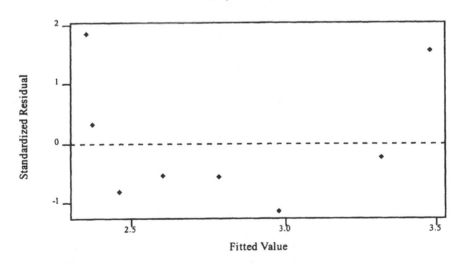

Residuals Versus the Fitted Values

(response is M)

Figure 8.19. A plot of *standardized residuals vs. fit* for the model $y = x^2$ for the metabolism data shown in Table 8.4. The U-shape indicates that an inappropriate model has been fitted to the data.

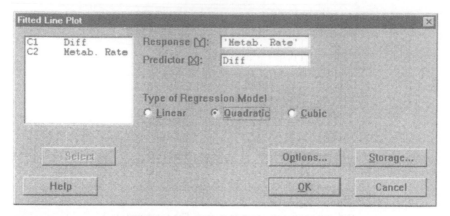

Exhibit 8.10. The Minitab dialog box to specify fitting a quadratic model to the mammal metabolism data in Table 8.4.

relationship involving both x and x^2, as described by Equation 8.11. below:

$$y = a + bx + cx^2 \qquad (8.11)$$

Equations including a single variable (in this case, x) raised to several different powers are referred to as *polynomials*. Equation 8.11, which includes only x and x^2, is known specifically as a *quadratic* equation, and one involving x, x^2, and x^3 would be referred to as *cubic*.

We can see how well a quadratic model describes the metabolism data by fitting a line using Minitab's fitted line function. In the dialog box (Exhibit 8.10.), there are options to fit quadratic or cubic lines through the data.* We need only to specify the dependent variable (in this example, *Metab. Rate*) and the independent variable (*Diff*; Minitab calculates the values of x^2 used in the calculations).

The plot of a quadratic model applied to the metabolism data (Figure 8.20.) appears to be a reasonable fit, with an R^2 of 98.4%. However, in order to confirm that the quadratic model is good, we must perform the full regression analysis, specifying the model using the dialog box as shown in Exhibit 8.11. In contrast to the fitted line plot, Minitab requires us to provide a column holding the data for each independent variable (even for polynomial regressions). Therefore, in this example we would have three columns representing *metabolic*

---

* The fitted line function will only work with a single variable or with quadratic or cubic polynomials.

Regression Plot

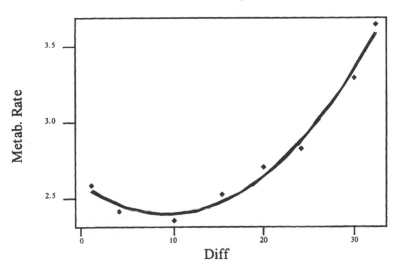

Y = 2.59825 - 4.33E-02X + 2.28E-03X**2
R-Sq = 98.4 %

Figure 8.20. A quadratic model fitted to the mammal metabolism data listed in Table 8.4.

rate, temperature difference, and *temperature difference*[2]. The squared values can be produced by arithmetic manipulation within Minitab.

The first stage of the process of determining if the quadratic model is appropriate is to examine the residuals plots. This particular data set is not very large and so the normality checks are not particularly informative, but the residuals vs. fits plot (Figure 8.21.) does show a considerable improvement over that of the simple $y = x^2$ model: there is no longer any evidence of the U-shape.

The Minitab output from the multiple regression analysis (Exhibit 8.12.) is an expansion of that of a simple regression with only a single independent variable. The regression equation includes extra terms because we have included more independent variables in the model, and each term has its own coefficient. We can see from the t-tests associated with the coefficients that both the difference and the difference[2] have had significant effects because their probability values are less than 0.05. Since the residuals plot (Figure 8.21.) gives no reason to doubt the appropriateness of the fit and the numerical results are statistically significant, this quadratic model would seem to be a perfectly adequate description of the relationship between

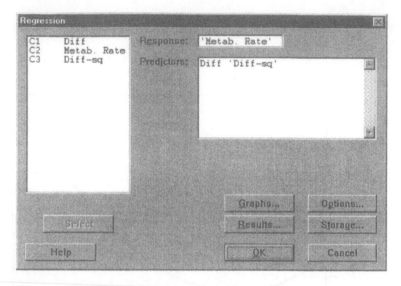

Exhibit 8.11. The Minitab dialog box to specify the quadratic model for the mammal metabolism example. Note the order in which the predictor variables are listed, the importance of the sequence is discussed in the relation to the interpretation of the results shown in Exhibits 8.12. and 8.13.

Residuals Versus the Fitted Values

(response is Metab. R)

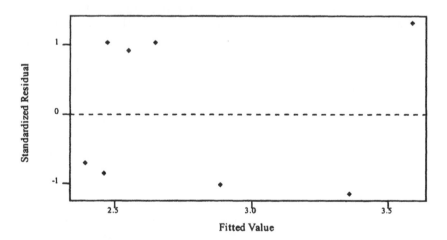

Figure 8.21. The residuals vs. fits plot of the quadratic model fitted to the mammal metabolism data. There does not seem to be any systematic pattern indicating that the fit may be reasonable.

metabolic rate and the difference between body and ambient temperature in mammals.

An additional feature of the output of multiple regression appears below the analysis of variance results. The regression sum of squares is partitioned into separate components for each independent variable, the *sequential sum of squares* (labeled *Seq SS* in Exhibit 8.12.); the total of this column will equal the regression SS in the ANOVA table above it. The sequential sum of squares is the contribution to

```
MTB > Regress 'Metab. Rate' 2 'Diff' 'Diff-sq';
SUBC>  GHistogram;
SUBC>  GNormalplot;
SUBC>  GFits;
SUBC>  RType 2;
SUBC>    Constant;
SUBC>    Brief 2.
```

Regression Analysis

```
The regression equation is
Metab. Rate = 2.60 - 0.0433 Diff + 0.00228 Diff-sq

Predictor        Coef        StDev          T        P
Constant      2.59825      0.06177      42.07    0.000
Diff         -0.043259     0.008867     -4.88    0.005
Diff-sq       0.0022849    0.0002554     8.94    0.000

S = 0.06785     R-Sq = 98.4%     R-Sq(adj) = 97.9%

Analysis of Variance

Source            DF          SS          MS        F        P
Regression         2     1.41643     0.70822   153.85    0.000
Residual Error     5     0.02302     0.00460
Total              7     1.43945

Source          DF     Seq SS
Diff             1    1.04814
Diff-sq          1    0.36829


MTB >
```

Exhibit 8.12. Minitab results of a multiple regression of metabolic rate on the difference between body and ambient temperature in mammals (the data are shown in Table 8.4.). A quadratic model has been fitted using the predictors difference and difference², including difference as the first term in the model. Note the sequential sum of squares is different from the results shown in Exhibit 8.13. where the predictor terms were included in the reverse order.

the regression sum of squares made by each variable as it is added into the regression model in the sequence specified in the *Predictors* box shown in Exhibit 8.11. In this particular example, the results show that the variable *difference* (labeled *Diff* in the exhibit) had a Seq SS of 1.04814 when it was the first variable included in the model (accounting for 72.8% of the total variation[*]) and the variable *difference*² (labeled *Diff-sq* in the exhibit) had a sequential sum of squares of 0.36829 when it was included in the model as the second variable. However, if we construct the model in the reverse order (*difference*² before *difference*, achieved by listing them in reverse order in the *Predictors* box of the Exhibit 8.11.), we can see from the results shown in Exhibit 8.13. that the sequential sums of squares have changed. The variable *difference*² now accounts for 90.8% of the total sum of squares, and *difference* accounts for only 7.6%. Why should this be so? What does it mean in terms of understanding the regression analysis?

The reason why the sequential sums of squares change is because the two variables (*difference* and *difference*²) are not independent of each other. Consequently, when *difference* was included first, a considerable proportion of the effect of *difference*² was also included. By comparing the sequential sum of squares in the two alternative sequences (as shown in Exhibits 8.12. and 8.13.), we can see that *difference*² by itself accounts for a larger proportion of the total than *difference* (1.30688 against 1.04814).

The conclusion that we can draw from this multiple regression analysis is that there is a statistically significant relationship between metabolic rates and the *difference* between body and ambient temperatures in mammals ($R^2 = 98.4\%$; $F = 153.85$; $df = 2.7$; $P < 0.001$) that can be described by the following equation:

$$Metabolic\ rate = 2.6 - 0.0433\ difference + 0.00228\ difference^2$$

In a report of this investigation, we would include Figure 8.20. to illustrate the relationship and confirm in a visual manner that the model described by the equation was a true reflection of the relationship between the biological variables.

In other investigations where the predictor variables are independent of each other (often described as *orthogonal*), the sequence in which the variables are included in the model would not affect the

---

[*] Seq SS (*difference*)/Total SS = 1.04184/1.43945 = 72.8%; effectively the R^2 for *difference*.

```
MTB > Regress 'Metab. Rate' 2 'Diff-sq' 'Diff';
SUBC>   GHistogram;
SUBC>   GNormalplot;
SUBC>   GFits;
SUBC>   RType 2;
SUBC>    Constant;
SUBC>    Brief 2.
```

Regression Analysis

```
The regression equation is
Metab. Rate = 2.60 + 0.00228 Diff-sq - 0.0433 Diff

Predictor          Coef       StDev          T       P
Constant        2.59825     0.06177      42.07   0.000
Diff-sq       0.0022849   0.0002554       8.94   0.000
Diff          -0.043259    0.008867      -4.88   0.005

S = 0.06785     R-Sq = 98.4%     R-Sq(adj) = 97.8%

Analysis of Variance

Source             DF          SS          MS         F       P
Regression          2     1.41643     0.70822    153.85   0.000
Residual Error      5     0.02302     0.00460
Total               7     1.43945

Source       DF     Seq SS
Diff-sq       1    1.30688
Diff          1    0.10955


MTB >
```

Exhibit 8.13. Minitab results of a multiple regression of metabolic rate on the difference between body and ambient temperature in mammals (the data are shown in Table 8.4.). A quadratic model has been fitted using the predictors difference and difference[2], including difference[2] as the first term in the model. Note the sequential sum of squares is different from the results shown in Exhibit 8.12. where the predictor terms were included in the reverse order.

sequential sums of squares. Where the effects of several predictor variables are being investigated, we can assess the relative importance of each predictor by repeatedly including different combinations and subsets of the variables and submitting them in different sequences. This trial and error procedure is not described here, but there are techniques (and Minitab commands, e.g., **STEPWISE** regression) that perform this process in a systematic and automated manner.

8.4   Exercises

1. In an investigation of the way in which molluscs cope with different levels of salinity in sea water, the following data were obtained that relate the salinity of sea water and the concentration of taurine (a product of the metabolism of cysteine) in mussels of the genus *Mytilus*. A hypothesis has been proposed that suggests that a high concentration of taurine might be expected under very saline conditions as a mechanism to avoid osmotic desiccation. Calculate the correlation coefficient, and comment on whether the results warrant further exploration of this hypothesis.

Sea water salinity (parts per thousand)	Taurine conc. (mM)
7	0
8	4
16	23
21	25
27	40
31	45
34	49

2. The data tabulated below were gathered to investigate the relationship between dominance of songbirds and their frequency of singing bouts. The data are ranked. Is there any correlation?

Dominance	Song frequency
1.0	4.0
2.5	5.0
2.5	3.0
4.0	1.0
5.0	9.0
6.0	2.0
7.0	6.5
8.0	6.5
9.0	11.0
10.5	8.0
10.5	10.0

3. Using the data in Exercise 7 of Chapter 2, regress brain weight on body weight. What transformation is necessary? Examine the residuals carefully, and comment biologically about any outliers.

4. The following data table relates to the velocity of an enzyme reaction that is described by the Michaelis-Menten equation. Transform the data appropriately and evaluate the Michaelis constant and the maximum velocity.

Substrate conc. (mol)	Velocity (mol per unit time)
0.208333	21.95
0.142857	21.15
0.068493	18.65
0.053763	17.55
0.044843	16.8
0.031847	15.15
0.025575	14.08
0.019194	12.40
0.015552	11.15
0.009597	8.35
0.009042	8.05
0.007541	7.15

5. Using the pulse rate data, examine the relationship between the resting pulse rate and pulse rate after exercise. What effect does adding body weight into the relationship have in addition to resting pulse rate?

6. The data tabulated below record the flow rates and depths of various streams. Construct a regression model that describes how flow rate depends upon stream depth.

Flow rate	Depth
0.34	0.636
0.29	0.319
0.28	0.734
0.42	1.327
0.29	0.487
0.41	0.924
0.76	7.350
0.73	5.890
0.46	1.979
0.40	1.124

8.5   Summary of Minitab commands

CORRELATE

RANK

REGRESS (including residuals plots and fitted line plot)

8.6   Summary

1. Correlation coefficients have been introduced to measure the degree of association between variables measured on ordinal, interval, or ratio scales.

2. A positive correlation is when a high value of one variable is associated with a high value of the other; a negative correlation occurs when a high value of one variable is associated with a low value of the other.

3. The Pearson product-moment correlation coefficient can be used when both variables are drawn from normal distributions. Spearman correlation coefficients are used when the data are not drawn from normal distributions (e.g., when they are in the form of ranks).

4. A significant correlation does *not* imply that a change in one variable *causes* a change in the other; merely that a high value in one variable is generally found with a high (positive correlation) or low (negative correlation) value of the other variable.

5. Regression analysis is used to characterise causal relationships between one dependent (response) and one or more independent (or predictor) variables.

6. The outcome of regression analysis is a mathematical model (equation) that describes the relationship under investigation.

7. All regression models, whether they involve a single predictor (simple regression) or several predictor variables (multiple regression), must be validated as statistically significant (assessed by examination of the statistical analysis) and appropriate (assessed by graphical methods involving residuals) before they can be used.

8. Valid regression models can be used predictively, but caution should be exercised in the range of values of the predictor variables; one should not extrapolate beyond the range used to construct the model.

PART IV

Questions of Association
and Agreement

CHAPTER 9

Tabular Relationships

9.1   Introduction

Associations between variables or characteristics of biological subjects are many and varied. In Chapter 8, we saw how correlation coefficients can be used to measure the association between variables that are measured on ordinal, interval, or ratio scales (e.g., bone lengths, dominance, and song activity), but how can we explore associations between characteristics that are described by nominal variables? What form might such associations take? And how can they be summarised? In order to illustrate the type of associations that we might wish to describe, consider the following examples.

In plant ecology, we may quite often only be able to record a species as either present or absent rather than in terms of numbers of individuals. We could do this by sampling habitats using quadrats, for example. If we wished to determine whether there is an association between two species (A and B), we could record the number of quadrats in which: both species are present; only Species A, but not Species B; only Species B, but not Species A; or neither species, and then construct a table like Table 9.1 to summarise the results.

Similarly, in epidemiology, we might wish to compare the susceptibilities of individuals with different blood groups to a particular disease by recording the number of diseased and unaffected individuals in each group. Again, we would summarise the data in a table by *cross-tabulating* blood group against disease status (diseased or unaffected).

In both of these examples, the measured data are the frequencies of subjects that share attributes (e.g., quadrats in which both Species A and B are present, or diseased individuals with blood group A). The examples are very similar (but we shall see later in this chapter that they require slightly different methods of analysis), and they illustrate an extremely common type of problem in which we attempt to answer the following question:

		Species A		
		Present	Absent	Total
	Present	16	40	56
Species B	Absent	35	15	50
	Total	51	55	106

Table 9.1. Presence/absence of two species measured as frequency of occurrence in a quadrat survey.

What is the frequency of sampling units that share a particular set of characteristics?

Or, in other words,

Are specific combinations of characteristics observed more often than would be expected by chance?

The data are summarised in tables that might have one or more columns and one or more rows. In this sort of question, we may have a fairly clear idea of the nature of the associations that the results might reveal, such as Plant Species A rarely being found in the presence of Plant Species B, but we would not always be in this position; we might pose this type of question out of pure curiosity. Furthermore, the exact frequencies of the subjects having the various combinations of characteristics would not normally be predictable. For example, although we might have expected that Species A was not often found with Species B as the data in Table 9.1. would appear to suggest, we would not have any means of predicting with any certainty the actual frequencies that would arise from the survey.

A closely-related type of question that can be analysed by a similar technique arises when we can use theoretical knowledge to predict the frequencies of different characteristics in a population. For example, Mendel developed his law of independent assortment through observing the offspring of crosses of peas that he had described in terms of two phenotypic characters: shape (round or wrinkled) and colour (yellow or green). His law predicts that if we cross individuals that are heterozygotic for two genes (e.g., RrYy) in which the alleles R (round) and Y (yellow) are dominant to alleles r (wrinkled) and y (green), the resulting offspring will exhibit the phenotypic characteristics, round/yellow, round/green, wrinkled/yellow, and wrinkled/green in the ratio of 9:3:3:1. We could test the hypothesis that phenotypic characteristics were controlled by such a pair of genes by performing a breeding experiment and comparing the frequencies of the various phenotypic characteristics in the resulting offspring with Mendel's predictions. In this type of question, we are not simply looking for a combination of particular characteristics, but actually examining the *ratio* of the characteristics in the sampled population. In other words, we are comparing the observed frequencies against predictions: this is a *goodness-of-fit question* or a *question of agreement*.

These two types of question (association and goodness-of-fit) are very similar and so, not surprisingly, the analytical techniques used to answer them share many common features, not least of which is that they both rely upon the chi-squared (χ^2) distribution, and it is convenient to discuss them together. However, because one question is simply asking whether or not there is an association, while the other is comparing observed frequencies against predictions, there are some differences in the detail of the analysis. The structure of this chapter reflects this distinction, and, in Section 9.2, I describe the way in which we answer questions of comparison. Following that, in Section 9.3, methods to investigate questions of goodness-of-fit are outlined.

9.2 The contingency table

The example of the association between disease susceptibility and blood groups is typical of many investigations in the biosciences. The researcher is looking for any evidence of non-random occurrences of the possible combinations of the different forms of the classification variables. In other words, we can characterise the activity of the researcher by the following question:

> Is there any evidence of one or more blood groups being associated with a particular disease status more than would be expected by chance?

This question leads directly to a testable null hypothesis:

> H_N: There is no association between blood group and disease status.

And the complementary, alternative hypothesis:

> H_A: There is an association between blood group(s) and disease status.

| | Blood Group | | | | |
	A	B	AB	O	Total
Diseased	11	21	32	9	73
Unaffected	35	29	12	42	118
Total	46	50	44	51	191

Table 9.2. The frequency of diseased and unaffected people classified according to blood group.

Having constructed a null hypothesis, how can we go about testing it? How can we actually determine whether or not there is an association? In fact, how are we to define an association?

The question preceding the null hypothesis provides a very good clue as to the way that we proceed. We shall generate a prediction of the frequencies of the various combinations that would be expected by chance, i.e., if there is no association. Then we shall compare the observed frequencies with those predicted under the null hypothesis of random assortment. The statistical test that we shall use will enable us to calculate the probability of our observations occurring if the null hypothesis is true. This type of analysis is known as a *contingency table* χ^2.

The first stage in this analytical process is to determine what frequencies would be expected by chance. Let us consider this in terms of the blood group/disease example, the data of which are tabulated in Table 9.2.

The calculation of any expected frequency must take into account several aspects of the data. To take an example, if we were trying to establish the number of diseased individuals one might expect with Blood Group A, we would have to consider the total number of subjects who were diseased irrespective of blood group, the total number with Blood Group A irrespective of disease status, and also the total sample size. The way that we actually calculate an expected value is given by Equation 9.1:

$$Expected\,frequency = \frac{(Row\,total)(Column\,total)}{Grand\,total} \qquad (9.1)$$

and, in the case of Blood Group A, we would expect

$$\frac{(73)(46)}{191} = 17.581\,diseased\,individuals$$

	Blood Group				
	A	B	AB	O	Total
Diseased	17.581	19.110	16.817	19.492	73
Unaffected	28.419	30.890	27.183	31.508	118
Total	46	50	44	51	191

Table 9.3. The expected frequency of diseased and unaffected people classified according to their blood group.

Calculate the remaining expected frequencies and compare your answers with those in Table 9.3.

Once we have established the frequencies that would be expected if there was no association, we can compare them with the observed frequencies (Table 9.2.). For this, we use the formula shown in Equation 9.2 to obtain the test statistic, χ^2.

$$\chi^2 = \Sigma \frac{(O-E)^2}{E} \qquad (9.2)$$

where O and E represent, respectively, the observed and expected frequencies.

The function of this test statistic is to assess the difference between the observed and expected frequencies relative to the expected frequency for each cell in the contingency table. An extremely important point to remember is that the only data to which this test may be applied are frequencies or counts; *it cannot be applied to proportions or percentages*. There are several other constraints that must be satisfied, and they are discussed later in this section. By way of illustration, the calculation for blood group *AB/non-diseased* is calculated thus:

$$\frac{(O-E)^2}{E} = \frac{(12-27.183)^2}{27.183} = 8.480 \qquad (9.3)$$

The results of the complete calculation are shown in Table 9.4. The test statistic, χ^2, is the sum of the values in each cell:

$$\chi^2 = \Sigma \frac{(O-E)^2}{E} = 35.62$$

In order to interpret this test statistic, we need to calculate the appropriate degrees of freedom, which are evaluated using the following expression:

$$degrees\ of\ freedom = (no.\ of\ rows - 1)(no.\ of\ columns - 1)\quad (9.4)$$

which, in the case of this example, is:

$$degrees\ of\ freedom = (2 - 1)(4 - 1) = 3$$

The probability of obtaining a chi-squared value of 35.62 with three degrees of freedom is determined using the critical values of the chi-squared distribution (Appendix A.10.). We can see in that table that the critical value of χ^2 at $P = 0.05$ is 7.8147, and the calculated value of 35.62 is much greater than this; in fact, the value of χ^2 that we have obtained is greater than the critical value for $P = 0.001$ (which is16.2662). Therefore, we can draw the conclusion that there is a statistically significant association between blood group and disease status ($\chi^2 = 35.62$; d.f. = 3; $P < 0.001$). But what does a significant association actually mean biologically (or even statistically)? In terms of the biology of the investigation, we need to ask the supplementary question. Between which blood group(s) and disease status(es) are there significant associations?

In some ways, this situation parallels very closely the position in which researchers find themselves after a carrying out analyses of variance; they may know that there is an effect, but they need to identify it using a Tukey test. With a contingency table χ^2, there is not a supplementary statistical test like that of Tukey, and so we have to devise an alternative methodology.

Intuitively, we might proceed by identifying the cells in which the difference between the observed and expected frequencies is greatest. The underlying logic of this would appear to be straightforward. If the null hypothesis of no association is true, we would expect the values of chi-squared for each cell in the table to be 0 (i.e., no difference between observed and expected frequencies). However, the difference will be influenced not only by any association, but also by the absolute magnitude of the observed and expected values. In order to illustrate this point, consider the following two scenarios. In the first situation, the observed frequency is 1050 and the expected is 1000; a difference of 50, with the observed frequency being 5% greater than the expected. Contrast that with the situation where the observed frequency is 20 and the expected is 10; it is a difference of only 10, but the observed frequency is 100% greater than the expected. In absolute terms, the difference is greater in the first example, but, if we express the difference relative to the magnitude of the expected value, then the difference is far greater in the second example. The simple approach of comparing observed and expected would appear to be ignoring an important element, namely the relative scale of the difference. In order to reflect this relative difference, we must take into consideration the magnitude of the expected frequencies when examining the difference between observed and expected frequencies. The way that we achieve this is to calculate the *standardized residuals* (sometimes referred to as *adjusted residuals*) for each cell in a contingency table. These are evaluated using the formula given by Equation 9.5.

$$standardized\,residual = \frac{(O-E)}{\sqrt{E}} \qquad (9.5)$$

In terms of magnitude, the standardized residual is the square root of the χ^2 value for each cell in the table (as calculated by Equation 9.3). However, it is calculated differently to $\sqrt{\chi^2}$ in order to preserve the direction or sign of the difference; standardized residuals can be positive or negative, whereas $\sqrt{\chi^2}$ can only give positive values. By having positive or negative standardized residuals, we can see the direction in which an observed frequency deviates from the expected. Having calculated standardized residuals for every cell in a contingency table, statistically significant associations are identified as those with residuals outside the range −1.96 to +1.96. This range can be regarded essentially as a confidence interval around a standardized residual of 0 that would occur if the null hypothesis is true. In terms of the blood group/disease investigation, the standardized residuals are shown in Table 9.5.

Now that we have completed an inspection of the standardized residuals, we are in a position to report the results of the analysis with a paragraph such as this:

A contingency table χ^2 analysis (*insert reference here*) identified a significant association between blood group and disease status ($\chi^2 = 35.62$; d.f. = 3; $P < 0.001$). Inspection of the standardized residuals revealed that individuals with Blood Group AB were more likely to suffer from the disease than be unaffected

	Blood Group			
	A	B	AB	O
Disease	−1.57	0.43	3.70	−2.38
Unaffected	1.23	−0.34	−2.91	1.87

Table 9.5. The standardized residuals (calculated using Equation 9.5) for each cell in the contingency table for the diseased and unaffected people classified according to their blood group. These results show that individuals of Blood Group AB tend to be more likely to suffer from the disease and those of Blood Group O tend not to suffer from the disease.

by it, and that subjects with Blood Group O had a significantly lower than expected incidence of the disease than would be expected by chance.

Having identified the specific associations, the researchers would be in a position to explore the medical reasons why such associations exist.

Minitab provides two different commands to perform this type of chi-squared analysis. The first that I shall describe is the **CHISQUARE** command that is found by the following menu route:

STAT > Tables > Chi-Square Test

When using this command, the data should be stored in columns within Minitab in the form of a table. In the context of the blood group data, we would store the data for each blood group in a separate column, with the data for diseased and unaffected subjects in rows 1 and 2, respectively (the data should appear like Table 9.2.). The Minitab dialog box to specify the detail is shown in Exhibit 9.1., with the data for blood groups A, B, AB, and O being stored in columns *C1–C4*, respectively.

The output from the **CHISQUARE** command is illustrated in Exhibit 9.2. Minitab provides the observed, expected, and chi-squared values for each cell. However, it does not include the standardized residuals; they must be calculated by finding $\sqrt{\chi^2}$ for each cell and then attaching the appropriate sign by comparing the relative size of the observed and expected frequencies. For example, the standardized

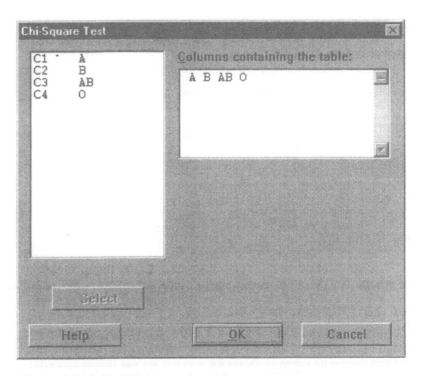

Exhibit 9.1. The Minitab dialog box to specify the chi-square command for the blood group/disease status example. The data are presented to Minitab in the form of a table: four columns (one for each blood group) and two rows (diseased and unaffected).

residuals for Group A/diseased will be $-(\sqrt{2.464})$ because the observed frequency (11) is less than the expected (17.58).

The **CHISQUARE** command described above can be applied to data that have already been organised into a table of frequencies. There is an alternative method of performing a contingency table χ^2 analysis available within the cross-tabulation commands that were introduced in Chapter 2. Not only does this latter method perform the analysis, but it also constructs the contingency table itself. The data are stored in two columns (for example, one for blood group and the other for disease status), with codes to denote the different categories within each variable; the codes can be integers or characters. In the blood group example illustrated in Exhibits 9.3. and 9.4., I have used *A, B, AB,* and *O* to represent the blood groups and *Dis* and *Not-D* for disease status. Each row of the two columns would relate to a single subject, defining his or her blood group and disease status. The Minitab

```
MTB > ChiSquare  'A' 'B' 'AB' 'O'.

Chi-Square Test


Expected counts are printed below observed counts

            A        B        AB        O      Total
   1       11       21        32        9        73
         17.58    19.11     16.82    19.49

   2       35       29        12       42       118
         28.42    30.89     27.18    31.51

Total     46       50        44       51       191

Chi-Sq =  2.464 +  0.187 + 13.708 +  5.648 +
          1.524 +  0.116 +  8.481 +  3.494 = 35.621
DF = 3, P-Value = 0.000

MTB >
```

Exhibit 9.2. The Minitab output from the chi-square command. The observed and expected frequencies are shown within each cell, and, underneath the chi-square table are the chi-square values for each cell (calculated using Equation 9.3). *N.B.* This command does not calculate the standardized residuals. Differences between values in this output and the results shown in Tables 9.3. and 9.4. are due to rounding errors.

dialog box illustrated in Exhibit 9.3. shows that this method also includes the facility to calculate and display the standardized residuals (selected in this example). The dialog box is accessed via the following menu route:

STAT > Tables > Cross Tabulation

The output of this method of performing the chi-squared analysis is shown in Exhibit 9.4. Included within each cell of the table, are the observed and expected frequencies along with the standardized residuals.

This method is particularly useful when the data set is very large because it reduces the scope for human error in determining all the observed frequencies. It is also the method that we would employ to handle the analysis of surveys or questionnaires where respondents are required to answer questions. We would code the data into Minitab with one row per subject and one column per question. At a very practical level, whenever we gather data by means of a questionnaire, we should

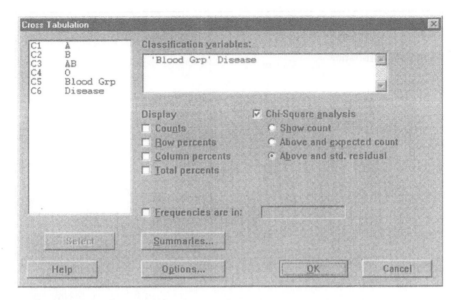

Exhibit 9.3. The Minitab dialog box to specify a contingency table χ^2 including standardized residuals in the output.

physically number or code the individual questionnaires and include the code in the computerised data as well. By doing so, one can always match hardcopy and electronic versions of the data to check for errors.

The contingency table χ^2 analysis technique is fairly simple in terms of arithmetic, and Minitab certainly makes it very easy to perform. However, just like many of the other techniques described in earlier chapters, there are some constraints on its use. There are five simple, general rules, and one specific rule that applies to the special case of a 2-row × 2-column table. First of all, I shall outline the general rules and then I will illustrate the special case in the context of an example.

1. Only frequencies or counts (not percentages or proportions) may be analysed by χ^2.
2. Each category must be mutually exclusive.
3. Each subject must be independent.
4. No *expected* values less than 1 are permitted.
5. No more than 10% of *expected* values in a contingency table can be less than 5.

I have already mentioned the first of these rules. The second rule should be self-explanatory; we cannot have an individual contributing

```
MTB > Table 'Blood Grp' 'Disease';
SUBC>    ChiSquare 3.
```

Tabulated Statistics

```
Rows: Blood Gr     Columns: Disease

         Dis     Not-D      All

A         11        35       46
        17.58     28.42    46.00
        -1.57      1.23      --

B         21        29       50
        19.11     30.89    50.00
         0.43     -0.34      --

AB        32        12       44
        16.82     27.18    44.00
         3.70     -2.91      --

O          9        42       51
        19.49     31.51    51.00
        -2.38      1.87      --

All       73       118      191
        73.00    118.00   191.00
         --        --       --

Chi-Square = 35.621, DF = 3, P-Value = 0.000


  Cell Contents --
                 Count
                 Exp Freq
                 St Resid

MTB >
```

Exhibit 9.4. The Minitab output of a contingency table χ^2 analysis performed by the cross-tabulation command (Exhibit 9.3.) for the blood group/disease example. The observed and expected frequencies along with the standardized residuals are displayed within each cell.

to more than one cell in a contingency table. But this does not preclude the categories being drawn from continuous variables. For example, if we wished to examine whether age was associated with disease status in humans, we could create several mutually exclusive age categories, such as <16, 16–25, 26–35, 36–45, and 46+. The importance of the third constraint cannot be overstated; clearly if two subjects are not independent, they will exert undue influence and bias the results. The remaining two constraints are closely related to each other, and it is extremely important to note they are framed in terms of the *expected* values, even though such problems normally arise because of low numbers of observations. If our data suffer from problems caused by either of these last two rules, we can generally resolve them by one of two methods.

First, we could remove all the data for the category in which the low expected frequencies are located. If in the blood group investigation, for example, we found that the expected value for group AB/diseased was less than 1, we could remove all subjects with this blood group and analyse the remaining data. As long as the remaining data satisfied the constraints, we would be able to draw conclusions about blood groups (excluding AB) and disease status.

The second option is to combine categories (often referred to as *collapsing categories*) in order to increase expected frequencies. With this technique, we have to recode subjects into broader categories. This method will reduce the detail of potential conclusions because we have fewer categories, but it should also enable us to carry out a valid analysis, which otherwise would not be possible. As an illustration of this technique, consider the age categories described above. Let us imagine that we have low expected frequencies in the <16 age group. We could reorganise the age groups into <26, 26–35, 36–45, and 46+, and continue with the analysis. This manipulation is perfectly admissible, but it should be used with care. We can only use it with variables for which a reorganisation makes sense. We could not, for example, reorganise blood groups in a sensible way biologically. The simple rule to follow is that you must have a strong *biological* argument to justify changing categories. Whichever method we use, we must never lose sight of the biological objectives. We cannot continue removing or collapsing categories without limit because, ultimately, the analysis would become biologically meaningless.

In the special case of a 2×2 table, we have to apply a correction factor in the calculation of the χ^2 statistic. This continuity correction,[*] named after its developer Yates, requires us to use an amended version

* This modification to the method is required because the χ^2 distribution is a continuous distribution that we are using with discrete (frequency) data. This only presents a problem when the tables are very small (two rows and two columns), hence the need for the continuity correction for this special case.

| | | Species A | | |
		Present	Absent	Total
	Present	16	40	56
		26.943	29.057	
		4.048	3.753	
Species B	Absent	35	15	50
		24.057	25.943	
		4.533	4.204	
	Total	51	55	106

Table 9.6. Presence/absence of two species measured as frequency of occurrence in a quadrat survey. Each cell contains the observed and expected frequencies along with the χ^2 value corrected using Yates' continuity correction.

of Equation 9.2 to calculate the χ^2 test statistic, which is given by Equation 9.6, below:

$$\chi^2 = \Sigma \frac{(|O - E| - 0.5)^2}{E} \qquad (9.6)$$

The vertical bars in the equation mean that we should ignore the sign of the difference between the observed and expected values (i.e., all values should be regarded as positive; the mathematical term describing this function is the *modulus*) and the magnitude of the difference is reduced by 0.5. Use of Yates' correction can be illustrated by considering the plant presence/absence example shown in Table 9.1. For quadrats where both Species A and B are present, the expected frequency will be 26.943, and the corrected value of χ^2 for that cell will be:

$$\chi^2 = \frac{(|16 - 26.943| - 0.5)^2}{26.943} = 4.048$$

The observed and expected frequencies as well as the χ^2 values for each cell are shown in Table 9.6.

The final χ^2 value for this investigation is the sum of the individual cell values:

$$\chi^2 = 4.048 + 4.533 + 3.753 + 4.204 = 16.538$$

with one degree of freedom (calculated using Equation 9.4) which has a probability of less than 0.001 according to the critical values of χ^2 (Appendix A.10.). Inspection of the standardized residuals reveals that the two species and are rarely found together, and are found separately more than would be expected by chance. Therefore, we would report these results as follows:

> A contingency table chi-squared analysis (with Yates' continuity correction) was performed that showed that Species A and B tended to be found in different habitats ($\chi^2 = 16.538$; d.f. = 1; $P < 0.001$).

The ecologists could then examine the differences between the habitats occupied by the two species and attempt to identify the specific habitat requirements of Species A and B.

Unfortunately, Minitab does not provide the facility to include Yates' correction automatically, so we must apply it manually. The simplest way to do this is to use the normal **Chisquare** or **Cross-Tabulation** commands to obtain the expected frequencies, and then calculate the χ^2 values for each cell manually using Equation 9.6 (the output that we would use to carry this out for the plant presence/absence investigation is shown in Exhibit 9.5).

9.3   Goodness-of-fit

I outlined two question types in the introduction to this chapter: questions of association and those concerning the degree to which observed frequencies agree with predictions of theoretical or independently-derived models. The latter *goodness-of-fit* or *agreement* questions are found in many diverse areas of biology. The example described in Section 9.1 was based upon the early genetics studies of Gregor Mendel, and we shall consider that example in some detail in this section. We shall also examine an ecological example in which we compare the spacing of limpets (*Patella vulgaris*) on a rocky shore with the predictions of the Poisson distribution (cf., Chapter 3) to determine if the limpets are distributed in a random or non-random manner.

The fundamental biological question underlying goodness-of-fit questions is:

> Do the frequencies of the subjects in each of the categories of interest conform to the predictions of a specified model?

```
MTB > ChiSquare C1 C2.

Chi-Square Test


Expected counts are printed below observed counts

          Present    Absent    Total
    1          16        40       56
            26.94     29.06

    2          35        15       50
            24.06     25.94

Total         51        55      106

Chi-Sq =   4.445 +   4.122 +
           4.978 +   4.616 = 18.161
DF = 1, P-Value = 0.000

MTB >
```

Exhibit 9.5. Minitab output for the plant presence/absence data. Since this is a 2×2 table, Yates' continuity correction must be applied. We would use Minitab to calculate the expected frequencies and then complete the analysis manually.

We can rephrase this question in terms of Mendel's dihybrid cross with peas as follows:

> Do the frequencies of the phenotypic characters round/yellow, round/green, wrinkled/yellow, and wrinkled/green occur in the ratio of 9:3:3:1 as predicted by Mendel's law of independent assortment?

This question gives rise to a null hypothesis of:

> H_N: The frequencies of the phenotypic characters round/yellow, round/green, wrinkled/yellow, and wrinkled/green are consistent with the predicted ratio of 9:3:3:1.

and the complementary, alternative hypothesis:

> H_A: The frequencies of the phenotypic characters round/yellow, round/green, wrinkled/yellow, and wrinkled/green are not consistent with the predicted ratio of 9:3:3:1.

	Round/ Yellow	Round/ Green	Wrinkled/ Yellow	Wrinkled/ Green
Predicted ratio	9	3	3	1
Proportion of offspring	9/16	3/16	3/16	1/16
Predicted frequency (proportion × total offspring)	28.125	9.375	9.375	9.125
Observed frequency	30	7	8	5

Table 9.7. Predicted and observed frequencies of the four phenotypic characteristics resulting from Mendel's dihybrid cross of peas (the total number of offspring = 50).

We can see how to test this null hypothesis using the following data that were obtained from a repeat of Mendel's experiment in which the results of 50 crosses were recorded. If the offspring of these crosses are to exhibit phenotypic characteristics in the same ratio as the predictions (9:3:3:1), then the results that would be expected are those tabulated along with the observed frequencies (Table 9.7.).

We test the null hypothesis simply by regarding the observed frequencies as a 4×1 contingency table, with the expected frequencies predicted by the model rather than being calculated using Equation 9.1. The test statistic, χ^2, for this goodness-of-fit question is evaluated thus:

$$\chi^2 = \frac{(30-28.125)^2}{28.125} + \frac{(7-9.375)^2}{9.375} + \frac{(8-9.375)^2}{9.375} + \frac{(5-3.125)^2}{3.125} = 2.053$$

and the degrees of freedom are given by the *number of categories* − 1. Therefore, in this example, there are 3 degrees of freedom. The critical value of χ^2 in Appendix A.10. for 3 degrees of freedom at $P = 0.05$ is 7.8147, so we can conclude that there is no evidence to reject the null hypothesis. These results are consistent with the predictions of Mendel's law of independent assortment. One should remember, however, that accepting a null hypothesis is not the same as confirming that it is true; it is simply stating that there is no evidence to doubt it.

Unlike the contingency table χ^2, Minitab does not provide a specific goodness-of-fit function, but, as Exhibit 9.6. illustrates, the calculator function within Minitab enables us to calculate the test statistic very easily. We store the observed and expected frequencies in two columns (*C1* and *C2*, respectively, in Exhibit 9.6.) and then enter the chi-

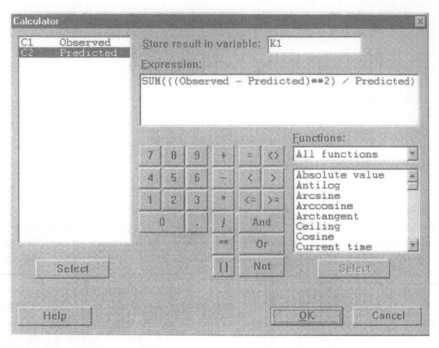

Exhibit 9.6. The Minitab Calculator dialog box with the formula for the chi-squared statistic (Equation 9.2) entered in the *Expression* box. The observed and expected frequencies are stored in columns *C1* and *C2*, respectively, and the test statistic will be stored in constant *K1*.

squared formula (Equation 9.2) into the *Expression* box in the calculator dialog box as shown in the exhibit.

The actual calculation of the goodness-of-fit χ^2 is not very difficult arithmetically, and the expression shown in the Minitab illustration above (Exhibit 9.6.) can be used for any goodness-of-fit question.

I conclude this final chapter with a second example of a goodness-of-fit question. The rationale behind its inclusion is obviously not because of any intrinsic complexity in the calculation of the test statistic, but to demonstrate how to generate predicted frequencies based upon a statistical distribution because this is a common task for some biologists. Many statistical distributions are used to model biological populations or processes (e.g., Poisson, log-normal, gamma, negative binomial, and many others), and so, as biologists, we should be aware of the general principles of generating expected frequencies based upon a statistical distribution. I shall illustrate this process in the context of the Poisson distribution, which, as we saw in Chapter 3, is the distribution that relates to random events.

The Poisson distribution is known to accurately describe events that occur randomly in time or space. Classic examples are the frequency of release of packets of the neurotransmitter acetyl choline at nerve synapses and the spatial dispersion of unassociated organisms. The particular example that I describe here relates to the dispersion of limpets over a rocky shore, in which we will test the following null hypothesis:

H_N: The spatial dispersion of limpets on a rocky shore (as measured by the number of limpets per quadrat) does not differ from the predictions of the Poisson distribution.

An area of shore was surveyed by recording the number of limpets found within each of 100 quadrats. The mean number of limpets per quadrat was found to be 2.05. We saw in Chapter 3, that Equation 3.11 (repeated below) could be used to calculate the probability of a quadrat containing 0, 1, 2, or r limpets:

$$P(r) = \frac{e^{-\mu}\mu^r}{r!}$$

where μ is the mean number of limpets per quadrat. We can use this formula to generate the predicted numbers of quadrats containing any number of limpets simply by multiplying the probability $P(r)$ by the number of quadrats (100 in this example), and then comparing those predictions against the observed frequencies in exactly the same way as we did for Mendel's peas. For example, the following expression predicts the number of quadrats that would contain 1 limpet:

No. of quadrats containing 1 *limpet* $= 100P(1)$

$$= 100\left(\frac{e^{-2.05}2.05^1}{1!}\right)$$

$$= 26.391\ quadrats$$

The actual data (frequencies) that were obtained and the predicted frequencies are shown in Table 9.8.

No. of limpets per quadrat	Observed frequency	Predicted frequency
0	12	12.874
1	29	26.391
2	24	27.050
3	21	18.485
4	7	9.473
5	5	3.884
6	2	1.327

Table 9.8. Observed and expected frequency of the number of limpets per quadrat. The predicted frequencies are based upon a Poisson distribution (mean = 2.05). Poisson probabilities are calculated using Equation 3.11 and then multiplied by the number of quadrats (100 in this example) to give the expected frequencies.

Repetitive manual calculations of the expected values would become quite tedious and prone to error so we should, wherever possible, use Minitab to perform the calculations. Exhibit 9.7. shows the dialog box in which we instruct Minitab to generate the expected probabilities. This dialog box is found by the following menu route:

Calc > Probability Distributions > Poisson

The three columns shown in the exhibit are the same three columns shown in Table 9.8. In this example, the mean number of limpets per quadrat (2.05) is entered, the number of limpets per quadrat for which we wish to calculate the probabilities (0, 1, 2, ..., 6 limpets per quadrat) is stored in $C1$, the observed frequencies in $C2$, and the resulting probabilities calculated by Minitab are stored in $C3$. The column of probabilities would be multiplied by the number of quadrats to give the expected frequencies, thus enabling us to use the expression shown in Exhibit 9.6. to calculate the goodness-of-fit χ^2. Calculate the test statistic and draw appropriate conclusions. Is the dispersion of limpets consistent with the Poisson distribution?* If it is, we would accept the null hypothesis, and conclude that the limpets are distributed randomly. But what would be the biological significance of a non-random pattern? If the limpets were regularly spaced, that would suggest that

* The goodness-of-fit χ^2 = 2.311 with 6 degrees of freedom, giving $P > 0.05$. We accept the null hypothesis that the limpets are dispersed randomly over the shore.

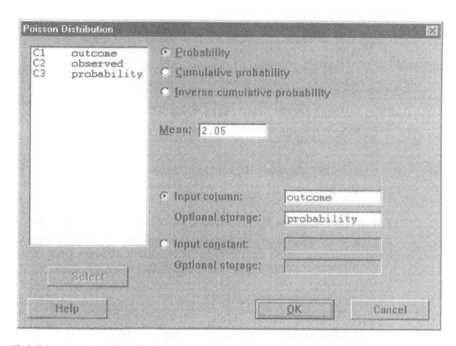

Exhibit 9.7. The Minitab dialog box to generate Poisson probabilities. The frequencies for which probabilities are to be calculated are stored in *C1* and the resulting probabilities are stored in *C3*. The observed frequencies are stored in *C2*, but they are not required for the purposes of calculating the probabilities.

competition was influencing their spacing, forcing individuals to spread themselves as far apart as possible. In contrast, if they were clumped, we could conclude that there was some degree of association between them.

The key question that arises in the case of a non-random pattern is: How would we determine whether the limpets are regularly spaced or aggregated? To answer this we must use a unique property of the Poisson distribution, the mean equals the variance. By comparing the mean with the variance, we can see whether the subjects are aggregated (mean < variance) or regularly spaced (mean > variance). The possible conclusions are summarised in Table 9.9.

This example illustrates the method for generating predicted frequencies from a statistical distribution. Minitab provides this function for several important distributions; however, like all statistical packages, it cannot provide a function for every possible distribution. In such circumstances, it will be necessary to calculate the probabilities using the Minitab calculator function.

Relationship between mean and variance	Spacing	Biological meaning
Mean = Variance	Random	No association
Mean > Variance	Regularly spaced	Competition
Mean < Variance	Aggregated	Association

Table 9.9. Summary of the conclusions that can be drawn from a goodness-of-fit χ^2 where the observed frequencies are compared against the predictions of the Poisson distribution.

9.4   Exercises

1.   The data below were obtained in a study to examine the sex-linked disorder of haemophilia. Analyse the data using a contingency table and conclude whether this sample of data show any evidence of association between gender and haemophilia.

	Male	Female	Total
Haemophiliac	469	35	504
Non-Haemophiliac	5250	6002	11252
Total	5719	6037	11756

2.   Analyse the data in Table 2.2. to determine if there is any association between blood groups and gender.

3.   Examine the data below to determine if there is any evidence for a link between malaria and sickle-cell anaemia. If there is, what biological conclusions can you draw?

	Sickle-cell sufferer	Sickle-cell non-sufferer	Total
Malaria	62	350	412
Non-sufferer	302	420	770
Total	364	770	1134

4.   Three phenotypic characters resulting from a cross of *Drosophila* were recorded with the following frequency: 32:66:29. Is this consistent with the ratio of 1:2:1 predicted by Mendelian genetics? Perform a goodness-of-fit χ^2 and draw appropriate conclusions.

5. If the dispersion of ground nesting birds is random, the distribution of the distance between each nest and its nearest neighbour will follow a Poisson distribution. Analyse the following data and draw appropriate biological conclusions.

Nearest neighbour distance (m)	Observed frequency
1	2
2	14
3	48
4	18
5	10
6	6
7	2

9.5 Summary of Minitab commands

CHISQUARE
TABLE (Cross-Tabulation with Chisquare subcommand)

9.6 Summary

1. Two types of questions have been identified: questions of association where the frequencies of occurrence of different combinations are not predictable in advance; and questions of goodness-of-fit or agreement where we can use independent information to predict expected frequencies.
2. Questions of association are analysed using contingency table χ^2; questions of goodness-of-fit are analysed by the goodness-of-fit χ^2.
3. Various constraints to the use of χ^2 were outlined.
4. When statistically significant results are obtained in questions of association, we use standardized residuals to identify significant associations.
5. Various methods dependent upon the biological and statistical context of the problem are used to follow up deviations from predictions in the goodness-of-fit χ^2.

Concluding Remarks

The objective of this book has been to provide a guide to those involved in answering biological questions, and I hope that I have succeeded in demonstrating that there is more to investigative science than just statistics. Inevitably, a considerable part of the text within this book has been devoted to describing how to carry out the appropriate analysis, but the structure has been designed so that the various techniques have been described within the context of the four main problem types that were defined in Chapter 1 (comparison, sequential relationship, association, and goodness-of-fit or agreement). My intention has been to ensure that your attention has not been drawn too much toward the arithmetic and away from the biological question, because without a biological context in which the answers can be discussed, the statistical results have no real relevance. In the biosciences, statistics are a very useful tool to help us get from a *biological* question to a *biological* conclusion.

In the introduction, I discussed the structure of an investigation and emphasised the planning stages prior to any physical data gathering; and this must include a definition of the objectives of the investigation. When will you know when to stop? When will you know if the investigation has been successful? I cannot stress enough the need to gather as much information about the biological material, organisms, and processes involved in a proposed investigation *prior* to commencing practical work. In biological investigations, statistical analysis alone cannot compensate for lack of biological knowledge.

The book has been structured around a categorisation of problem types in an effort to help identify the sorts of questions that we might have to ask. Categorisations are always instructive at the outset; they help to get our minds in order. But, inevitably, we find that there are always grey areas where individuals do not seem to fit into any classification system or perhaps seem to fit into more than one. For example, we saw problems (such as, the *Drosophila* egg-laying in Chapter 7) that appeared to fall into the grey area between analysis of variance and regression. Another ambiguity to consider is whether correlation is a sequential relationship or a question of association. I hope that by identifying the flaws in these initially neat-looking groupings, I have focussed your mind on defining the question carefully, because, if you cannot specify the question ...

Finally, a colleague of mine once said that if the results of the analysis don't make biological sense, there are two possible reasons: you have discovered a very interesting biological phenomenon, or you have made a mistake in the analysis!

Appendix

A.1. Table of normal probabilities

z	0.00	0.01	0.02	0.03	0.04	0.05	0.06	0.07	0.08	0.09
0.0	0.5000	0.5040	0.5080	0.5120	0.5160	0.5199	0.5239	0.5279	0.5319	0.5359
0.1	0.5398	0.5438	0.5478	0.5517	0.5557	0.5596	0.5636	0.5675	0.5714	0.5753
0.2	0.5793	0.5832	0.5871	0.5910	0.5948	0.5987	0.6026	0.6064	0.6103	0.6141
0.3	0.6179	0.6217	0.6255	0.6293	0.6331	0.6368	0.6406	0.6443	0.6480	0.6517
0.4	0.6554	0.6591	0.6628	0.6664	0.6700	0.6736	0.6772	0.6808	0.6844	0.6879
0.5	0.6915	0.6950	0.6985	0.7019	0.7054	0.7088	0.7123	0.7157	0.7190	0.7224
0.6	0.7257	0.7291	0.7324	0.7357	0.7389	0.7422	0.7454	0.7486	0.7517	0.7549
0.7	0.7580	0.7611	0.7642	0.7673	0.7704	0.7734	0.7764	0.7794	0.7823	0.7852
0.8	0.7881	0.7910	0.7939	0.7967	0.7995	0.8023	0.8051	0.8078	0.8106	0.8133
0.9	0.8159	0.8186	0.8212	0.8238	0.8264	0.8289	0.8315	0.8340	0.8365	0.8389
1.0	0.8413	0.8438	0.8461	0.8485	0.8508	0.8531	0.8554	0.8577	0.8599	0.8621
1.1	0.8643	0.8665	0.8686	0.8708	0.8729	0.8749	0.8770	0.8790	0.8810	0.8830
1.2	0.8849	0.8869	0.8888	0.8907	0.8925	0.8944	0.8962	0.8980	0.8997	0.9015
1.3	0.9032	0.9049	0.9066	0.9082	0.9099	0.9115	0.9131	0.9147	0.9162	0.9177
1.4	0.9192	0.9207	0.9222	0.9236	0.9251	0.9265	0.9279	0.9292	0.9306	0.9319
1.5	0.9332	0.9345	0.9357	0.9370	0.9382	0.9394	0.9406	0.9418	0.9429	0.9441
1.6	0.9452	0.9463	0.9474	0.9484	0.9495	0.9505	0.9515	0.9525	0.9535	0.9545
1.7	0.9554	0.9564	0.9573	0.9582	0.9591	0.9599	0.9608	0.9616	0.9625	0.9633
1.8	0.9641	0.9649	0.9656	0.9664	0.9671	0.9678	0.9686	0.9693	0.9699	0.9706
1.9	0.9713	0.9719	0.9726	0.9732	0.9738	0.9744	0.9750	0.9756	0.9761	0.9767
2.0	0.9772	0.9778	0.9783	0.9788	0.9793	0.9798	0.9803	0.9808	0.9812	0.9817
2.1	0.9821	0.9826	0.9830	0.9834	0.9838	0.9842	0.9846	0.9850	0.9854	0.9857
2.2	0.9861	0.9864	0.9868	0.9871	0.9875	0.9878	0.9881	0.9884	0.9887	0.9890
2.3	0.9893	0.9896	0.9898	0.9901	0.9904	0.9906	0.9909	0.9911	0.9913	0.9916
2.4	0.9918	0.9920	0.9922	0.9925	0.9927	0.9929	0.9931	0.9932	0.9934	0.9936
2.5	0.9938	0.9940	0.9941	0.9943	0.9945	0.9946	0.9948	0.9949	0.9951	0.9952
2.6	0.9953	0.9955	0.9956	0.9957	0.9959	0.9960	0.9961	0.9962	0.9963	0.9964
2.7	0.9965	0.9966	0.9967	0.9968	0.9969	0.9970	0.9971	0.9972	0.9973	0.9974
2.8	0.9974	0.9975	0.9976	0.9977	0.9977	0.9978	0.9979	0.9979	0.9980	0.9981
2.9	0.9981	0.9982	0.9982	0.9983	0.9984	0.9984	0.9985	0.9985	0.9986	0.9986

z	0.00	0.01	0.02	0.03	0.04	0.05	0.06	0.07	0.08	0.09
3.0	0.9987	0.9987	0.9987	0.9988	0.9988	0.9989	0.9989	0.9989	0.9990	0.9990
3.1	0.9990	0.9991	0.9991	0.9991	0.9992	0.9992	0.9992	0.9992	0.9993	0.9993
3.2	0.9993	0.9993	0.9994	0.9994	0.9994	0.9994	0.9994	0.9995	0.9995	0.9995
3.3	0.9995	0.9995	0.9995	0.9996	0.9996	0.9996	0.9996	0.9996	0.9996	0.9997
3.4	0.9997	0.9997	0.9997	0.9997	0.9997	0.9997	0.9997	0.9997	0.9997	0.9998
3.5	0.9998	0.9998	0.9998	0.9998	0.9998	0.9998	0.9998	0.9998	0.9998	0.9998
3.6	0.9998	0.9998	0.9999	0.9999	0.9999	0.9999	0.9999	0.9999	0.9999	0.9999
3.7	0.9999	0.9999	0.9999	0.9999	0.9999	0.9999	0.9999	0.9999	0.9999	0.9999
3.8	0.9999	0.9999	0.9999	0.9999	0.9999	0.9999	0.9999	0.9999	0.9999	0.9999
3.9	1.0000	1.0000	1.0000	1.0000	1.0000	1.0000	1.0000	1.0000	1.0000	1.0000

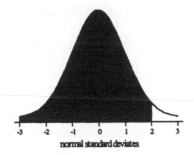

normal standard deviates

Figure A.1.

A.2.   Student's t-distribution

df	0.1	0.05	0.02	0.01	0.001
1	6.3138	12.7062	31.8206	63.6570	636.6072
2	2.9200	4.3027	6.9646	9.9248	31.5983
3	2.3534	3.1824	4.5407	5.8409	12.9238
4	2.1319	2.7764	3.7470	4.6041	8.6102
5	2.0150	2.5706	3.3649	4.0322	6.8688
6	1.9432	2.4469	3.1427	3.7075	5.9588
7	1.8946	2.3646	2.9980	3.4995	5.4079
8	1.8595	2.3060	2.8965	3.3554	5.0413
9	1.8331	2.2622	2.8214	3.2499	4.7809
10	1.8125	2.2281	2.7638	3.1693	4.5869
11	1.7959	2.2010	2.7181	3.1058	4.4370
12	1.7823	2.1788	2.6810	3.0546	4.3178
13	1.7709	2.1604	2.6503	3.0123	4.2208
14	1.7613	2.1448	2.6245	2.9769	4.1404
15	1.7531	2.1315	2.6025	2.9467	4.0727
16	1.7459	2.1199	2.5835	2.9208	4.0150
17	1.7396	2.1098	2.5669	2.8982	3.9651
18	1.7341	2.1009	2.5524	2.8784	3.9216
19	1.7291	2.0930	2.5395	2.8610	3.8834
20	1.7247	2.0860	2.5280	2.8453	3.8495
21	1.7208	2.0796	2.5176	2.8314	3.8193
22	1.7172	2.0739	2.5083	2.8188	3.7921
23	1.7139	2.0687	2.4999	2.8073	3.7676
24	1.7109	2.0639	2.4922	2.7969	3.7454
25	1.7081	2.0595	2.4851	2.7874	3.7251
26	1.7056	2.0555	2.4786	2.7787	3.7066
27	1.7033	2.0518	2.4727	2.7707	3.6896
28	1.7011	2.0484	2.4671	2.7633	3.6739
29	1.6991	2.0452	2.4620	2.7564	3.6594
30	1.6973	2.0423	2.4573	2.7500	3.6459
40	1.6839	2.0211	2.4233	2.7045	3.5509
50	1.6759	2.0086	2.4033	2.6778	3.4960
60	1.6707	2.0003	2.3902	2.6604	3.4604
70	1.6669	1.9944	2.3808	2.6480	3.4351
80	1.6641	1.9901	2.3739	2.6387	3.4164

df	0.1	0.05	0.02	0.01	0.001
90	1.6620	1.9867	2.3685	2.6316	3.4020
100	1.6602	1.9840	2.3642	2.6259	3.3905
∞	1.6450	1.9600	2.3260	2.5760	3.2910

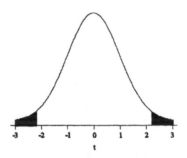

Figure A.2.

A.3. Two thousand five hundred random numbers

	1	2	3	4	5	6	7	8	9	10	
1	49452	78423	50347	08383	25418	16917	48276	75101	06997	50465	1
2	55376	31660	50016	17546	55047	93356	03106	37394	34093	87392	2
3	86205	25179	85053	61300	66532	36308	24179	47654	59963	77377	3
4	62719	77982	15816	06336	19212	88181	82723	64455	73562	02390	4
5	69245	88742	69626	24780	39507	65625	35632	98879	91384	55528	5
6	60578	52501	72438	28730	04463	96069	24928	25021	87795	18333	6
7	95113	60550	96725	76448	29523	82553	99501	89929	85851	76039	7
8	46252	92473	53605	85869	77219	95369	08301	55862	00643	15082	8
9	79853	98159	56662	26172	87414	47557	66340	76819	77212	03229	9
10	37020	60735	67472	93557	04023	15657	68563	86650	69446	41573	10
11	06368	03205	23034	60676	83375	25397	87729	12153	43262	62298	11
12	96699	34359	48870	87247	23592	83599	90524	20813	25598	04209	12
13	75363	42175	26533	09516	68350	93344	41277	82867	94281	75622	13
14	52478	10883	58501	53315	60952	76339	27920	12372	22028	99599	14
15	90100	97954	74984	71031	93096	67277	28255	97015	36814	06408	15
16	35062	55427	28114	22043	25443	87687	12742	60775	02449	69959	16
17	19753	21933	65831	63331	69771	49184	63523	47825	57016	10849	17
18	46287	32491	88651	22666	80375	07585	12269	28208	79338	78246	18
19	94864	62124	79805	49623	62221	83681	85043	22607	56953	53252	19
20	98982	21984	84121	22667	29790	54031	39748	41526	47483	45327	20
21	79252	98501	98689	76133	15811	03684	66599	29604	26314	03940	21
22	91368	18249	80123	37834	32573	81406	46233	63796	94310	06621	22
23	15025	23984	84757	96455	23196	19902	64838	64173	21074	20471	23
24	57234	90957	81099	01462	73976	83339	14839	36338	43944	05726	24
25	95227	22489	94917	20567	75375	69249	40424	71660	98422	19549	25
26	09254	47034	63229	69045	92422	28664	86809	40513	34647	17079	26
27	68821	35240	47011	15170	29120	90589	92229	13736	66388	50144	27
28	26140	82308	92973	21116	66684	00759	41405	66045	61428	76602	28
29	75988	34076	19720	09044	22413	13947	88883	86459	03967	09473	29
30	88554	73134	37292	41502	82572	90501	31140	99841	16988	10785	30
31	03247	37720	76745	27155	38765	04399	39746	55112	68237	38432	31
32	34290	35459	48345	66391	23244	16328	57841	89421	77404	21617	32
33	90672	35472	01159	81075	01052	38621	99387	68454	72011	38954	33
34	88659	58406	07482	66240	45285	52776	45946	23039	06591	50153	34
35	49075	94987	10911	13616	39258	48999	28779	52536	88585	69084	35
36	55076	48698	71427	66082	72590	34372	16235	99474	12402	97243	36
37	42392	55315	89345	70207	83102	79709	00756	12478	29646	43832	37
38	40893	02106	49474	91895	08078	28621	42764	81881	25806	63736	38
39	34774	68991	41803	30223	54834	54480	51648	00684	18416	21220	39
40	31335	62580	52154	63589	20731	57559	16007	38125	63468	09116	40

	1	2	3	4	5	6	7	8	9	10	
41	26472	84906	20851	97074	78885	88051	59891	67402	52875	40007	41
42	56714	09677	02852	68981	65324	30403	32872	50544	81107	96149	42
43	67106	86419	14485	72436	78278	12177	73706	66529	99797	31105	43
44	71509	78636	48121	99080	54271	32961	33035	20172	77786	29596	44
45	33666	73355	15203	40265	46896	72926	65855	62083	89515	30092	45
46	84237	58947	68166	26180	40408	28696	64290	74175	66011	94846	46
47	62104	53965	24895	95223	19697	50338	86545	91867	53447	77780	47
48	45950	24108	42879	67034	13406	70454	74418	89723	48690	21471	48
49	50411	24893	76178	13990	42428	71187	75021	89348	51497	48572	49
50	58696	40780	57572	23033	86284	14101	34014	29346	69886	47061	50

A.4. Binomial probabilities providing critical values for a two-tailed sign test (sample sizes less than 5 cannot distinguish a difference)

n	0.2	0.1	0.05	0.02	0.01	0.002	0.001
5	5	5					
6	6	6	6				
7	6	7	7	7			
8	7	7	8	8	8		
9	7	8	8	9	9		
10	8	9	9	10	10	10	
11	9	9	10	10	11	11	11
12	9	10	10	11	11	12	12
13	10	10	11	12	12	13	13
14	10	11	12	12	13	13	14
15	11	12	12	13	13	14	14
16	12	12	13	14	14	15	15
17	12	13	13	14	15	16	16
18	13	13	14	15	15	16	17
19	13	14	15	15	16	17	17
20	14	15	15	16	17	18	18
21	14	15	16	17	17	18	19
22	15	16	17	17	18	19	19
23	16	16	17	18	19	20	20
24	16	17	18	19	19	20	21
25	17	18	18	19	20	21	21
26	17	18	19	20	20	22	22
27	18	19	20	20	21	22	23
28	18	19	20	21	22	23	23
29	19	20	21	22	22	24	24
30	20	20	21	22	23	24	25
31	20	21	22	23	24	25	25
32	21	22	23	24	24	26	26
33	21	22	23	24	25	26	27
34	22	23	24	25	25	27	27
35	22	23	24	25	26	27	28

A.5.   Critical values for the Wilcoxon test (two-tailed)

N	0.2	0.1	0.05	0.02	0.01	0.005	0.001
5	13	15					
6	18	19	21				
7	23	25	26	28			
8	28	31	33	35	36		
9	35	37	40	42	44	45	
10	41	45	47	50	52	54	
11	49	53	56	59	61	63	66
12	57	61	65	69	71	73	77
13	65	70	74	79	82	84	89
14	74	80	84	90	93	96	101
15	84	90	95	101	105	108	114
16	94	101	107	113	117	121	128
17	105	112	119	126	130	134	142
18	116	124	131	139	144	148	157
19	128	137	144	153	158	163	172
20	141	150	158	167	173	178	189

A.6.   Critical values for the Mann-Whitney test

n_1	n_2	0.1 0.05	0.05 0.025	0.02 0.01	0.01 0.005	two-tailed one-tailed
2	4	0	0	0	0	
	5	1	0	0	0	
	6	1	0	0	0	
	7	1	0	0	0	
	8	2	1	0	0	
	9	2	1	0	0	
	10	2	1	0	0	
	11	2	1	0	0	
	12	3	2	0	0	
	13	3	2	1	0	
	14	4	2	1	0	
	15	4	2	1	0	
	16	4	2	1	0	
	17	4	3	1	0	
	18	5	3	1	0	
	19	5	3	2	1	
	20	5	3	2	1	
3	2	0	0	0	0	
	3	1	0	0	0	
	4	1	0	0	0	
	5	2	1	0	0	
	6	3	2	0	0	
	7	3	2	1	0	
	8	4	3	1	0	
	9	5	3	2	1	
	10	5	4	2	1	
	11	6	4	2	1	
	12	6	5	3	2	
	13	7	5	3	2	
	14	8	6	3	2	
	15	8	6	4	3	
	16	9	7	4	3	
	17	10	7	5	3	
	18	10	8	5	3	
	19	11	8	5	4	
	20	12	9	6	4	
4	2	0	0	0	0	
	3	1	0	0	0	

n_1	n_2	0.1 / 0.05	0.05 / 0.025	0.02 / 0.01	0.01 / 0.005	
						two-tailed
						one-tailed
4	4	2	1	0	0	
	5	3	2	1	0	
	6	4	3	2	1	
	7	5	4	2	1	
	8	6	5	3	2	
	9	7	5	4	2	
	10	8	6	4	3	
	11	9	7	5	3	
	12	10	8	6	4	
	13	11	9	6	4	
	14	12	10	7	5	
	15	13	11	8	6	
	16	15	12	8	6	
	17	16	12	9	7	
	18	17	13	10	7	
	19	18	14	10	8	
	20	19	15	11	9	
5	2	1	0	0	0	
	3	2	1	0	0	
	4	3	2	1	0	
	5	5	3	2	1	
	6	6	4	3	2	
	7	7	6	4	2	
	8	9	7	5	3	
	9	10	8	6	4	
	10	12	9	7	5	
	11	13	10	8	6	
	12	14	12	9	7	
	13	16	13	10	8	
	14	17	14	11	8	
	15	19	15	12	9	
	16	20	16	13	10	
	17	21	18	14	11	
	18	23	19	15	12	
	19	24	20	16	13	
	20	26	21	17	14	
6	2	1	0	0	0	
	3	3	2	0	0	
	4	4	3	2	1	
	5	6	4	3	2	

n_1	n_2	0.1 / 0.05	0.05 / 0.025	0.02 / 0.01	0.01 / 0.005	
		0.1	0.05	0.02	0.01	*two-tailed*
n_1	n_2	0.05	0.025	0.01	0.005	*one-tailed*
6	6	8	6	4	3	
	7	9	7	5	4	
	8	11	9	7	5	
	9	13	11	8	6	
	10	15	12	9	7	
	11	17	14	10	8	
	12	18	15	12	10	
	13	20	17	13	11	
	14	22	18	14	12	
	15	24	20	16	13	
	16	26	22	17	14	
	17	27	23	19	16	
	18	29	25	20	17	
	19	31	26	21	18	
	20	33	28	23	19	
7	2	1	0	0	0	
	3	3	2	1	0	
	4	5	4	2	1	
	5	7	6	4	2	
	6	9	7	5	4	
	7	12	9	7	5	
	8	14	11	8	7	
	9	16	13	10	8	
	10	18	15	12	10	
	11	20	17	13	11	
	12	22	19	15	13	
	13	25	21	17	14	
	14	27	23	18	16	
	15	29	25	20	17	
	16	31	27	22	19	
	17	34	29	24	20	
	18	36	31	25	22	
	19	38	33	27	23	
	20	40	35	29	25	
8	2	2	1	0	0	
	3	4	3	1	0	
	4	6	5	3	2	
	5	9	7	5	3	
	6	11	9	7	5	
	7	14	11	8	7	

n_1	n_2	0.1 / 0.05	0.05 / 0.025	0.02 / 0.01	0.01 / 0.005	*two-tailed* / *one-tailed*
8	8	16	14	10	8	
	9	19	16	12	10	
	10	21	18	14	12	
	11	24	20	16	14	
	12	27	23	18	16	
	13	29	25	21	18	
	14	32	27	23	19	
	15	34	30	25	21	
	16	37	32	27	23	
	17	40	35	29	25	
	18	42	37	31	27	
	19	45	39	33	29	
	20	48	42	35	31	
9	2	2	1	0	0	
	3	5	3	2	1	
	4	7	5	4	2	
	5	10	8	6	4	
	6	13	11	8	6	
	7	16	13	10	8	
	8	19	16	12	10	
	9	22	18	15	12	
	10	25	21	17	14	
	11	28	24	19	17	
	12	31	27	22	19	
	13	34	29	24	21	
	14	37	32	27	23	
	15	40	35	29	25	
	16	43	38	32	28	
	17	46	40	34	30	
	18	49	43	37	32	
	19	52	46	39	34	
	20	55	49	41	37	
10	2	2	1	0	0	
	3	5	4	2	1	
	4	8	6	4	3	
	5	12	9	7	5	
	6	15	12	9	7	
	7	18	15	12	10	
	8	21	18	14	12	
	9	25	21	17	14	

n_1	n_2	0.1 0.05	0.05 0.025	0.02 0.01	0.01 0.005	*two-tailed* *one-tailed*
10	10	28	24	20	17	
	11	32	27	23	19	
	12	35	30	25	22	
	13	38	34	28	25	
	14	42	37	31	27	
	15	45	40	34	30	
	16	49	43	37	32	
	17	52	46	39	35	
	18	56	49	42	38	
	19	59	53	45	40	
	20	63	56	48	43	
11	2	2	1	0	0	
	3	6	4	2	1	
	4	9	7	5	3	
	5	13	10	8	6	
	6	17	14	10	8	
	7	20	17	13	11	
	8	24	20	16	14	
	9	28	24	19	17	
	10	32	27	23	19	
	11	35	31	26	22	
	12	39	34	29	25	
	13	43	38	32	28	
	14	47	41	35	31	
	15	51	45	38	34	
	16	55	48	42	37	
	17	58	52	45	40	
	18	62	56	48	43	
	19	66	59	51	46	
	20	70	63	54	49	
12	2	3	2	0	0	
	3	6	5	3	2	
	4	10	8	6	4	
	5	14	12	9	7	
	6	18	15	12	10	
	7	22	19	15	13	
	8	27	23	18	16	
	9	31	27	22	19	
	10	35	30	25	22	
	11	39	34	29	25	

n_1	n_2	0.1 0.05	0.05 0.025	0.02 0.01	0.01 0.005	two-tailed one-tailed
12	12	43	38	32	28	
	13	48	42	36	32	
	14	52	46	39	35	
	15	56	50	43	38	
	16	61	54	47	42	
	17	65	58	50	45	
	18	69	62	54	48	
	19	73	66	57	52	
	20	79	70	61	55	
13	2	3	2	1	0	
	3	7	5	3	2	
	4	11	9	6	4	
	5	16	13	10	8	
	6	20	17	13	11	
	7	25	21	17	14	
	8	29	25	21	18	
	9	34	29	24	21	
	10	38	34	28	25	
	11	43	38	32	28	
	12	48	42	36	32	
	13	52	46	40	35	
	14	57	51	44	39	
	15	62	55	48	43	
	16	66	60	52	46	
	17	71	64	56	50	
	18	76	68	60	54	
	19	81	73	64	58	
	20	85	77	68	61	
14	2	4	2	1	0	
	3	8	6	3	2	
	4	12	10	7	5	
	5	17	14	11	8	
	6	22	18	14	12	
	7	27	23	18	16	
	8	32	27	23	19	
	9	37	32	27	23	
	10	42	37	31	27	
	11	47	41	35	31	
	12	52	46	39	35	
	13	57	51	44	39	

n_1	n_2	0.1 / 0.05	0.05 / 0.025	0.02 / 0.01	0.01 / 0.005	*two-tailed* / *one-tailed*
14	14	62	56	48	43	
	15	67	60	52	47	
	16	72	65	57	51	
	17	78	70	61	55	
	18	83	75	66	59	
	19	88	79	70	64	
	20	93	84	74	68	
15	2	4	2	1	0	
	3	8	6	4	3	
	4	13	11	8	6	
	5	19	15	12	9	
	6	24	20	16	13	
	7	29	25	20	17	
	8	34	30	25	21	
	9	40	35	29	25	
	10	45	40	34	30	
	11	51	45	38	34	
	12	56	50	43	38	
	13	62	55	48	43	
	14	67	60	52	47	
	15	73	65	57	52	
	16	78	71	62	58	
	17	84	76	67	61	
	18	89	81	71	65	
	19	95	86	76	70	
	20	101	91	81	74	
16	2	4	2	1	0	
	3	9	7	4	3	
	4	15	12	8	6	
	5	20	16	13	10	
	6	26	22	17	14	
	7	31	27	22	19	
	8	37	32	27	23	
	9	43	38	32	28	
	10	49	43	37	32	
	11	55	48	42	37	
	12	61	54	47	42	
	13	66	60	52	46	
	14	72	65	57	51	
	15	78	71	62	56	

		0.1	0.05	0.02	0.01	*two-tailed*
n_1	n_2	0.05	0.025	0.01	0.005	*one-tailed*
16	16	84	76	67	61	
	17	90	82	72	66	
	18	96	87	77	71	
	19	102	93	83	75	
	20	108	99	88	80	
17	2	4	3	1	0	
	3	10	7	5	3	
	4	16	12	9	7	
	5	21	18	14	11	
	6	27	23	19	16	
	7	34	29	24	20	
	8	40	35	29	25	
	9	46	40	34	30	
	10	52	46	39	35	
	11	58	52	45	40	
	12	65	58	50	45	
	13	71	64	56	50	
	14	78	70	61	55	
	15	84	76	67	61	
	16	90	82	72	66	
	17	97	88	78	71	
	18	103	94	83	76	
	19	110	100	89	82	
	20	116	106	94	87	
18	2	5	3	1	0	
	3	10	8	5	3	
	4	17	13	10	7	
	5	23	19	15	12	
	6	29	25	20	17	
	7	36	31	25	22	
	8	42	37	31	27	
	9	49	43	37	32	
	10	56	49	42	38	
	11	62	56	48	43	
	12	69	62	54	48	
	13	76	68	60	54	
	14	83	75	66	59	
	15	89	81	71	65	
	16	96	87	77	71	

n_1	n_2	0.1 0.05	0.05 0.025	0.02 0.01	0.01 0.005	*two-tailed* *one-tailed*
18	17	103	94	83	76	
	18	110	100	89	82	
	19	117	107	95	98	
	20	124	113	10	93	
19	2	5	3	2	1	
	3	11	8	5	4	
	4	18	14	10	8	
	5	24	20	16	13	
	6	31	26	21	18	
	7	38	33	27	23	
	8	45	39	33	29	
	9	52	46	39	34	
	10	59	53	45	40	
	11	66	59	51	46	
	12	73	66	57	52	
	13	81	73	64	58	
	14	88	79	70	64	
	15	95	86	76	70	
	16	102	93	83	75	
	17	110	100	89	82	
	18	117	107	95	88	
	19	124	114	102	94	
	20	131	120	108	100	
20	2	5	3	2	1	
	3	12	9	6	4	
	4	19	15	11	9	
	5	26	21	17	14	
	6	33	28	23	19	
	7	40	35	29	25	
	8	48	42	35	31	
	9	55	49	41	37	
	10	63	56	48	43	
	11	70	63	54	49	
	12	78	70	61	55	
	13	85	77	68	61	
	14	93	84	74	68	
	15	101	91	81	74	
	16	108	99	88	80	
	17	116	106	94	87	

n_1	n_2	0.1 0.05	0.05 0.025	0.02 0.01	0.01 0.005	two-tailed one-tailed
20	18	124	113	101	93	
	19	131	120	108	100	
	20	139	128	115	106	

A.7a. Variance Ratio (F) Tables P = 0.05

Treatment d.f.

Error df	1	2	3	4	5	6	7	8	9	10	11	12	13	14	15	20	30	60	120	∞
1	161	199	216	225	230	234	237	239	241	242	243	244	245	245	246	248	250	252	253	254
2	18.51	19.00	19.16	19.25	19.30	19.33	19.35	19.37	19.38	19.40	19.41	19.41	19.42	19.42	19.43	19.45	19.46	19.48	19.49	19.50
3	10.13	9.55	9.28	9.12	9.01	8.94	8.89	8.85	8.81	8.79	8.76	8.74	8.73	8.71	8.70	8.66	8.62	8.57	8.55	8.53
4	7.71	6.94	6.59	6.39	6.26	6.16	6.09	6.04	6.00	5.96	5.94	5.91	5.89	5.87	5.86	5.80	5.75	5.69	5.66	5.63
5	6.61	5.79	5.41	5.19	5.05	4.95	4.88	4.82	4.77	4.74	4.70	4.68	4.66	4.64	4.62	4.56	4.50	4.43	4.40	4.37
6	5.99	5.14	4.76	4.53	4.39	4.28	4.21	4.15	4.10	4.06	4.03	4.00	3.98	3.96	3.94	3.87	3.81	3.74	3.70	3.67
7	5.59	4.74	4.35	4.12	3.97	3.87	3.79	3.73	3.68	3.64	3.60	3.57	3.55	3.53	3.51	3.44	3.38	3.30	3.27	3.23
8	5.32	4.46	4.07	3.84	3.69	3.58	3.50	3.44	3.39	3.35	3.31	3.28	3.26	3.24	3.22	3.15	3.08	3.01	2.97	2.93
9	5.12	4.26	3.86	3.63	3.48	3.37	3.29	3.23	3.18	3.14	3.10	3.07	3.05	3.03	3.01	2.94	2.86	2.79	2.75	2.71
10	4.96	4.10	3.71	3.48	3.33	3.22	3.14	3.07	3.02	2.98	2.94	2.91	2.89	2.86	2.85	2.77	2.70	2.62	2.58	2.54
11	4.84	3.98	3.59	3.36	3.20	3.09	3.01	2.95	2.90	2.85	2.82	2.79	2.76	2.74	2.72	2.65	2.57	2.49	2.45	2.40
12	4.75	3.89	3.49	3.26	3.11	3.00	2.91	2.85	2.80	2.75	2.72	2.69	2.66	2.64	2.62	2.54	2.47	2.38	2.34	2.30
13	4.67	3.81	3.41	3.18	3.03	2.92	2.83	2.77	2.71	2.67	2.63	2.60	2.58	2.55	2.53	2.46	2.38	2.30	2.25	2.21
14	4.60	3.74	3.34	3.11	2.96	2.85	2.76	2.70	2.65	2.60	2.57	2.53	2.51	2.48	2.46	2.39	2.31	2.22	2.18	2.13
15	4.54	3.68	3.29	3.06	2.90	2.79	2.71	2.64	2.59	2.54	2.51	2.48	2.45	2.42	2.40	2.33	2.25	2.16	2.11	2.07
16	4.49	3.63	3.24	3.01	2.85	2.74	2.66	2.59	2.54	2.49	2.46	2.42	2.40	2.37	2.35	2.28	2.19	2.11	2.06	2.01
17	4.45	3.59	3.20	2.96	2.81	2.70	2.61	2.55	2.49	2.45	2.41	2.38	2.35	2.33	2.31	2.23	2.15	2.06	2.01	1.96
18	4.41	3.55	3.16	2.93	2.77	2.66	2.58	2.51	2.46	2.41	2.37	2.34	2.31	2.29	2.27	2.19	2.11	2.02	1.97	1.92
19	4.38	3.52	3.13	2.90	2.74	2.63	2.54	2.48	2.42	2.38	2.34	2.31	2.28	2.26	2.23	2.16	2.07	1.98	1.93	1.88

A.7a. Variance Ratio (F) Tables P = 0.05 (continued)

Treatment d.f.

Error df	1	2	3	4	5	6	7	8	9	10	11	12	13	14	15	20	30	60	120	∞
20	4.35	3.49	3.10	2.87	2.71	2.60	2.51	2.45	2.39	2.35	2.31	2.28	2.25	2.23	2.20	2.12	2.04	1.95	1.90	1.84
30	4.17	3.32	2.92	2.69	2.53	2.42	2.33	2.27	2.21	2.16	2.13	2.09	2.06	2.04	2.01	1.93	1.84	1.74	1.68	1.62
40	4.08	3.23	2.84	2.61	2.45	2.34	2.25	2.18	2.12	2.08	2.04	2.00	1.97	1.95	1.92	1.84	1.74	1.64	1.58	1.51
50	4.03	3.18	2.79	2.56	2.40	2.29	2.20	2.13	2.07	2.03	1.99	1.95	1.92	1.89	1.87	1.78	1.69	1.58	1.51	1.44
60	4.00	3.15	2.76	2.53	2.37	2.25	2.17	2.10	2.04	1.99	1.95	1.92	1.89	1.86	1.84	1.75	1.65	1.53	1.47	1.39
70	3.98	3.13	2.74	2.50	2.35	2.23	2.14	2.07	2.02	1.97	1.93	1.89	1.86	1.84	1.81	1.72	1.62	1.50	1.44	1.35
80	3.96	3.11	2.72	2.49	2.33	2.21	2.13	2.06	2.00	1.95	1.91	1.88	1.84	1.82	1.79	1.70	1.60	1.48	1.41	1.32
90	3.95	3.10	2.71	2.47	2.32	2.20	2.11	2.04	1.99	1.94	1.90	1.86	1.83	1.80	1.78	1.69	1.59	1.46	1.39	1.30
100	3.94	3.09	2.70	2.46	2.31	2.19	2.10	2.03	1.97	1.93	1.89	1.85	1.82	1.79	1.77	1.68	1.57	1.45	1.38	1.28
120	3.92	3.07	2.68	2.45	2.29	2.18	2.09	2.02	1.96	1.91	1.87	1.83	1.80	1.78	1.75	1.66	1.55	1.43	1.35	1.25
∞	3.84	3.00	2.61	2.37	2.21	2.10	2.01	1.94	1.88	1.83	1.79	1.75	1.72	1.69	1.67	1.57	1.46	1.32	1.22	1.00

A.7b. Variance Ratio (*F*) Tables *P* = 0.01

Treatment d.f.

Error df	1	2	3	4	5	6	7	8	9	10	11	12	13	14	15	20	30	60	120	∞
1	4052	4999	5403	5625	5764	5859	5928	5981	6022	6056	6083	6106	6126	6143	6157	6209	6261	6313	6339	6370
2	98.50	99.00	99.17	99.25	99.30	99.33	99.36	99.37	99.39	99.40	99.41	99.42	99.42	99.43	99.43	99.45	99.46	99.48	99.49	99.50
3	34.12	30.82	29.46	28.71	28.24	27.91	27.67	27.49	27.35	27.23	27.13	27.05	26.98	26.92	26.87	26.69	26.50	26.32	26.22	26.10
4	21.20	18.00	16.69	15.98	15.52	15.21	14.98	14.80	14.66	14.55	14.45	14.37	14.31	14.25	14.20	14.02	13.84	13.65	13.56	13.50
5	16.26	13.27	12.06	11.39	10.97	10.67	10.46	10.29	10.16	10.05	9.96	9.89	9.82	9.77	9.72	9.55	9.38	9.20	9.11	9.02
6	13.75	10.92	9.78	9.15	8.75	8.47	8.26	8.10	7.98	7.87	7.79	7.72	7.66	7.61	7.56	7.40	7.23	7.06	6.97	6.88
7	12.25	9.55	8.45	7.85	7.46	7.19	6.99	6.84	6.72	6.62	6.54	6.47	6.41	6.36	6.31	6.16	5.99	5.82	5.74	5.65
8	11.26	8.65	7.59	7.01	6.63	6.37	6.18	6.03	5.91	5.81	5.73	5.67	5.61	5.56	5.52	5.36	5.20	5.03	4.95	4.86
9	10.56	8.02	6.99	6.42	6.06	5.80	5.61	5.47	5.35	5.26	5.18	5.11	5.05	5.01	4.96	4.81	4.65	4.48	4.40	4.31
10	10.04	7.56	6.55	5.99	5.64	5.39	5.20	5.06	4.94	4.85	4.77	4.71	4.65	4.60	4.56	4.41	4.25	4.08	4.00	3.91
11	9.65	7.21	6.22	5.67	5.32	5.07	4.89	4.74	4.63	4.54	4.46	4.40	4.34	4.29	4.25	4.10	3.94	3.78	3.69	3.60
12	9.33	6.93	5.95	5.41	5.06	4.82	4.64	4.50	4.39	4.30	4.22	4.16	4.10	4.05	4.01	3.86	3.70	3.54	3.45	3.36
13	9.07	6.70	5.74	5.21	4.86	4.62	4.44	4.30	4.19	4.10	4.02	3.96	3.91	3.86	3.82	3.66	3.51	3.34	3.25	3.17
14	8.86	6.51	5.56	5.04	4.70	4.46	4.28	4.14	4.03	3.94	3.86	3.80	3.75	3.70	3.66	3.51	3.35	3.18	3.09	3.00
15	8.68	6.36	5.42	4.89	4.56	4.32	4.14	4.00	3.89	3.80	3.73	3.67	3.61	3.56	3.52	3.37	3.21	3.05	2.96	2.87
16	8.53	6.23	5.29	4.77	4.44	4.20	4.03	3.89	3.78	3.69	3.62	3.55	3.50	3.45	3.41	3.26	3.10	2.93	2.84	2.75
17	8.40	6.11	5.19	4.67	4.34	4.10	3.93	3.79	3.68	3.59	3.52	3.46	3.40	3.35	3.31	3.16	3.00	2.83	2.75	2.65
18	8.29	6.01	5.09	4.58	4.25	4.01	3.84	3.71	3.60	3.51	3.43	3.37	3.32	3.27	3.23	3.08	2.92	2.75	2.66	2.57
19	8.19	5.93	5.01	4.50	4.17	3.94	3.77	3.63	3.52	3.43	3.36	3.30	3.24	3.19	3.15	3.00	2.84	2.67	2.58	2.49

A.7b. Variance Ratio (F) Tables $P = 0.01$ (*continued*)

Treatment d.f.

Error df	1	2	3	4	5	6	7	8	9	10	11	12	13	14	15	20	30	60	120	∞
20	8.10	5.85	4.94	4.43	4.10	3.87	3.70	3.56	3.46	3.37	3.29	3.23	3.18	3.13	3.09	2.94	2.78	2.61	2.52	2.42
30	7.56	5.39	4.51	4.02	3.70	3.47	3.30	3.17	3.07	2.98	2.91	2.84	2.79	2.74	2.70	2.55	2.39	2.21	2.11	2.01
40	7.31	5.18	4.31	3.83	3.51	3.29	3.12	2.99	2.89	2.80	2.73	2.66	2.61	2.56	2.52	2.37	2.20	2.02	1.92	1.80
50	7.17	5.06	4.20	3.72	3.41	3.19	3.02	2.89	2.79	2.70	2.63	2.56	2.51	2.46	2.42	2.27	2.10	1.91	1.80	1.68
60	7.08	4.98	4.13	3.65	3.34	3.12	2.95	2.82	2.72	2.63	2.56	2.50	2.44	2.39	2.35	2.20	2.03	1.84	1.73	1.60
70	7.01	4.92	4.07	3.60	3.29	3.07	2.91	2.78	2.67	2.59	2.51	2.45	2.40	2.35	2.31	2.15	1.98	1.78	1.67	1.54
80	6.96	4.88	4.04	3.56	3.26	3.04	2.87	2.74	2.64	2.55	2.48	2.42	2.36	2.31	2.27	2.12	1.94	1.75	1.63	1.49
90	6.93	4.85	4.01	3.54	3.23	3.01	2.84	2.72	2.61	2.52	2.45	2.39	2.33	2.29	2.24	2.09	1.92	1.72	1.60	1.46
100	6.90	4.82	3.98	3.51	3.21	2.99	2.82	2.69	2.59	2.50	2.43	2.37	2.31	2.27	2.22	2.07	1.89	1.69	1.57	1.43
120	6.85	4.79	3.95	3.48	3.17	2.96	2.79	2.66	2.56	2.47	2.40	2.34	2.28	2.23	2.19	2.03	1.86	1.66	1.53	1.38
∞	6.64	4.61	3.78	3.32	3.02	2.80	2.64	2.51	2.41	2.32	2.25	2.18	2.13	2.08	2.04	1.88	1.70	1.47	1.32	1.00

A.8. Studentised Range (*q*) Tables *P* = 0.05

No. of treatment means (k)

Error df	2	3	4	5	6	7	8	9	10	11	12	13	14	15	20	30	60
1	17.97	26.98	32.82	37.08	40.41	43.12	45.40	47.36	49.07	50.59	51.96	53.20	54.33	55.36	59.56	65.15	73.97
2	6.09	8.33	9.80	10.88	11.74	12.44	13.03	13.54	13.99	14.39	14.75	15.08	15.38	15.65	16.77	18.27	20.66
3	4.50	5.91	6.83	7.50	8.04	8.48	8.85	9.12	9.46	9.72	9.95	10.15	10.35	10.53	11.24	12.21	13.76
4	3.93	5.04	5.78	6.29	6.71	7.05	7.35	7.60	7.83	8.03	8.21	8.37	8.53	8.66	9.23	10.00	11.24
5	3.64	4.60	5.22	5.67	6.03	6.33	6.58	6.80	7.00	7.17	7.32	7.47	7.60	7.72	8.21	8.88	9.95
6	3.46	4.34	4.90	5.31	5.63	5.90	6.12	6.32	6.49	6.65	6.79	6.92	7.03	7.14	7.59	8.19	9.16
7	3.34	4.17	4.68	5.06	5.36	5.61	5.82	6.00	6.16	6.30	6.43	6.55	6.66	6.76	7.17	7.73	8.63
8	3.26	4.04	4.53	4.89	5.17	5.40	5.60	5.77	5.92	6.05	6.18	6.28	6.40	6.48	6.87	7.40	8.25
9	3.20	3.95	4.42	4.76	5.02	5.24	5.43	5.60	5.74	5.87	5.98	6.09	6.19	6.28	6.64	7.15	7.96
10	3.15	3.88	4.33	4.65	4.91	5.12	5.31	5.46	5.60	5.72	5.83	5.94	6.03	6.11	6.47	6.95	7.73
11	3.11	3.82	4.26	4.57	4.82	5.03	5.20	5.35	5.49	5.61	5.71	5.81	5.90	5.98	6.33	6.79	7.55
12	3.08	3.77	4.20	4.51	4.75	4.95	5.12	5.27	5.40	5.51	5.62	5.71	5.80	5.88	6.21	6.66	7.39
13	3.06	3.74	4.15	4.45	4.69	4.89	5.05	5.19	5.32	5.43	5.53	5.63	5.71	5.79	6.11	6.55	7.28
14	3.03	3.70	4.11	4.41	4.64	4.83	4.99	5.13	5.25	5.36	5.46	5.55	5.64	5.71	6.03	6.46	7.16
15	3.01	3.67	4.08	4.37	4.60	4.78	4.94	5.08	5.20	5.31	5.40	5.49	5.57	5.65	5.96	6.38	7.07
16	2.99	3.65	4.05	4.33	4.56	4.74	4.90	5.03	5.15	5.26	5.35	5.44	5.52	5.59	5.90	6.31	6.98
17	2.98	3.63	4.02	4.30	4.52	4.71	4.86	4.99	5.11	5.21	5.31	5.39	5.47	5.54	5.84	6.25	6.91
18	2.97	3.61	4.00	4.28	4.50	4.67	4.82	4.96	5.07	5.17	5.27	5.35	5.43	5.50	5.79	6.20	6.85

A.8.   Studentised Range (q) Tables $P = 0.05$ (continued)

No. of treatment means (k)

Error df	2	3	4	5	6	7	8	9	10	11	12	13	14	15	20	30	60
19	2.96	3.59	3.98	4.25	4.47	4.65	4.79	4.92	5.04	5.14	5.23	5.32	5.39	5.46	5.75	6.15	6.79
20	2.95	3.58	3.96	4.23	4.45	4.62	4.77	4.90	5.01	5.11	5.20	5.28	5.36	5.43	5.71	6.10	6.74
30	2.89	3.49	3.85	4.10	4.30	4.46	4.60	4.72	4.82	4.92	5.00	5.08	5.15	5.21	5.48	5.83	6.42
40	2.86	3.44	3.79	4.04	4.23	4.39	4.52	4.64	4.74	4.82	4.90	4.98	5.04	5.11	5.36	5.70	6.26
60	2.83	3.40	3.74	3.98	4.16	4.31	4.44	4.55	4.65	4.73	4.81	4.88	4.94	5.00	5.24	5.57	6.09
120	2.80	3.36	3.69	3.92	4.10	4.24	4.36	4.47	4.56	4.64	4.71	4.78	4.84	4.90	5.13	5.43	5.93
∞	2.77	3.31	3.63	3.86	4.03	4.17	4.29	4.39	4.47	4.55	4.62	4.69	4.74	4.80	5.01	5.30	5.76

A.9. Critical values for the Kruskal-Wallis non-parametric ANOVA (H)

n_1	n_2	n_3	0.05	0.01	0.001
5	5	5	5.729	8.028	10.271
5	5	4	5.661	7.936	9.960
5	4	4	5.657	7.760	9.168
5	4	3	5.656	7.445	8.795
5	3	3	5.648	7.079	8.727
4	4	4	5.692	7.654	9.269
4	4	3	5.598	7.144	8.909
4	3	3	5.791	6.745	
3	3	3	5.600		

Where n_1, n_2, and n_3 represent the sample sizes.

A.10. Critical values of the chi-square (χ^2) distribution

df	0.05	0.01	0.001
1	3.8415	6.6349	10.8276
2	5.9915	9.2103	13.8155
3	7.8147	11.3449	16.2662
4	9.4877	13.2767	18.4668
5	11.0705	15.0863	20.5150
6	12.5916	16.8119	22.4578
7	14.0671	18.4753	24.3219
8	15.5073	20.0902	26.1245
9	16.9190	21.6660	27.8772
10	18.3070	23.2093	29.5883
11	19.6751	24.7250	31.2641
12	21.0261	26.2170	32.9095
13	22.3620	27.6883	34.5282
14	23.6848	29.1412	36.1233
15	24.9958	30.5779	37.6973
16	26.2962	31.9999	39.2524
17	27.5871	33.4087	40.7902
18	28.8693	34.8053	42.3124
19	30.1435	36.1909	43.8202
20	31.4104	37.5662	45.3148
21	32.6706	38.9322	46.7971
22	33.9244	40.2894	48.2680
23	35.1725	41.6384	49.7283
24	36.4150	42.9798	51.1786
25	37.6525	44.3141	52.6197
26	38.8851	45.6417	54.0520
27	40.1133	46.9629	55.4760
28	41.3371	48.2782	56.8923
29	42.5570	49.5879	58.3012
30	43.7730	50.8922	59.7031
40	55.7585	63.6907	73.4020
50	67.5048	76.1539	86.6608
60	79.0819	88.3794	99.6073
70	90.5312	100.4252	112.3170
80	101.8795	112.3288	124.8393
90	113.1453	124.1163	137.2084
100	124.3421	135.8067	149.4493

A.11. Critical values of Q for nonparametric comparisons ($P = 0.05$)

k treatment means	Probability 0.05
2	1.960
3	2.394
4	2.639
5	2.807
6	2.936
7	3.038
8	3.124
9	3.197
10	3.261

A.12. Critical values of the correlation coefficient

df	0.05	0.01	0.001
1	0.9969	0.9999	0.9999
2	0.9500	0.9900	0.9990
3	0.8783	0.9587	0.9911
4	0.8114	0.9172	0.9741
5	0.7545	0.8745	0.9509
6	0.7067	0.8343	0.9249
7	0.6664	0.7977	0.8983
8	0.6319	0.7646	0.8721
9	0.6021	0.7348	0.8470
10	0.5760	0.7079	0.8233
11	0.5529	0.6835	0.8010
12	0.5324	0.6614	0.7800
13	0.5140	0.6411	0.7603
14	0.4973	0.6226	0.7419
15	0.4821	0.6055	0.7247
16	0.4683	0.5897	0.7084
17	0.4555	0.5751	0.6932
18	0.4438	0.5614	0.6788
19	0.4329	0.5487	0.6652
20	0.4227	0.5368	0.6524
21	0.4133	0.5256	0.6402
22	0.4044	0.5151	0.6287
23	0.3961	0.5052	0.6178
24	0.3882	0.4958	0.6074
25	0.3809	0.4869	0.5974
26	0.3739	0.4785	0.5880
27	0.3673	0.4705	0.5790
28	0.3610	0.4629	0.5703
29	0.3550	0.4556	0.5620
30	0.3494	0.4487	0.5541
35	0.3246	0.4182	0.5189

df	0.05	0.01	0.001
40	0.3044	0.3932	0.4896
45	0.2876	0.3721	0.4647
50	0.2732	0.3542	0.4432
55	0.2609	0.3385	0.4244
60	0.2500	0.3248	0.4079
65	0.2404	0.3126	0.3931
70	0.2319	0.3017	0.3798
75	0.2242	0.2919	0.3678
80	0.2172	0.2830	0.3568
85	0.2108	0.2748	0.3468
90	0.2050	0.2673	0.3376
95	0.1996	0.2604	0.3290
100	0.1946	0.2540	0.3211

Solutions

Chapter 2

4.
```
LET K1 = (SSQ(C1) - ((SUM(C1))**2)/COUNT(C1))/COUNT (C1)
LET K2 = SQRT(C1)
PRINT K1 K2
```

Data Display

```
K1 = 119.874
K2 = 10.9487
```

Mean ± 1 s.d.   72.87 ± 10.9487
Mean ± 2 s.d.   72.87 ± 21.8574
67.3913% of data within 1 s.d. of mean; 95.6522 of data within 2 s.d. of mean
46.74% of data within inter-quartile range (i.e., greater than 64 and less than 80)

5. In the following example, columns 9–11 have been named 'Pulse After,' 'Active?' and 'Smoker?' by entering the names at the head of each of the columns in the worksheet.

```
MTB > Copy 'Pulse2' 'Activity' 'Smokes' C9-c11;
SUBC> Use 'Ran' = 1.
MTB > Table 'Active?' 'Smoker?';
SUBC> Means 'Pulse After';
SUBC> Medians 'Pulse After';
SUBC> Minimums 'Pulse After';
SUBC> Maximums 'Pulse After';
SUBC> StDev 'Pulse After'.
```

6.   **Tabulated Statistics**

Rows: Active?   Columns: Smoker?

	1	2	All
1	86.00	84.00	85.33
	86.00	84.00	--
	78.00	84.00	78.00
	94.00	84.00	94.00
	11.31	--	8.08
2	97.00	98.50	98.08
	98.00	100.00	--
	76.00	70.00	70.00
	115.00	140.00	140.00
	14.25	20.94	19.02
3	77.33	74.50	75.71
	76.00	78.00	--
	72.00	58.00	58.00
	84.00	84.00	84.00
	6.11	11.47	8.98
All	90.25	93.70	92.51
	--	--	--
	72.00	58.00	58.00
	115.00	140.00	140.00
	14.39	21.14	18.94

Cell Contents --
            Pulse Af:Mean
                    Median
                    Minimum
                    Maximum
                    StDev

7.   The range of measurements is very large: fractions of kg to tens of thousands. Therefore, use a logarithmic scale. The obvious outliers are the three prehistoric reptiles, all the rest are mammals.

8.   MTB > desc c1-c4

Descriptive Statistics

Variable	N	Mean	Median	TrMean	StDev	SE Mean
A	13	208.69	208.00	207.91	28.96	8.03
B	13	263.54	263.00	264.55	34.43	9.55

C	13	240.54	236.00	240.73	26.52	7.36
D	13	297.92	303.00	297.45	26.20	7.27

Variable	Minimum	Maximum	Q1	Q3
A	160.00	266.00	187.50	227.50
B	197.00	319.00	244.50	285.50
C	196.00	283.00	222.50	260.50
D	263.00	338.00	272.00	321.50

Use a histogram or scatter plot to present the data (but also see Interval Plot which is described in Chapter 3).

9.

Variable	N	Mean	Median	TrMean	StDev	SE Mean
C1	23	1.913	2.000	1.810	1.621	0.338

Variable	Minimum	Maximum	Q1	Q3
C1	0.000	6.000	1.000	3.000

Variance = 2.6276; CV = 84.74%. Use median and inter-quartile range to represent the data because the histogram shows a skewed distribution.

10 CV at day 1 = 6.07%; CV at day 30 = 3.88%; CV at day 64 = 2.81%. Use scatter plot or histogram (but see Interval Plot described in Chapter 3).

Chapter 3

1.

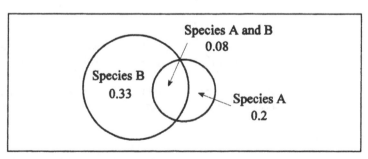

(a) P (neither sp A nor B) = 1 − (0.2 + 0.33 − 0.08) = 0.55
(b) P (sp. A or sp. B or both) = P(A) + P(B) − P(both)
      = 0.2 + 0.33 − 0.08 = 0.45

(c) P (sp. B but not sp. A) = P(B) – P(both)
     = 0.33 – 0.08 = 0.25
(d) P (sp. A or sp. B but not both) = P(A) + P(B) – 2(P(both))
     = 0.2 + 0.33 – 2(0.08) = 0.37

2.

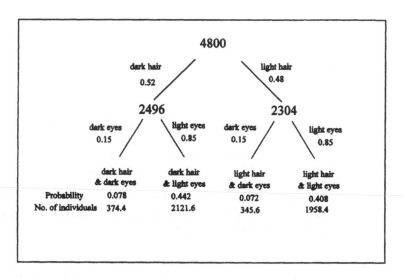

3.

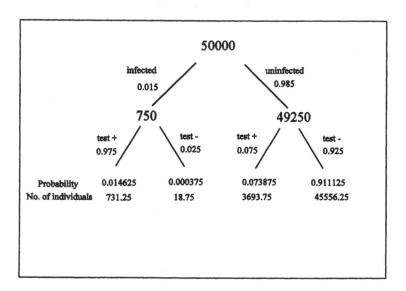

a) 731.25 + 3693.75 = 4425
b) 3693.75. Since one cannot have 0.75 of an individual, round the answer to the nearest integer, 3694.
c) P(uninfected | test negative)
   = 0.911125 ÷ (0.911125 + 0.000375) = 0.9115

4. Binomial probabilities: $n = 8$, $p = 0.5$

$$P(X = 6) = \frac{8!}{6!\,2!}(0.5^6)(0.5^2) = 0.1094$$

Minitab: **PDF 6;**
          BINOMIAL 8 0.5.

5. Binomial probability: $n = 20$, $p = 0.37$

$$P(X > 4) = 1 - (X \le 4) = 0.9141$$

Minitab: **CDF 4 K1;**
          BINOMIAL 20 0.37.
          LET K2=1-K1

6. Binomial probabilities: $n = 12$, $p = 0.45$

(a) $P(X = 3) = \dfrac{12!}{3!\,9!}(0.45^3)(0.55^9) = 0.09233$

Minitab: **PDF 3;**
          BINOMIAL 12 0.45.

(b) $P(X < 5) = P(X = 4) + P(X = 3) + P(X = 2) + P(X = 1) + P(X = 0) = 0.3044$

Minitab: **CDF 4;**
          BINOMIAL 12 0.45.

(c) $P(6 \le X \le 8) = P(X \le 8) - P(X \le 5) = 0.4375$

Minitab: **CDF 8 K1;**
          BINOMIAL 12 0.45.
          CDF 5 K2;

BINOMIAL 12 0.45.
LET K3=K1-K2

7.   Poisson probabilities: $\mu = 2.33$

(a) $P(X = 3) = \dfrac{e^{-2.33}\mu^3}{3!} = 0.2051$

Minitab: **PDF 3;**
            POISSON 2.33.

(b) $P(X \geq 7) = 1 - P(X \leq 6) = 1 - 0.9900 = 0.01$

Minitab: **CDF 6 K1;**
            POISSON 2.33.
            LET K2=1-K1

(c) $P(3 \leq X \leq 5) = P(X \leq 5) - (X \leq 2) = 0.9683 - 0.5880$
       $= 0.3803$

Minitab: **CDF 5 K1;**
            POISSON 2.33.
            CDF 2 K2;
            POISSON 2.33.
            LET K3=K1-K2

(d) $P(0 \leq X < 4) = P(X = 0) + P(X = 1) + P(X = 2)$
       $+ P(X = 3) = 0.7932$

Minitab: **CDF 3;**
            POISSON 2.33.

8.   Poisson probabilities $\mu = 1.913$

(a) $P(2 \leq X \leq 4) = P(X = 2) + P(X = 3) + P(X = 4) = 0.5248$

Minitab: **CDF 4 K1;**
            POISSON 1.913.
            CDF 1 K2;
            POISSON 1.913.
            LET K3=K1-K2

(b) $P(X > 4) = 1 - P(X \leq 4) = 1 - 0.9549 = 0.0451$

Minitab: **CDF 4 K1;**
POISSON 1.913.
LET K2=1-K1

9.   95% confidence intervals: *Species A*   2.8685 – 2.9986 cm
Species B   3.7122 – 3.9304 cm

Minitab: **DESCRIBE C1 C2**
ZINTERVAL 0.1847 C1
ZINTERVAL 0.3098 C2

(a)

$$P(X > 3.0) = 1 - P\left(z \le \frac{3.0 - \mu}{\sigma}\right) = 1 - P(z \le 0.3600) = 1 - 0.6406 = 0.3594$$

Minitab: **CDF 3.0 K1;**
NORMAL 2.9335 0.1847.
LET K2=1-K1

(b)

$$P(X < 3.4) = P\left(z < \frac{3.4 - \mu}{\sigma}\right) = P(z < -1.3599) = 1 - P(z \ge 1.3599)$$

$$= 1 - 0.9131 = 0.0869$$

Minitab: **CDF 3.4;**
NORMAL 3.8213 0.3098.

10.

$$P(X > 23.55) = 1 - P\left(z \le \frac{23.55 - \mu}{\sigma}\right) = 1 - P(z \le 1.0576) = 1 - 0.8554 = 0.1446$$

Minitab: **CDF 23.55 K1;**
NORMAL 22.412 1.076.
LET K2=1-K1

95% confidence interval: length: 22.28 – 22.55 mm
breadth: 16.46 – 16.63 mm

11. 99% confidence interval: 0.1236 – 0.14348 g

   Minitab: **DESCRIBE C1**
   ZINTERVAL 0.99 0.02112 C1

   $P(0.15 < X < 0.16) = P(0.7794 < z < 1.2528) = 0.1121$

   Minitab: **CDF 0.16 K1;**
   NORMAL 0.13354 0.02112.
   CDF 0.15 K2;
   NORMAL 0.13354 0.02112.
   LET K3=K1-K2

12. fertile: mean = 245.32; s.d. = 12.74;
   95% c.i.: 237.22 – 253.42;
   sterile: mean = 230.93; s.d. = 13.68; 95%
   c.i.: 222.24 – 239.62;

   Minitab: **DESCRIBE C1 C2**
   TINTERVAL C1
   TINTERVAL C2

13. Binomial probabilities for a proportion: $p = 0.65$ $n = 40$.
   95% confidence interval: 0.5022 – 0.7978

Chapter 4

4. Assume an initial value of $t = 2$:

$$n = \frac{(0.01)^2(2)^2}{(0.005)^2} = 16$$

$$n = \frac{(0.01)^2(2.1315)^2}{(0.005)^2} = 18.1732 \approx 18$$

$$n = \frac{(0.01)^2(2.1098)^2}{(0.005)^2} = 17.8050$$

Answer: $n = 18$

5.   Assume an initial value of $t = 2$:

$$n = \frac{(10.03)^2(2)^2}{(5)^2} = 16.096 \approx 16$$

$$n = \frac{(10.03)^2(2.9467)^2}{(5)^2} = 34.9409 \approx 35$$

$$n = \frac{(10.03)^2(2.75)^2}{(5)^2} = 30.43 \approx 30$$

$$n = \frac{(10.03)^2(2.7564)^2}{(5)^2} = 30.57$$

Answer: $n = 31$

6.

$$n = (0.13)(0.87)\frac{(2.576)^2}{(0.1)^2} = 75.05$$

Recommend sample size of 76.

7.

$$n = \frac{(6.5)^2(2.093)^2}{2^2} = 46.27$$

$$n = \frac{(6.5)^2(2.0211)^2}{2^2} = 43.14$$

Recommend a sample size of 44.

8.

$$n = (0.175)(0.825)\frac{(1.96)^2}{(0.05)^2} = 221.85 \approx 222$$

Using conservative estimate,

$$n = (0.25)\frac{(1.96)^2}{(0.05)^2} = 384.16 \approx 385$$

Chapter 5

1. (a) One sample t-test: $n \leq 30$, the data are normally distributed and the standard deviation is unknown.
    (b) Binomial or sign tests: frequencies of nominal data, of which there are two categories and that are predicted to have equal probabilities of occurrence.
    (c) One sample z-test: $n > 30$ and the data are distributed normally.
    (d) Binomial test: frequencies of nominal data, of which there are two categories that have known probabilities of occurrence (0.75 and 0.25).
    (e) Wilcoxon one sample test: measurements on an ordinal scale, so the magnitude of the difference between each measurement and the hypothetical median can be measured to a certain extent.

2. Summary of statistical results:

    Test of H_N: = 250.00; H_A: μ not = 250.00

Variety	N	Mean	StDev	SE Mean	t	P
A	13	208.69	28.96	8.03	-5.14	0.0002
B	13	263.54	34.43	9.55	1.42	0.18
C	13	240.54	26.52	7.36	-1.29	0.22
D	13	297.92	26.20	7.27	6.60	0.0000

    The number of degrees of freedom is the same for all four varieties: 12.
    The total leaf areas of Varieties B and C do not differ significantly from 250 cm²; the leaf area of Variety A is significantly smaller than 250 cm²; and Variety D has a significantly greater leaf area than 250 cm².

    The t-test is appropriate because the sample sizes are smaller than 30 and the data are assumed to be distributed normally.

3. (a) Test of H_N: $\mu \geq 3.1$ cms; H_A: $\mu < 3.1$ cm
    The population standard deviation is unknown, so that of the sample is used, $s = 0.1847$. The sample size is more than 30, so the one sample z-test is used.

```
Variable N    Mean   StDev SE Mean     T      P
Species  31  2.9335  0.1847  0.0332  -5.02  0.0000
```

The mean toe length of *Species A* is significantly smaller than 3.1 cm.

(b) Test of H_N: $\mu = 3.9$ cm; H_A: $\mu \neq 3.9$ cm
The population standard deviation is unknown, so that of the sample is used, $s = 0.3098$. The sample size is more than 30, so the one sample z-test is used.

```
Variable N    Mean   StDev SE Mean     Z      P
Species  31  3.8213  0.3098  0.0556  -1.41   0.16
```

The mean toe length of *Species B* is not significantly different from 3.9 cm.

4. Test of H_N: $\mu = 0.2$ kg.day^{-1}; H_A: $\mu \neq 0.2$ kg.day^{-1}

```
Variable      N    Mean    StDev  SE Mean     t      P
growth rate   9  0.18889  0.01316  0.00439  -2.53  0.035
```

The number of degrees of freedom is 8.
The growth rate of the pigs is significantly slower than 0.2 kg.day^{-1}.

If the data are reanalysed using a sign test, the null and alternative hypotheses become:
H_N: $\eta = 0.2$ kg.day^{-1}; H_A: $\eta \neq 0.2$ kg.day^{-1}

because the median, not the mean is compared.

```
             N Below Equal Above    P  Median
growth rate  9    6     1     2  0.2891  0.1810
```

No evidence of difference from the expected median growth rate is found by the sign test.

If the data are reanalysed using a Wilcoxon test, the null and alternative hypotheses are similar to those used with the sign test:
H_N: $\eta = 0.2$ kg.day^{-1}; H_A: $\eta \neq 0.2$ kg.day^{-1}

```
                        N for  Wilcoxon              Estimated
                   N    Test   Statistic      P       Median
growth rate        9     8        3.5      0.050      0.1895
```

Using the Wilcoxon test, the result is on the borderline between accepting or rejecting the null hypothesis. The relative power of the three tests is demonstrated, the t-test being the most powerful and the sign test having the least power.

5. Since the data are quantified measurements but are censored, the Wilcoxon test is the most appropriate test of a two-tailed hypothesis.
    Test of H_N: $\eta = 11.00$: H_A: $\eta \neq 11.00$

```
                        N for  Wilcoxon              Estimated
                   N    Test   Statistic      P       Median
fledging          15    13       79.5      0.019       12.50
```

The results are not very consistent with the known fledging period of 11 days: there is a small probability of obtaining these results if the null hypothesis is true, and so we reject the null hypothesis: the median fledging period appears to be longer than 11 days.

6. Since the probability of yellow flowers is known (0.75) and the data are nominal variables (yellow or green flowers), this problem can be analysed by a binomial test. The number of trials, n is 18; the probability of a success, p is 0.75. The question asks whether 16 yellow flowers is more than would be expected: a question that leads to a one-tailed hypothesis, requiring us to calculate the probability of observing 16, 17, or 18 yellow flowers, which is the sum of the three individual probabilities. These are calculated using Equation 3.6:

$$P(16) = \frac{18!}{16!\ 2!}(0.75)^{16}(0.25)^2 = 0.09584$$

$$P(17) = \frac{18!}{17!\ 1!}(0.75)^{17}(0.25)^1 = 0.03383$$

$$P(18) = \frac{18!}{18!\ 0!}(0.75)^{18}(0.25)^0 = 0.00564$$

The probability of obtaining 16 yellow flowers (or more) from an experiment is 0.13531, a result that is not inconsistent with the hypothetical ratio of 3:1. The ease with which this can be done using Minitab is shown below by calculating the cumulative probability of 15 or less yellow flowers:

```
CDF 15 K1;
BINO 18 0.75.
LET K2=1-K1
```

Chapter 6

1. Assume equal variances.

    H_N: There is no difference in mean aldosterone concentration between rabbit populations in coastal region of Australia and mountainous region.
    H_A: There is a difference in mean aldosterone concentration between rabbit populations in coastal region of Australia and mountainous region.

    $$t = \frac{64.2 - 20.5}{10.61\sqrt{\dfrac{1}{32} + \dfrac{1}{32}}} = 16.48$$

    degrees of freedom $= 32 + 32 - 2 = 62$; $P < 0.001$; reject H_N: the aldosterone concentration of rabbits in the sodium-scarce mountainous regions is much higher than in rabbits from the coastal region. This difference may be an adaptation to the scarcity of sodium in the mountainous area.

2. The data comprise independent measures of leaf area from 2 distinct samples: only 1 measurement of leaf area is derived from each sampling unit. Therefore, an unpaired t-test is appropriate.

    H_N: There is no difference in leaf area between the two varieties of soya.

H_A: There is a difference in leaf area between the two varieties of soya.

$$t = \frac{263.5 - 240.5}{30.7\sqrt{\frac{1}{13} + \frac{1}{13}}} = 1.91$$

degrees of freedom = 13 + 13 – 2 = 24; $P > 0.05$; accept H_N: there is no evidence to contradict the null hypothesis.

3. H_N: Methods B and C make no difference in active life of the drug.
H_A: There is a difference in active life of the drug due to differing effects of Methods B and C.

$W_1 = 371$; $W_2 = 224$

$$T = 371 - \frac{(17)(18)}{2} = 218$$

$$or \quad T = 244 - \frac{(17)(18)}{2} = 71$$

$n_1 = n_2 = 17$; $P = 0.01$: the active life of the drug depends upon the method of administration. Method B = 12.2 h; Method C = 8.7 h: Method B leads to a longer active life than Method A.

4. H_N: There is no difference in the median no. of ticks found on sheep of two areas.
H_A: There is a difference in the median no. of ticks found on sheep of two areas.

Since $n_1 = n_2 = 70$, use the normal equivalent rather than the Mann-Whitney test for small samples.

$$z = \frac{4458 - 70(141)/2}{\sqrt{(70^2)(141)/12}} = -1.988$$

Since this is a two-tailed test, $P = (1 - 0.9767)2 = 0.0466$. Reject H_N, the median no. of ticks on sheep in area B is greater than on sheep in area A ($\eta_A = 4$; $\eta_B = 6.5$).

5. The number of visits to each of the feeders each day are not independent, therefore, a paired t-test is appropriate.
H_N: The mean daily difference in number of visits to the feeders is zero.
H_A: The mean daily difference in number of visits to the feeders each day is not zero.

$$t = \frac{-1.333}{1.543/\sqrt{15}} = -3.35$$

$P < 0.01$. Therefore, reject H_N: there is a difference in usage of the two feeders, Feeder B is used more than Feeder A (the mean difference in number of visits is 1.33 visits.day^{-1}).

6. H_N: The mean difference in growth per pair of seedlings is 0.
H_A: The mean difference in growth per pair of seedlings is not 0.

$$t = \frac{-1.86}{4.19/\sqrt{12}} = -1.54$$

$P > 0.05$: do not reject H_N.

7. H_N: There is no change in blood concentration of β-endorphin in cats after surgery relative to the concentration before surgery.
H_A: There is an elevation in blood concentration of β-endorphin in cats after surgery compared to the pre-surgery blood concentration.

This is a one-tailed test (H_A includes the word *elevation*, i.e., a change in one direction). Twelve cats showed an elevation, five cats showed a decrease, and two cats showed no change. N for test $= 17$, $P = 0.1$; therefore accept H_N.

8. For the thorax weights:

$$t = \frac{-0.01}{1.004 / \sqrt{24}} = -0.0488$$

degrees of freedom = 24 − 1 = 23; $P > 0.05$; there is no evidence of any difference in body weight between winners and losers.

For the fat weights:

$$t = \frac{0.465}{0.834 / \sqrt{24}} = 2.73$$

degrees of freedom = 24 − 1 = 23; $P < 0.02$; the probability of these results occurring if the null hypothesis is true is such that we reject H_N and accept that the fat content of damsel flies that win contests is greater than those who lose.

Taking the thorax and fat results together we conclude that it is fat (energy) reserves and not body weight *per se* that determine the outcome of contests between damsel flies.

9. Normal approximation because N > 20. A 2-tailed test. The null and alternative hypotheses are the same as Exercise 8, except that the median differences are compared (not the mean).

Fat: The sum of the negative ranks for the fat differences = 48.5
The sum of the positive ranks for the fat differences = 227.5
$z = 2.71$; $P = (1 − 0.9966)2 = 0.0068$; Reject H_N

Thorax: The sum of the negative ranks for the thorax differences = 129.5
The sum of the positive ranks for the thorax differences = 123.5
$z = 0.081$; $P = (1 − 0.5319)2 = 0.9362$; Accept H_N.

Chapter 7

1. The variances appear to be proportional to the means. Application of square root transformation seems to rectify the problem. Oneway analysis shows a significant treatment effect.

```
Analysis of Variance for Response

Source    DF        SS        MS        F         P
Group      3    14.2480    4.7493    178.72    0.000
Error     15     0.3986    0.0266
Total     18    14.6466

Means

Group    N    Response
A        5       7.778
B        5       8.364
C        4      10.102
D        5       9.312
```

Tukey tests show the following differences:

Comparison	q	Comparison	q
A vs. B	8.03	B vs. C	20.47
A vs. C	30.04	B vs. D	13.00
A vs. D	21.03	C vs. D	10.21

Critical of q value in tables: 4.08; therefore all treatments differ from each other A < B < D < C.

2. Result of comparing blood retention times based upon data in Table 6.2. Data are censored (24+ hours), hence need to use Kruskal-Wallis test.

Kruskal-Wallis Test

```
Kruskal-Wallis Test on Data

Method    N    Median    Ave Rank      Z
A        17    17.000      35.1      3.10
B        17    12.200      26.9      0.30
C        17     8.700      16.0     -3.40
Overall  51                26.0
```

```
H = 14.15 DF = 2 P = 0.001
H = 14.16 DF = 2 P = 0.001 (adjusted for ties)
```

Comparisons of the three administration methods demonstrate that:
$\Sigma t = 120$; SE = 5.0969

Comparison	Q	Comparison	Q
A vs. B	1.61	B vs. C	2.14
A vs. C	3.75		

Critical value in Q tables: 2.394; therefore, *Method A* causes drug to be retained in blood longer than *Method C*, no other differences were detected.

3.   A oneway ANOVA does not reveal any treatment effect, but the randomized block design does.

```
Analysis of Variance for Response

Source      DF       SS       MS       F       P
Treatment    3   10.6762   3.5587   28.94   0.000
block        4   37.5512   9.3878   76.33   0.000
Error       12    1.4758   0.1230
Total       19   49.7033

Means

Treatment   N   Response
A           5    5.2514
B           5    5.9290
C           5    6.5824
D           5    7.2116
```

Tukey tests reveal the following differences:

Compariso n	q	Compariso n	q
A vs. B	4.321	B vs. C	4.167
A vs. C	8.489	B vs. D	8.180
A vs. D	12.501	C vs. D	4.013

Critical value of q from tables: 4.20. We can conclude that *A* is less than all of the other treatments; *B* is less than *D*; and there are no other differences.

4. Re-analysis of the data from Exercise 3 using Friedman nonparametric randomized block test.

Friedman Test

Friedman test for Response by Variety blocked by block

$\chi^2 = 14.04$ DF $= 3$ P $= 0.003$

Variety	N	Est Median	Sum of Ranks
A	5	5.5206	5.0
B	5	6.2466	10.0
C	5	6.8214	16.0
D	5	7.3659	19.0

Grand median = 6.4886

Comparison of rank sums show the following differences:

Comparison	q	Comparison	q
A vs. B	1.732	B vs. C	2.078
A vs. C	3.811	B vs. D	3.118
A vs. D	4.850	C vs. D	1.039

Critical value from q tables: 3.63; therefore A < C, A < D, no other significant differences. The Friedman test has not been as powerful as the parametric randomized block ANOVA.

5. Results of the analysis of the two factor factorial design investigation of the effect of spacing on three varieties of strawberry.

Analysis of Variance (Balanced Designs)

Factor	Type	Levels	Values		
Variety	fixed	3	A	B	C
Spacing	fixed	3	1	2	3

Analysis of Variance for Kg

Source	DF	SS	MS	F	P
Variety	2	55.257	27.629	42.22	0.000

```
Spacing              2     6.265   3.133   4.79   0.014
Variety*Spacing      4    22.073   5.518   8.43   0.000
Error               36    23.556   0.654
Total               44   107.152
```

```
Means

Variety   N       Kg
A        15     9.680
B        15    11.567
C        15    12.313

Spacing   N       Kg
1        15    10.700
2        15    11.607
3        15    11.253

Variety   Spacing   N       Kg
A         1         5     9.580
A         2         5     8.880
A         3         5    10.580
B         1         5    10.340
B         2         5    12.940
B         3         5    11.420
C         1         5    12.180
C         2         5    13.000
C         3         5    11.760
```

Calculation of Tukey tests: SE = 0.3617
Critical value of q from studentized range tables ($k = 9$; error d.f. = 36, but use 30): 4.72.

		Spacing			
		1	2	3	Overall
Variety	A	9.58[a]	8.88[c,d]	10.58	9.68
	B	10.34[b,e]	12.94[c,e]	11.42	11.57
	C	12.18[a,b]	13.00[d]	11.76	12.31
Overall		10.7	11.61	11.25	

Table of means; values with similar subscripts within the same row or column differ $P < 0.05$.

6.   Re-analysis of the strawberry data with the replicates blocked.

Analysis of Variance (Balanced Designs)

```
Factor      Type   Levels  Values
Variety     fixed      3      A  B  C
Spacing     fixed      3      1  2  3
Block       random     5      1  2  3  4  5
```

Analysis of Variance for Kg

Source	DF	SS	MS	F	P
Variety	2	55.2573	27.6287	34.43	0.000
Spacing	2	6.2653	3.1327	5.05	0.038
Variety*Spacing	4	22.0733	5.5183	9.37	0.000
Block	4	2.7476	0.6869		
Variety*Block	8	6.4204	0.8026		
Spacing*Block	8	4.9658	0.6207		
Error	16	9.4222	0.5889		
Total	44	107.1520			

Means

Variety	N	Kg
A	15	9.680
B	15	11.567
C	15	12.313

Spacing	N	Kg
1	15	10.700
2	15	11.607
3	15	11.253

Variety	Spacing	N	Kg
A	1	5	9.580
A	2	5	8.880
A	3	5	10.580
B	1	5	10.340
B	2	5	12.940
B	3	5	11.420
C	1	5	12.180
C	2	5	13.000
C	3	5	11.760

Calculation of Tukey test results: SE = 0.343
Critical value of q from studentized range tables ($k = 9$; error d.f. = 16): 5.03.

		Spacing			
		1	2	3	Overall
Variety	A	9.58ᵃ	8.88ᵇ,ᶜ	10.58ᶜ	9.68
	B	10.34ᵈ	12.94ᵇ,ᵈ	11.42	11.57
	C	12.18ᵃ	13.00ᵇ	11.76	12.31
Overall		10.70	11.61	11.25	

Table of means; values with similar subscripts within the same row or column differ $P < 0.05$. This experimental design has identified a difference within variety 1 between spacings 2 and 3 that was not identified without blocking.

Chapter 8

1. The result of the correlation: $r = 0.991$; $P < 0.001$. There is a positive correlation between salinity and taurine concentration: the relationship should be examined further experimentally.

2. Spearman rank correlation, $r_S = 0.682$, $P = 0.021$. There is a positive correlation between dominance and song activity.

3. A logarithmic transformation is necessary. Three outliers (standardized residuals >2): Brachiosaurus, Diplodocus and Triceratops (all reptiles) whereas all the other animals are mammals. The regression equation is:

   log (brain wt.) = 1.11 + 0.496 log (body wt.) $R^2 = 60.8\%$

   If those dinosaurs are removed, the new equation is:

   log (brain wt.) = 0.934 + 0.752 log (body wt.) $R^2 = 92.2\%$

   or

   brain wt. = $10^{0.934}$(body wt.)$^{0.752}$ = 8.59 (body wt.)$^{0.752}$

   With the new relationship, the residuals associated with humans and rhesus monkeys are large, and should be examined further.

4.   The regression equation is:

$$\frac{1}{V} = 0.0426007 + 0.00073630 \, \frac{1}{S}$$

Therefore, $V_{max} = 23.47379$ and $K_M = 0.017284$

5.   Using resting pulse rate alone to predict after-exercise pulse rate, the regression equation is a good linear fit with $R^2 = 36.9\%$

after $= 18.5 + 1.01$ resting

If weight is included, the $R^2 = 55.6\%$ and the equation is a good fit:

after $= 96.0 + 0.747$ resting $- 0.385$ weight

Both predictor variables are statistically significant. Note the counter-intuitive effect of weight: it appears to have a negative influence on post-exercise pulse rate (i.e., heavier people have lower pulse rates after exercise).

6.   A quadratic model gives a very good fit (better than *flow* on $depth^2$ by itself), $R^2 = 99.0\%$ and the equation is:

Flow $= 1.68 + 23.5$ Depth$^2 - 10.9$ Depth

It should come as no surprise that the *flow* will be related to the square of the *depth* since this will be related to the cross-sectional area of the stream.

Chapter 9

1.   This is a 2×2 table, therefore we must use Yates' continuity correction. $\chi^2 = 413.840$, d.f. $= 1$; $P < 0.001$. Examination of standardized residuals shows that haemophilia is significantly associated with males.

2. $\chi^2 = 3.279$; d.f. = 3; $P = 0.351$. There is no evidence to suggest that there is an association between gender and blood group.

3.   This is a 2×2 table, therefore we must use Yates' continuity correction. $\chi^2 = 85.094$, d.f. = 1; $P < 0.001$. Examination of standardized residuals shows that sickle cell anaemia is significantly associated with non-sufferers of malaria. There is evidence to suggest that sickle-cell anaemia may confer some protection against malaria.

4. $\chi^2 = 0.338583$; d.f. = 2; $P > 0.05$. The offspring of this cross exhibit the phenotypic characters in a ratio that is consistent with Mendelian genetics.

5. $\chi^2 = 42.155$; d.f. = 6; $P < 0.001$. The nearest neighbour distances do not conform to a Poisson distribution, the birds are nesting in a non-random pattern. The mean nearest neighbour distance is 3.46 m and the variance is 1.46 m^2. In terms of magnitude, the mean is larger than the variance indicating that the birds are regularly spaced, possibly indicating competition for nesting sites.

References

Altmann, J. (1974) Observational study of behavior: sampling methods. *Behav.*, 49, 227–267.

Barnett, V. (1991) *Sample Survey Principles and Methods.* 2nd ed., Edward Arnold, London.

Chatfield, C. (1970) *Statistics for Technology*, Penguin Books, Harmondsworth.

Chatfield, C. (1988) *Problem Solving: A Statistician's Guide*, Chapman & Hall, London.

Clarke, G.M. and Cooke, D. (1992) *A Basic Course in Statistics*, 3rd ed., Edward Arnold, London.

Cochran, W.G. (1977) *Sampling Techniques*, 3rd ed., John Wiley & Sons, New York.

Cox, D.R. (1958) *Planning of Experiments*, John Wiley & Sons, New York.

Crowder, M.J. and Hand, D.J. (1990) *Analysis of Repeated Measures*, Chapman & Hall, London.

Darwin, C. (1876) *The Effect of Cross- and Self-fertilisation in the Vegetable Kingdom*, 2nd ed., John Murray, London.

Hull, D.L. (1974) *Philosophy of Biological Science*, Prentice-Hall, Engelwood Cliffs.

Jerison, H.J. (1973) *Evolution of the Brain and Intelligence*, Academic Press, New York.

Krebs, C.J. (1989) *Ecological Methodology*, Harper Collins, New York.

Kuhn, T.S. (1970) *The Structure of Scientific Revolutions*, 2nd ed., The University of Chicago, Chicago.

Latter, O.H. (1902) The egg of *Cuculus canorus*, *Biometrika*, 1, 164–176.

Martin, P. and Bateson, P. (1993) *Measuring Behaviour: An Introductory Guide*, 2nd ed., Cambridge University Press, Cambridge.

Morris, R.F. (1955) The development of sampling techniques for forest insect defoliators, with particular reference to the spruce budworm, *Can. J. Zool.*, 33, 225–294.

Popper, K.R. (1959) *The Logic of Scientific Discovery*, Hutchinson, London.

Siegel, S. and Castellan, N.J., jnr (1988) *Nonparametric Statistics for the Behavioral Sciences*, 2nd ed., McGraw-Hill, New York.

Snedecor, G.W. and Cochran, W.G. (1956) *Statistical Methods Applied to Experiments in Agriculture and Biology,* 5th ed., Iowa State University Press, Ames.

Southwood, T.R.E. (1978) *Ecological Methods with Particular Reference to Insect Populations,* 2nd ed., Chapman & Hall, London.

Sprent, P. (1993) *Applied Nonparametric Statistical Methods,* 2nd ed., Chapman & Hall, London.

Tinbergen, N. (1963) On aims and methods in ethology, *Zeitschrift für Tierpsychologie,* 20, 410–433.

Tukey, J.W. (1977) *Exploratory Data Analysis,* Addison-Wesley, Reading.

Velleman, P.F. and Hoaglin, D.C. (1981) *Applications, Basics, and Computing of Exploratory Data Analysis,* Duxbury Press, Boston.

Zar, J.H. (1996) *Biostatistical Analysis,* 3rd ed., Prentice-Hall, Upper Saddle River.

Zelen, M. and Severo, N.C. (1964) Probability functions. In: *Handbook of Mathematical Functions,* ed. M. Abramowitz and I. Styun, National Bureau of Standards, Washington, D.C., 925–995.

Index